Lecture Notes in Mathematics 1992

Alberto Parmeggiani

Spectral Theory of Non-Commutative Harmonic Oscillators: An Introduction

Springer

Alberto Parmeggiani
Department of Mathematics
University of Bologna
Piazza di Porta San Donato, 5
40126 Bologna
Italy
parmeggi@dm.unibo.it

ISBN: 978-3-642-11921-7 e-ISBN: 978-3-642-11922-4
DOI: 10.1007/978-3-642-11922-4
Springer Heidelberg Dordrecht London New York

Lecture Notes in Mathematics ISSN print edition: 0075-8434
ISSN electronic edition: 1617-9692

Library of Congress Control Number: 2010924182

Mathematics Subject Classification (2000): 35P20, 35J45, 35P15, 35S05, 47A10, 47G30, 47N50, 58J52, 81Q20

Cover design: SPi Publisher Services

Printed on acid-free paper

springer.com

To Serena & Lorenzo, Luisa & Pier Luigi

Preface

This book grew out of a series of lectures given at the Mathematics Department of Kyushu University in the Fall 2006, within the support of the 21st Century COE Program (2003–2007) "Development of Dynamical Mathematics with High Functionality" (Program Leader: prof. Mitsuhiro Nakao).

It was initially published as the Kyushu University COE Lecture Note number 8 (COE Lecture Note, 8. Kyushu University, The 21st Century COE Program "DMHF", Fukuoka, 2008. vi+234 pp.), and in the present form is an extended version of it (in particular, I have added a section dedicated to the Maslov index).

The book is intended as a rapid (though not so straightforward) *pseudodifferential* introduction to the spectral theory of certain systems, mainly of the form $a_2 + a_0$ where the entries of a_2 are homogeneous polynomials of degree 2 in the (x,ξ)-variables, $(x,\xi) \in \mathbb{R}^n \times \mathbb{R}^n$, and a_0 is a constant matrix, the so-called *non-commutative harmonic oscillators*, with particular emphasis on a class of systems introduced by M. Wakayama and myself about ten years ago. The class of non-commutative harmonic oscillators is very rich, and many problems are still open, and worth of being pursued.

I wish to thank Masato Wakayama, dearest friend and collaborator, and my friends and colleagues Nicola Arcozzi, Sandro Coriasco, Sandro Graffi, Frédéric Hérau, Takashi Ichinose, Miyuki Kuze, Lidia Maniccia, Luca Migliorini, Cesare Parenti and Cosimo Senni for their invaluable help in giving these notes a (hopefully) decent shape.

Bologna
December 2009

Alberto Parmeggiani

Contents

Chapter 1
Introduction

A **non-commutative harmonic oscillator** (NCHO for short) is the Weyl-quantization $a^{\mathrm{w}}(x,D)$ of an $N \times N$ system of the form $a(x,\xi) = a_2(x,\xi) + a_0$, $(x,\xi) \in \mathbb{R}^n \times \mathbb{R}^n = T^*\mathbb{R}^n$, where $a_2(x,\xi)$ is an $N \times N$ matrix whose entries are *homogeneous polynomials of degree* 2 *in the* (x,ξ) *variables*, and a_0 is a constant $N \times N$ matrix. In other words, a NCHO is the Weyl-quantization of a matrix-valued quadratic form in (x,ξ), plus a constant matrix term.

The name NCHO (which was given by M. Wakayama and myself) originates on the one hand from the fact that a scalar harmonic oscillator is a single quadratic form in (x,ξ), and on the other from the two levels of non-commutativity that one has to deal with when studying these systems: the level due to the matrix-valued nature of the symbol of the system, and the level due to the Weyl-quantization rule $x_k\xi_j \leftrightarrow (x_k D_{x_j} + D_{x_j} x_k)/2$ (where $D = -i\partial$, as usual), reflected through symplectic geometry by the Poisson-bracket relations $\{\xi_j, x_k\} = \delta_{jk}$, $1 \leq j,k \leq n$.

I shall say that a NCHO $a^{\mathrm{w}}(x,D)$ is elliptic when $\det a_2(x,\xi)$ behaves exactly like $(|x|^2 + |\xi|^2)^N$ when $|(x,\xi)|$ is large. When a_2 and a_0 are Hermitian matrices, the operator $a^{\mathrm{w}}(x,D)$ is "formally self-adjoint" (that is, symmetric on $\mathscr{S}(\mathbb{R}^n;\mathbb{C}^N)$), and when in addition it is positive elliptic (that is, the matrix $a_2(x,\xi)$ is positive definite for $(x,\xi) \neq (0,0)$), then it is self-adjoint as an unbounded operator in $L^2(\mathbb{R}^n;\mathbb{C}^N)$ with a discrete real spectrum.

It is particularly striking to realize that, while (almost) everything is known for scalar harmonic oscillators, very little is known about the spectral properties of self-adjoint elliptic systems, even in the basic (and seemingly simple) case of NCHOs.

This book is mainly intended as a *pseudodifferential* introduction to the spectral theory of elliptic NCHOs, although it contains also results for more general elliptic differential systems with polynomial coefficients, and is a first account of the properties of such systems.

A particularly important example of NCHO, is given by the system

$$Q^{\mathrm{w}}_{(\alpha,\beta)}(x,D) = \begin{bmatrix} \alpha\Big(-\dfrac{\partial_x^2}{2} + \dfrac{x^2}{2}\Big) & -\Big(x\partial_x + \dfrac{1}{2}\Big) \\ x\partial_x + \dfrac{1}{2} & \beta\Big(-\dfrac{\partial_x^2}{2} + \dfrac{x^2}{2}\Big) \end{bmatrix}, \quad x \in \mathbb{R},\ \alpha,\beta \in \mathbb{C},$$

A. Parmeggiani, *Spectral Theory of Non-Commutative Harmonic Oscillators: An Introduction*, Lecture Notes in Mathematics 1992,
DOI 10.1007/978-3-642-11922-4_1,

which is the Weyl-quantization of the matrix

$$Q_{(\alpha,\beta)}(x,\xi)=\begin{bmatrix}\alpha\Big(\dfrac{\xi^2+x^2}{2}\Big) & -ix\xi\\ ix\xi & \beta\Big(\dfrac{\xi^2+x^2}{2}\Big)\end{bmatrix},\quad (x,\xi)\in\mathbb{R}\times\mathbb{R},$$

that was introduced by M. Wakayama and myself in a series of papers (see Parmeggiani-Wakayama [58, 59]). When $\alpha,\beta>0$ with $\alpha\beta>1$, the system is positive elliptic, self-adjoint, and so it has a discrete spectrum in $L^2(\mathbb{R};\mathbb{C}^2)$, and a very rich and remarkable structure. It is worth remarking that in [59] (see also [58]) the eigenvalues are described in terms of a *scalar three-term* recurrence, and hence in terms of a continued fraction (however, it is very difficult to have a direct and explicit expression of them). It is also worth mentioning that when $\alpha=\beta>1$, one has (see Parmeggiani-Wakayama [58, 59]) that $Q^{\mathrm{w}}_{(\alpha,\alpha)}(x,D)$ is *unitarily* equivalent to a scalar harmonic oscillator times the identity 2×2 matrix, so that its spectral properties are governed by the tensor product of the oscillator representation and the 2-dimensional trivial representation of $\mathfrak{sl}_2(\mathbb{R})$ (see e.g. Howe-Tan [25]), i.e. one has matrix-valued creation/annihilation operators that can be used to "construct" the spectrum (see Parmeggiani-Wakayama [60] for further examples of NCHOs whose spectrum can be studied through the tensor product of the oscillator representation and the finite dimensional representation of $\mathfrak{sl}_2(\mathbb{R})$ and its perturbations). Hence, when $\alpha,\beta>0$ and $\alpha\beta>1$, $Q^{\mathrm{w}}_{(\alpha,\beta)}(x,D)$ can be thought of as a matrix-valued deformation of the scalar harmonic oscillator (it is not yet known whether $Q^{\mathrm{w}}_{(\alpha,\beta)}(x,D)$ when $\alpha\neq\beta$ admits creation/annihilation operators or not). System $Q^{\mathrm{w}}_{(\alpha,\beta)}(x,D)$ for α, $\beta>0$ and $\alpha\beta>1$, will be the main example to bear in mind in this exploration of the basic spectral properties of elliptic NCHOs.

Of course, it is a natural and important problem to understand the spectral behavior of $Q^{\mathrm{w}}_{(\alpha,\beta)}(x,D)$ as α and β vary in $\mathbb{C}\setminus\{\alpha\beta=0\}$. I won't be dealing with this aspect here, and concentrate mainly only on the theory of self-adjoint elliptic and positive NCHOs.

My motivation for considering systems like $Q^{\mathrm{w}}_{(\alpha,\beta)}(x,D)$ comes from PDEs, namely from the study of a-priori lower bound estimates, such as Melin's or Hörmander's or Fefferman-Phong's , in the case of pseudodifferential systems (see Parenti-Parmeggiani [49], and also Parmeggiani [53, 54], and references therein). A basic tool for studying such inequalities is the use of the *localized operator*, introduced by L. Boutet de Monvel (and further developed by L. Boutet de Monvel, A. Grigis and B. Helffer in their study of the hypoellipticity with loss of derivatives; see [49] and the references therein). The localized operator is, loosely speaking, the Weyl quantization of the "meaningful" part of the Taylor-expansion of the symbol at characteristic points, whose spectral properties become relevant to the existence of the lower-bound estimate. In the case of systems, the localized operator is a differential system with polynomial coefficients, an example of which is given by $Q^{\mathrm{w}}_{(\alpha,\beta)}(x,D)$. Another motivation comes from the search of vector-valued deformations of the scalar, fundamental, harmonic oscillator. In some sense, as the reader will see, the class of polynomial differential systems studied here is by all means a remarkable vector-valued deformation of the harmonic oscillator which inherits many beautiful spectral properties.

From the point of view of Mathematical Physics, it is interesting to mention here the recent results of Hirokawa [22] on the Dicke-type crossings among the eigenvalues (of basic importance in many physical problems) for certain NCHOs.

It is also conceivable that NCHOs (and more general differental systems with polynomial coefficients) appear naturally in questions related to the Born-Oppenheimer approximation or to Solid State Physics. For instance,[1] when dealing with the Flux $1/2$ problem relative to the spectral analysis of the Harper model, B. Helffer and J. Sjöstrand [21] are led to work with a semiclassical pseudodifferental system whose symbol is $\begin{bmatrix} \cos\xi & \cos x \\ \cos x & -\cos\xi \end{bmatrix}$, and one is concerned with the analysis for energies near 0 and phase-space points (x,ξ) near $(\pi/2,\pi/2)$. After a phase-space translation, one is therefore forced to consider the system $\begin{bmatrix} \sin\xi & \sin x \\ \sin x & -\sin\xi \end{bmatrix}$, whose harmonic approximation is the system $\begin{bmatrix} \xi & x \\ x & -\xi \end{bmatrix}$. The square of the latter is the NCHO $\begin{bmatrix} D^2+x^2 & [D,x] \\ -[D,x] & D^2+x^2 \end{bmatrix}$.

It is interesting to remark that the three-term recurrence equations I have mentioned above, appearing in the study of the spectrum of $Q^{\mathrm{w}}_{(\alpha,\beta)}(x,D)$, are similar to certain ones appearing in Lay and Slavyanov's book [39] about special functions. It is thus plausible that some physical applications of NCHOs may also be found by considering, in analogy, those given in [39].

The main results contained in this book are:

- the meromorphic continuation of the spectral zeta function ζ_A of a general $N\times N$ elliptic NCHO A in $\mathbb{R}^n$, and the very elegant (and precise) result by Ichinose and Wakayama on the spectral zeta function of an elliptic $Q^{\mathrm{w}}_{(\alpha,\beta)}(x,D)$;
- some precise upper and lower bounds on the lowest eigenvalue of $Q^{\mathrm{w}}_{(\alpha,\beta)}(x,D)$ (due to Ichinose and Wakayama, and to myself);
- the "clustering" and "multiplicity" phenomena of eigenvalues in dependence on dynamical quantities, such as period and action (and Maslov index), related to the closed integral trajectories of the Hamilton vector-fields associated with the eigenvalues of the symbol of the system for a class of n-dimensional 2×2 positive elliptic NCHOs of order 2, that generalizes the class of positive elliptic $Q^{\mathrm{w}}_{(\alpha,\beta)}(x,D)$ (these results are due to myself).

In the course of achieving the afore-mentioned results, I have also developed some general tools, used here, that I hope will also prove useful to attack further spectral problems, beyond those treated in these notes.

[1] I am indebted with B. Helffer for explaining this to me.

Of some basic material I have not given a thoroughly self-contained presentation, at times preferring instead a kind of overall survey with the aim of giving a "user-guide" of it, my choice being due to the existence of beautiful (and important) books and papers to which I address the reader.

Here is now an account on the content of this book.

In Chapter 2, I introduce the Weyl-quantization of the Hamiltonian of the harmonic oscillator and describe its spectral properties. This is the basic *scalar* model which will serve as a guide throughout this study.

In Chapter 3, I give a rapid introduction to the general Weyl-Hörmander calculus, to be then specialized to the case of the "global metric". This is the setting where I shall develop the spectral theory of NCHOs.

In Chapter 4, I introduce the function that "counts" the eigenvalues of an operator with discrete spectrum (the spectral counting function), and give some of its elementary properties. I also describe the spectral counting function associated with a general harmonic oscillator, which is then used to derive a first crude information about the behavior of the spectral counting function associated with a general NCHO (information that, however, is already sufficient to obtain a first convergence region of the associated spectral zeta function).

In Chapter 5, I give an account of the functional calculus of the operators I am interested in, and introduce the notion of "trace" of an operator.

In Chapter 6, I decribe the construction of an approximation to e^{-tA}, based on the parametrix construction for $d/dt+A$, where A is a positive elliptic global polynomial differential system. This is then used to compute the leading coefficient of the asymptotics of the spectral counting function for large eigenvalues, in terms of the symbol of the system. To achieve this, I use (as is classical) the Karamata Tauberian theorem, of which I give a complete proof.

In Chapter 7, I recall Robert's theorem on the meromorphic continuation of the spectral zeta function, and then give a proof of it for a positive elliptic $N\times N$ NCHO in $\mathbb{R}^n$ by using the approximation to e^{-tA} constructed in Chapter 6, and finally recall (and sketch the proof of) the Ichinose-Wakayama theorem about the spectral zeta function associated with a positive elliptic $Q^{\mathrm{w}}_{(\alpha,\beta)}(x,D)$. (The Ichinose-Wakayama result is a consequence of the more general result I give for an elliptic $N\times N$ NCHO, except for their very precise description of the "trivial zeros". That is why I have decided to sketch their proof.)

In Chapter 8, I recall Ichinose-Wakayama's lower and upper bounds to the lowest eigenvalue of a positive elliptic $Q^{\mathrm{w}}_{(\alpha,\beta)}(x,D)$, and then show a refinement (which is new) of the upper bound, whose proof uses only some elementary symplectic linear algebra.

In Chapter 9, I review and construct some tools from the calculus of Semiclassical Pseudodifferential Operators (i.e. the Weyl-quantization of symbols of the kind $p(x,h\xi;h)$, where $h\in(0,1]$ is the semiclassical parameter), proving in particular the decoupling construction (modulo operators that have a norm which is $O(h^N)$ for every $N\in\mathbb{Z}_+$) of a given system whose principal symbol can be blockwise diagonalized.

Then in Chapter 10, I introduce a useful approach to study large eigenvalues of self-adjoint (unbounded) operators (in the usual and in the semiclassical setting), consisting of a "reduction" of the operator by "cutting off low energies".

In Chapter 11, I recall some basic facts about the dynamics of the integral trajectories of the Hamilton vector-field associated with the symbol of a pseudodifferential operator (the *bicharacteristics*), with special interest in the periodic trajectories. I will then introduce the Maslov index of such trajectories, and give an effective way of computing it, following the approach and results of Robbin-Salamon [62] (see also McDuff-Salamon [43]). In the notes to this chapter I rapidly recall some motivations for the introduction of the Maslov index, and give the appropriate references.

In the final Chapter 12, I describe in the first place the clustering and multiplicity properties of the large eigenvalues of a scalar semiclassical pseudodifferential operator, which are described through the dynamical quantities introduced in Chapter 11. This is afterwards used to study, through the decoupling seen in Chapter 9 and the methods developed in Chapter 10, the large eigenvalues of an elliptic 2×2 NCHO in $\mathbb{R}^n$ whose symbol has distinct eigenvalues. Since the eigenvalues of a semiclassical NCHO $a_2^{\mathrm{w}}(x,hD)$ (i.e. with $a_0 = 0$) are h times the eigenvalues of $a_2^{\mathrm{w}}(x,D)$, this gives us precise clustering and multiplicity properties of the large eigenvalues of a NCHO, with no lower-order part a_0, in terms of the dynamical quantities associated with the eigenvalues of the symbol $a_2(x,\xi)$. This part is new, and generalizes the results obtained in Parmeggiani [55] for an elliptic $Q_{(\alpha,\beta)}^{\mathrm{w}}(x,D)$ to n-dimensional 2×2 NCHOs.

In the Appendix, I show a few properties of the almost-analytic extension to the complex domain of compactly supported functions in $\mathbb{R}$, and (following Dimassi-Sjöstrand [7]) recall a proof of the Dyn'kin-Helffer-Sjöstrand formula.

Finally, a list of the main notation used throughout the book is given right before the Bibliography.

I end this Introduction by saying that in this book there was no space for including other issues related to an elliptic $Q_{(\alpha,\beta)}^{\mathrm{w}}(x,D)$, such as:

- the representation-theoretic approach, due to H. Ochiai, which establishes the deep relation of the spectral theory of these systems to the existence of particular holomorphic solutions of Heun differential operators in the complex domain (see Ochiai [46, 47]);
- the numerical study of the spectrum started by K. Nagatou, M.T. Nakao and M. Wakayama (see Nagatou-Nakao-Wakayama [45]);
- the number-theoretic investigation of the special values of the spectral zeta function associated with the spectrum initiated by T. Ichinose and M. Wakayama (see Ichinose-Wakayama [32, 33], Kimoto-Wakayama [36, 37], Kimoto-Yamasaki [38], and Ochiai [48]);
- the Poisson-like formulas on the spectral density of eigenvalues (see Parmeggiani [52]);
- the properties of the lowest eigenvalue and some perturbation theory in the parameters (α,β) (see Ichinose-Wakayama [33], Parmeggiani [51, 52]).

They will be part of a wider book, that M. Wakayama and I are planning to write.

Chapter 2
The Harmonic Oscillator

In this chapter we shall review the spectral properties of the harmonic oscillator, i.e. of the operator given by the Weyl-quantization of the Hamiltonian $(|\xi|^2 + |x|^2)/2$.

2.1 From the Hamiltonian to the Operator Acting on L^2

Consider the **phase-space** $\mathbb{R}^n \times \mathbb{R}^n = T^*\mathbb{R}^n = \mathbb{R}^{2n}$ (the **cotangent bundle** of $\mathbb{R}^n$), that we shall always consider endowed with the **canonical symplectic form**

$$\sigma = \sum_{j=1}^{n} d\xi_j \wedge dx_j, \tag{2.1}$$

where (x,ξ) are symplectic coordinates in $\mathbb{R}^n_x \times \mathbb{R}^n_\xi$. Equivalently, for any given tangent vector $v = \begin{bmatrix} \delta x \\ \delta \xi \end{bmatrix}, v' = \begin{bmatrix} \delta x' \\ \delta \xi' \end{bmatrix} \in \mathbb{R}^n \times \mathbb{R}^n$ one has

$$\sigma(v,v') = \langle \delta\xi, \delta x' \rangle - \langle \delta\xi', \delta x \rangle, \tag{2.2}$$

where $\langle \cdot, \cdot \rangle$ denotes the usual inner-product in $\mathbb{R}^n$. Remark that $\sigma = d(\sum_{j=1}^n \xi_j dx_j)$, where $\sum_{j=1}^n \xi_j dx_j$ is called the **canonical** 1**-form**.

Next, consider in $\mathbb{R}^{2n}$ the Hamiltonian

$$p_{0,n}(x,\xi) = \frac{1}{2}(|\xi|^2 + |x|^2) = \sum_{j=1}^{n} p_0(x_j, \xi_j),$$

where

$$p_0(x_j,\xi_j) = p_{0,1}(x_j,\xi_j) = \frac{1}{2}(\xi_j^2 + x_j^2),$$

A. Parmeggiani, *Spectral Theory of Non-Commutative Harmonic Oscillators: An Introduction*, Lecture Notes in Mathematics 1992,
DOI 10.1007/978-3-642-11922-4_2, © Springer-Verlag Berlin Heidelberg 2010

is the one-dimensional **harmonic oscillator** in the variables $(x_j, \xi_j) \in \mathbb{R}^2$. Its Weyl-quantization is the operator (we shall throughout use D_{x_j} or D_j for $-i\partial_{x_j}$ or $-i\partial_j$, respectively)

$$p_{0,n}^{\mathrm{w}}(x,D) = \frac{1}{2}(-\Delta_x + |x|^2) = \frac{1}{2}\sum_{j=1}^{n}(-\partial_{x_j}^2 + x_j^2) = \sum_{j=1}^{n} p_0^{\mathrm{w}}(x_j, D_{x_j}).$$

It is readily seen that $p_{0,n}^{\mathrm{w}}(x,D)$ is **continuous** when thought of as an operator $p_{0,n}^{\mathrm{w}}(x,D)\colon \mathscr{S}(\mathbb{R}^n) \longrightarrow \mathscr{S}(\mathbb{R}^n)$ and as an operator $p_{0,n}^{\mathrm{w}}(x,D)\colon \mathscr{S}'(\mathbb{R}^n) \longrightarrow \mathscr{S}'(\mathbb{R}^n)$, where $\mathscr{S}(\mathbb{R}^n)$ and $\mathscr{S}'(\mathbb{R}^n)$ are, respectively, the space of Schwartz functions and temperate distributions on $\mathbb{R}^n$.

The operator $p_{0,n}^{\mathrm{w}}(x,D)$ is obtained from $p_{0,n}(x,\xi)$ by replacing ξ_j by D_j, x_j by x_j and $x_j\xi_k$ by $(x_jD_k + x_kD_j)/2$, that is to say, it is obtained as

$$p_{0,n}^{\mathrm{w}}(x,D)u(x) = (2\pi)^{-n}\iint e^{i\langle x-y,\xi\rangle} p_{0,n}(\frac{x+y}{2},\xi)u(y)dyd\xi,\ \ u \in \mathscr{S}(\mathbb{R}^n), \tag{2.3}$$

where it is understood that the integration with respect to y is effected first and the one with respect to ξ last.

Exercise 2.1.1. Show the formula (2.3) indeed holds. △

When studying the spectral properties of $p_{0,n}^{\mathrm{w}}(x,D)$ it is also convenient to realize it as an unbounded operator in $L^2(\mathbb{R}^n)$, that we will denote by H regardless the dimension n, $H\colon D(H) \subset L^2(\mathbb{R}^n) \longrightarrow L^2(\mathbb{R}^n)$, with domain

$$D(H) = \{u \in L^2(\mathbb{R}^n);\ Hu = p_{0,n}^{\mathrm{w}}(x,D)u \in L^2(\mathbb{R}^n)\} =: B^2(\mathbb{R}^n) \tag{2.4}$$

(the so-called **maximal operator**; we shall come back to this later on), where $p_{0,n}^{\mathrm{w}}(x,D)u$ is understood in the sense of distributions. Note that one clearly has $\mathscr{S}(\mathbb{R}^n) \subset D(H)$, whence H is **densely** defined. It is also a **closed** operator. To see this, we have to show that

$$\left.\begin{array}{l} D(H) \ni u_j \xrightarrow{L^2} u,\ \text{as } j \to +\infty \\ \\ Hu_j \xrightarrow{L^2} v,\ \text{as } j \to +\infty \end{array}\right\} \Longrightarrow u \in D(H) \text{ and } Hu = v.$$

Since

$$u_j \xrightarrow{L^2} u \Longrightarrow u_j \xrightarrow{\mathscr{S}'} u,$$

we have on the one hand

$$p_{0,n}^{\mathrm{w}}(x,D)u_j \xrightarrow{\mathscr{S}'} p_{0,n}^{\mathrm{w}}(x,D)u,\ \text{as } j \to +\infty,$$

and on the other,

$$p_{0,n}^{\mathrm{w}}(x,D)u_j = Hu_j \xrightarrow{L^2} v, \text{ as } j \to +\infty,$$

whence

$$p_{0,n}^{\mathrm{w}}(x,D)u = v \in L^2(\mathbb{R}^n),$$

i.e. $u \in D(H)$ and $Hu = v$, which is what we wanted to prove.

Let us restrict our attention, for the time being, to the 1-dimensional case $n = 1$. We have

$$p_0(x,\xi) = \frac{x^2+\xi^2}{2} = \frac{x+i\xi}{\sqrt{2}}\frac{x-i\xi}{\sqrt{2}}, \quad (x,\xi) \in \mathbb{R}\times\mathbb{R}.$$

Set hence

$$\psi_{\pm}(x,\xi) := \frac{x \mp i\xi}{\sqrt{2}}.$$

Note that $|\psi_{\pm}(x,\xi)| \approx p_0(x,\xi)^{1/2}$, where, for positive A and B, $A \approx B$ means that there are universal constants $C_1, C_2 > 0$ such that $C_1 A \le B \le C_2 A$.

Notation. *Let $A, B > 0$. We shall always write $A \lesssim B$ (or $B \gtrsim A$) when there is a universal constant $C > 0$ such that $A \le CB$. We hence have $A \approx B$ whenever $A \lesssim B$ and $B \lesssim A$.*

Define next the **Poisson bracket** of two C^1 functions f and g by

$$\{f,g\} = \sum_{j=1}^{n}\Big(\frac{\partial f}{\partial \xi_j}\frac{\partial g}{\partial x_j} - \frac{\partial f}{\partial x_j}\frac{\partial g}{\partial \xi_j}\Big).$$

Then, on the one hand

$$p_0(x,\xi) = \prod_{\pm}\psi_{\pm}(x,\xi)\psi_{\mp}(x,\xi), \text{ and } \{\psi_+,\psi_-\} = -i,$$

and on the other hand, upon setting

$$\Psi_{\pm} := \psi_{\pm}^{\mathrm{w}}(x,D) = \frac{x \mp \partial_x}{\sqrt{2}},$$

we have, by a direct computation,

$$\Psi_+\Psi_- = H - \frac{1}{2}, \quad \Psi_-\Psi_+ = H + \frac{1}{2}, \quad [\Psi_+,\Psi_-] = -1. \tag{2.5}$$

The operators $\Psi_{\pm}$ are the celebrated **creation** (Ψ_+) and **annihilation** (Ψ_-) operators.

We now define, for $a, b \colon \mathbb{R}^2 \longrightarrow \mathbb{C}$ **affine** functions in x, ξ, the operation $a\sharp b$ by

$$a\sharp b = ab - \frac{i}{2}\{a,b\},$$

so that

$$\psi_+ \sharp \psi_- = \psi_+ \psi_- - \frac{i}{2}(-i) = p_0 - \frac{1}{2},$$

and

$$\psi_- \sharp \psi_+ = \psi_- \psi_+ - \frac{i}{2} i = p_0 + \frac{1}{2}.$$

We have therefore discovered the correspondence

$$\Psi_\pm \Psi_\mp = (\psi_\pm \sharp \psi_\mp)^{\mathrm{w}}(x,D).$$

Notice, moreover, that on $\mathscr{S}(\mathbb{R})$ and $\mathscr{S}'(\mathbb{R})$

$$\Psi_\pm^* = \Psi_\mp, \quad \Psi_\pm^* = (\overline{\psi_\pm})^{\mathrm{w}}(x,D),$$

and

$$[\Psi_+, \Psi_-] = (\psi_+ \sharp \psi_-)^{\mathrm{w}}(x,D) - (\psi_- \sharp \psi_+)^{\mathrm{w}}(x,D) = \frac{1}{i}\{\psi_+, \psi_-\}^{\mathrm{w}}(x,D). \quad (2.6)$$

Hence the fact that the operators Ψ_+ and Ψ_- do not commute may be explained by the fact that $\{\psi_+, \psi_-\} \neq 0$, which is hence a first level of non-commutativity when dealing with operators, due to the non-vanishing of the Poisson brackets $\{\xi_j, x_j\} = 1$.

Moreover, the fact that a commutator is a "lower order" operator will no longer hold true in general for matrix-valued symbols (see Remark 3.1.9 below). This is the second level at which non-commutativity manifests itself when considering systems of operators.

2.2 The Spectrum of the Harmonic Oscillator

We have now the following fundamental theorem about the spectrum of the harmonic oscillator when $n = 1$. The higher dimensional case will easily follow.

Theorem 2.2.1. *Let $H\colon D(H) \subset L^2(\mathbb{R}) \longrightarrow L^2(\mathbb{R})$, i.e. we think of H as an unbounded operator on L^2 with dense domain $D(H)$ given in (2.4), $Hu = p_0^{\mathrm{w}}(x,D)u$ in the sense of distributions. Then we have the following facts.*

1. *Self-adjointness: $H = H^*$, that is, $D(H) = D(H^*)$ and $H^*u = p_0^{\mathrm{w}}(x,D)u$ for all $u \in D(H)$.*
2. *$\mathrm{Spec}(H) = \{n + 1/2;\ n \in \mathbb{Z}_+\}$ (where $\mathbb{Z}_+ = \{0,1,2,\dots\}$), with multiplicity 1.*
3. *Define $v_0 = e^{-x^2/2}$ and $v_n = \Psi_+^n v_0$, $n \in \mathbb{Z}_+$. Then $\Psi_- v_0 = 0$, all the functions v_n belong to $\mathscr{S}$ and are of the form $v_n = f_n e^{-x^2/2}$, where $f_n \in \mathbb{R}[x]$ is a polynomial of degree n (i.e., of the form $a_n x^n + a_{n-1}x^{n-1} + \dots + a_0$, with non-zero leading coefficient a_n). Moreover, the v_n form a* **complete orthogonal** *system of L^2.*

Hence, upon defining $u_n = v_n/\|v_n\|$, the system $\{u_n\}_{n\in\mathbb{Z}_+}$ is an **orthonormal** *basis of L^2. The functions v_n are called* **Hermite functions.**

This theorem is well-known. However, we give the proof for the sake of completeness.

Proof. We shall denote throughout by $(\cdot,\cdot)$ the canonical inner product in L^2, and by $\|\cdot\|_0$ the induced L^2-norm.

Note, in the first place, that for any given $u_1, u_2 \in \mathscr{S}$ we have $(Hu_1, u_2) = (u_1, Hu_2)$ (i.e. H is symmetric, or, equivalently, $p_0^{\mathrm{w}}(x,D)$ is formally self-adjoint), and

$$(Hu,u) = \Big((\Psi_+\Psi_- + \frac{1}{2})u,u\Big) = \|\Psi_- u\|_0^2 + \frac{1}{2}\|u\|_0^2 \geq \frac{1}{2}\|u\|_0^2, \ \forall u \in \mathscr{S}.$$

On the other hand, since $\Psi_- v_0 = 0$ and $v_0 \in \mathscr{S}$, we indeed have that

$$\min_{u\in\mathscr{S}\setminus\{0\}} \frac{(Hu,u)}{\|u\|_0^2} = \frac{1}{2}.$$

Next, it is clear that the functions $v_n = \Psi_+^n v_0 \in \mathscr{S}$, for all $n \in \mathbb{Z}_+$. Hence we may compute by induction

$$\begin{aligned}
Hv_n &= (\Psi_+\Psi_- + \frac{1}{2})v_n = (\Psi_+\Psi_- + \frac{1}{2})\Psi_+ v_{n-1}\\
&= \Psi_+(\Psi_-\Psi_+ - \frac{1}{2})v_{n-1} + \Psi_+ v_{n-1}\\
&\quad \text{(using (2.5) and the induction hypothesis } Hv_{n-1} = (n-1+1/2)v_{n-1})\\
&= \Psi_+ Hv_{n-1} + \Psi_+ v_{n-1} = \Psi_+(n-1+\frac{1}{2})v_{n-1} + \Psi_+ v_{n-1}\\
&= (n+\frac{1}{2})\Psi_+ v_{n-1} = (n+\frac{1}{2})v_n.
\end{aligned}$$

Of course, we have to make sure that $v_n \not\equiv 0$ for all n. Hence, we compute by induction

$$\begin{aligned}
\|v_n\|_0^2 &= (v_n,v_n) = (\Psi_+^n v_0, \Psi_+^n v_0) = (\Psi_-^n \Psi_+^n v_0, v_0)\\
&= \Big(\Psi_-^{n-1}(\Psi_+\Psi_- + [\Psi_-,\Psi_+])\Psi_+^{n-1} v_0, v_0\Big) = \Big(\Psi_-^{n-1}(\Psi_+\Psi_- + 1)\Psi_+^{n-1} v_0, v_0\Big)\\
&= n(\Psi_-^{n-1}\Psi_+^{n-1} v_0, v_0) = n!\|v_0\|_0^2 > 0,
\end{aligned}$$

whence $v_n \not\equiv 0$ for all $n \in \mathbb{Z}_+$.

Now, since $\Psi_- v_0 = 0$, a similar computation also shows that

$$m > n \Longrightarrow (v_n, v_m) = 0,$$

that yields the orthogonality of the system $\{v_n\}_{n\in\mathbb{Z}_+}$. To prove its completeness, we note in the first place that it is clear that $v_n = f_n v_0$, for a real polynomial f_n of degree n (i.e., $f_n = a_n x^n + a_{n-1}x^{n-1} + \ldots + a_0$, with the $a_j \in \mathbb{R}$ and non-zero leading coefficient a_n). Hence, let $g \in L^2(\mathbb{R})$ be such that

$$\int_{-\infty}^{+\infty} g(x)v_n(x)dx = 0, \ \forall n \in \mathbb{Z}_+.$$

It then follows that, since any polynomial $f \in \mathbb{R}[x]$ can be written as a linear combination of the f_n (by virtue of the fact that their leading coefficients are not zero), we have

$$\int_{-\infty}^{+\infty} g(x)f(x)e^{-x^2/2}dx = 0, \ \forall f \in \mathbb{R}[x]. \tag{2.7}$$

Using $e^{-ix\xi} = \sum_{j\geq 0} \frac{(-ix\xi)^j}{j!}$ and (2.7) yields

$$\int_{-\infty}^{+\infty} e^{-ix\xi} g(x)e^{-x^2/2}dx = \mathscr{F}_{x\to\xi}(ge^{-x^2/2})(\xi) = 0, \ \forall \xi \in \mathbb{R}.$$

But an L^2-function which has zero Fourier-transform must be zero, that is

$$ge^{-x^2/2} \equiv 0 \Longrightarrow g \equiv 0 \ \text{in} \ L^2(\mathbb{R}).$$

Hence $\{v_n\}_{n\in\mathbb{Z}_+}$ is an **orthogonal basis** of $L^2(\mathbb{R})$ and $\{u_n\}_{n\in\mathbb{Z}_+}$, where $u_n := v_n/\|v_n\|_0$, is an **orthonormal basis** of $L^2(\mathbb{R})$.

Using this, one then sees that

$$\mathrm{Im}(H \pm i) = L^2(\mathbb{R}),$$

whence, by general arguments (see, for instance, Dunford-Schwartz [13] or Kato [35]), $H = H^*$.

This therefore shows that

$$\mathrm{Spec}(H) = \{n + \frac{1}{2};\ n \in \mathbb{Z}_+\},$$

with multiplicity 1, and concludes the proof. □

It is worth noting that the *non-commutativity* of the operators $\Psi_\pm$ allowed the operator H to be **positive** (i.e. the quadratic form $(H\cdot,\cdot)$ in L^2 is bounded from below by a positive constant) although its *symbol*, the function $p_0(x,\xi)$, is merely ≥ 0. We have this by virtue of the **uncertainty principle**, or equivalently, putting things geometrically, by virtue of the presence of symplectic variables (the pair $(x,\xi) \in \mathbb{R}^2$ satisfies $\{\xi, x\} = 1$) in the symbol, that makes $p_0^{-1}(0) (= \{0\}$ in this case) too small to support an eigenfunction.

How about higher dimensions? We start by putting

$$\Psi_{\pm}^{(j)} := \psi_{\pm}^{(j)\mathrm{w}}(x,D), \text{ where } \psi_{\pm}^{(j)}(x,\xi) = \frac{x_j \mp i\xi_j}{\sqrt{2}}, \ 1 \le j \le n$$

(note that when $n > 1$ we only have $|\psi_{\pm}^{(j)}(x,\xi)| \approx p_0(x_j,\xi_j)^{1/2}$ and no longer that $|\psi_{\pm}^{(j)}(x,\xi)| \approx p_{0,n}(x,\xi)^{1/2}$). Then, on $\mathscr{S}(\mathbb{R}^n)$ or $\mathscr{S}'(\mathbb{R}^n)$,

$$H = \sum_{j=1}^{n} (\Psi_{+}^{(j)}\Psi_{-}^{(j)} + \frac{1}{2}) = \sum_{j=1}^{n} (\Psi_{-}^{(j)}\Psi_{+}^{(j)} - \frac{1}{2}),$$

$$[\Psi_{+}^{(j)}, \Psi_{+}^{(j')}] = [\Psi_{-}^{(j)}, \Psi_{-}^{(j')}] = 0, \text{ and } [\Psi_{+}^{(j)}, \Psi_{-}^{(j')}] = -\delta_{jj'}.$$

Hence, if we define for $\alpha \in \mathbb{Z}_{+}^{n}$

$$u_\alpha(x) := \prod_{j=1}^{n} u_{\alpha_j}(x_j),$$

where u_{α_j} is given by Theorem 2.2.1, we have the following theorem.

Theorem 2.2.2.

1. *As an unbounded operator in $L^2(\mathbb{R}^n)$ with domain $B^2(\mathbb{R}^n)$ the operator H, the (maximal) realization of $p_{0,n}^{\mathrm{w}}(x,D)$, is self-adjoint;*
2. $\{u_\alpha\}_{\alpha \in \mathbb{Z}_{+}^{n}}$ *is a* **complete orthonormal** *system of $L^2(\mathbb{R}^n)$, made of eigenfunctions of H;*
3. $\mathrm{Spec}(H) = \{|\alpha| + n/2;\ \alpha \in \mathbb{Z}_{+}^{n}\}$ *(no longer with multiplicity 1 for $n > 1$; we shall examine the multiplicity later on, see Section 4.3).*

Chapter 3
The Weyl–Hörmander Calculus

In this chapter we shall review the Weyl-Hörmander pseudodifferential calculus (see Hörmander [29]; see also Hörmander [27]), in which we shall "embed" the "global" one (see Helffer [17] and Shubin [67]), to be recalled in Section 3.2. In section 3.3 we shall describe a few spectral properties of globally elliptic pseudodifferential operators.

3.1 Review of the Weyl–Hörmander Calculus

We shall denote by $X=(x,\xi)$, $Y=(y,\eta)$ and $Z=(z,\zeta)$ the points of $\mathbb{R}^{2n}=\mathbb{R}^n\times\mathbb{R}^n$.

Definition 3.1.1. An **admissible metric** in $\mathbb{R}^{2n}$ is a function $\mathbb{R}^{2n}\ni X\longmapsto g_X$ where g_X is a positive-definite quadratic form on $\mathbb{R}^{2n}$ such that:

- **Slowness:** There exists $C_0>0$ (the constant of **slowness**) such that for any given $X, Y\in\mathbb{R}^{2n}$ one has

$$g_X(Y-X)\le C_0^{-1}\Longrightarrow C_0^{-1}g_Y\le g_X\le C_0 g_Y;$$

- **Uncertainty:** For any given $X\in\mathbb{R}^{2n}$ one has

$$g_X\le g_X^\sigma,$$

  where g_X^σ is the **dual** metric (with respect to the symplectic form σ, see (2.1)) defined by

$$g_X^\sigma(Y)=\sup_{Z\neq 0}\frac{\sigma(Y,Z)^2}{g_X(Z)};$$

- **Temperateness:** There exists $C_1>0$ and $N_0\in\mathbb{Z}_+$ such that for all $X, Y\in\mathbb{R}^{2n}$ one has

$$g_X\le C_1 g_Y\Big(1+g_X^\sigma(X-Y)\Big)^{N_0}.$$

A. Parmeggiani, *Spectral Theory of Non-Commutative Harmonic Oscillators: An Introduction*, Lecture Notes in Mathematics 1992,
DOI 10.1007/978-3-642-11922-4_3,

The **Planck function** associated with g is by definition

$$h(X)^2 = \sup_{Z \neq 0} \frac{g_X(Z)}{g_X^\sigma(Z)}.$$

It is important to note that the uncertainty of g yields that $h(X) \leq 1$ for all $X \in \mathbb{R}^{2n}$.

Remark 3.1.2. One easily sees (exercise for the reader) that the metric

$$g_{x,\xi} = |dx|^2 + \frac{|d\xi|^2}{1+|\xi|^2}, \quad (x,\xi) \in \mathbb{R}^{2n},$$

is an admissible metric. In this case one has $h(x,\xi)^{-1} = (1+|\xi|^2)^{1/2}$, which is the usual pseudodifferential weight (which is then g-admissible; see Definition 3.1.4 below). △

Remark 3.1.3. The dual metric g^σ appears naturally to keep into account localization of Fourier transforms when dealing with the composition of symbols (see (3.5) below).

Notice that the uncertainty principle may also be rephrased as the basic observation that for the function $u_A(x) = e^{-\langle Ax,x\rangle}$, with A symmetric $n \times n$ and positive, the Fourier transform is $\hat{u}_A(\xi) = c(\det A)^{-1/2} e^{-\langle A^{-1}\xi,\xi\rangle}$, whence u_A is as much "concentrated" within the ellipsoid

$$\{x;\ \langle Ax,x\rangle \leq 1\},$$

as $\hat{u}_A(\xi)$ is "concentrated" within the ellipsoid

$$\{\xi;\ \langle A^{-1}\xi,\xi\rangle \leq 1\}.$$

As an example of computation of g^σ and h, consider the constant metric

$$g = \sum_{j=1}^{n} (\lambda_j dx_j^2 + \mu_j d\xi_j^2), \ \ \lambda_j, \mu_j > 0, \ 1 \leq j \leq n.$$

Then (exercise for the reader)

$$g^\sigma = \sum_{j=1}^{n} \Big(\frac{dx_j^2}{\mu_j} + \frac{d\xi_j^2}{\lambda_j}\Big),$$

and

$$h(X) = \max_{1 \leq j \leq n} (\lambda_j \mu_j)^{1/2}, \ \ \forall X \in \mathbb{R}^{2n}.$$

△

Definition 3.1.4. Given an admissible metric g, a g**-admissible weight** is a **positive** function m on $\mathbb{R}^{2n}$ (i.e. $m(X) > 0$ for all $X \in \mathbb{R}^{2n}$) for which there exist constants $c, C, C' > 0$ and $N_1 \in \mathbb{Z}_+$ such that for all $X, Y \in \mathbb{R}^{2n}$,

$$g_X(X - Y) \leq c \Longrightarrow C^{-1} \leq \frac{m(X)}{m(Y)} \leq C,$$

and

$$\frac{m(X)}{m(Y)} \leq C'\left(1 + g_X^\sigma(Y - X)\right)^{N_1}.$$

Remark 3.1.5. It is then seen (exercise for the reader) that the Planck function h is a g-admissible weight. △

Definition 3.1.6. Let g be an admissible metric and m be a g-admissible weight. Let $a \in C^\infty(\mathbb{R}^{2n})$. Denote by $a^{(k)}(X; v_1, \ldots, v_k)$ the k-th differential of a at X in the directions $v_1, \ldots, v_k$ of $\mathbb{R}^{2n}$ (thought of as tangential directions in $T_X\mathbb{R}^{2n}$, the tangent space of $\mathbb{R}^{2n}$ at the point X). Define

$$|a|_k^g(X) := \sup_{0 \neq v_1, \ldots, v_k \in \mathbb{R}^{2n}} \frac{|a^{(k)}(X; v_1, \ldots, v_k)|}{\prod_{j=1}^k g_X(v_j)^{1/2}}.$$

We say that $a \in S(m, g)$ if for any given integer $k \in \mathbb{Z}_+$ the following seminorms are finite:

$$\|a\|_{k;S(m,g)} := \sup_{\ell \leq k,\, X \in \mathbb{R}^{2n}} \frac{|a|_\ell^g(X)}{m(X)} < +\infty. \tag{3.1}$$

Given $\mu \in \mathbb{R}$, we shall say that $a \in S^\mu(g)$ if $a \in S(h^{-\mu}, g)$ (that is, we use the Planck function h as a weight for measuring the growth of the symbols).

With $B^g_{X_0,r} = \{X;\ g_{X_0}(X - X_0) < r^2\}$, following Bony and Lerner [3] we say that $a \in C^\infty(\mathbb{R}^{2n})$ is a **symbol (of weight** m**) confined to the ball** $B^g_{X_0,r}$, and write $a \in \mathrm{Conf}(m, g, X_0, r)$, if for all $k \in \mathbb{Z}_+$

$$\|a\|_{k,\mathrm{Conf}(m,g,X_0,r)} := \sup_{\ell \leq k,\, X \in \mathbb{R}^{2n}} \frac{|a|_\ell^{g_{X_0}}(X)}{m(X_0)}\left(1 + g_{X_0}^\sigma(X - B_{X_0,r})\right)^{k/2} < +\infty, \tag{3.2}$$

where $g_Y^\sigma(X - B) = \inf_{Z \in B} g_Y^\sigma(X - Z)$. Hence the space of symbols confined to the ball $B^g_{X_0,r}$ coincides with $\mathscr{S}(\mathbb{R}^{2n})$ endowed with the seminorms (3.2). Any given $\varphi \in C_0^\infty(B^g_{X_0,r})$ is automatically confined (of weight 1) to the ball $B^g_{X_0,r}$.

Remark 3.1.7. For the standard pseudodifferential metric

$$g_{x,\xi} = |dx|^2 + \frac{|d\xi|^2}{1 + |\xi|^2}, \quad (x, \xi) \in \mathbb{R}^{2n},$$

with $h(x, \xi)^{-1} = (1 + |\xi|^2)^{1/2}$, one has $S^\mu(g) = S^\mu_{1,0}(\mathbb{R}^n \times \mathbb{R}^n)$, the familiar class of symbols of order μ. △

Given a symbol $a \in S(m,g)$ one then defines its Weyl-quantization by the formula

$$a^{\mathrm{w}}(x,D)u(x) = (2\pi)^{-n} \iint e^{i\langle x-y,\xi\rangle} a(\frac{x+y}{2},\xi)u(y)dyd\xi, \;\; u \in \mathscr{S}(\mathbb{R}^n).$$

It is then seen (Hörmander [29]) that the expression makes sense, and that

$$A = a^{\mathrm{w}}(x,D)\colon \mathscr{S}(\mathbb{R}^n) \longrightarrow \mathscr{S}(\mathbb{R}^n)$$

and

$$A = a^{\mathrm{w}}(x,D)\colon \mathscr{S}'(\mathbb{R}^n) \longrightarrow \mathscr{S}'(\mathbb{R}^n)$$

is **continuous**, with Schwartz kernel given by the tempered distribution

$$\mathsf{K}_A(x,y) = (2\pi)^{-n} \int e^{i\langle x-y,\xi\rangle} a(\frac{x+y}{2},\xi)d\xi,$$

that is,

$$\mathsf{K}_A(x+t/2,x-t/2) = (2\pi)^{-n} \int e^{i\langle t,\xi\rangle} a(x,\xi)d\xi$$

is the inverse Fourier transform of a with respect to ξ, whence

$$a(x,\xi) = \int e^{-i\langle t,\xi\rangle} \mathsf{K}_A(x+t/2,x-t/2)dt.$$

For example, in the case of $a \in S^{\mu}(g)$, $g_{x,\xi} = |dx|^2 + |d\xi|^2/(1+|\xi|^2)$, the kernel K_A is interpreted as an **oscillatory integral**, that is

$$\mathsf{K}_A(x,y) = \mathscr{S}'\text{-}\lim_{\varepsilon\to 0+} (2\pi)^{-n} \int e^{i\langle x-y,\xi\rangle} a(\frac{x+y}{2},\xi)\chi(\varepsilon\xi)d\xi,$$

where χ is any given Schwartz function with $\chi(0) = 1$, the limit being independent of χ.

Recall that the expression for the Schwartz kernel comes from considering the "weak" form

$$\begin{aligned}(a^{\mathrm{w}}(x,D)u,v) &= (2\pi)^{-n} \iiint e^{i\langle x-y,\xi\rangle} a(\frac{x+y}{2},\xi)u(y)\overline{v(x)}dyd\xi dx \\ &= \langle \mathsf{K}_A | \bar{v} \otimes u\rangle_{\mathscr{S}',\mathscr{S}}, \;\; \forall u,v \in \mathscr{S}(\mathbb{R}^n),\end{aligned}$$

where $(v \otimes u)(x,y) = v(x)u(y)$ and $\langle\cdot|\cdot\rangle_{\mathscr{S}',\mathscr{S}}$ denotes the $\mathscr{S}'$-$\mathscr{S}$ duality.

One has the following composition theorem (see Hörmander [29] or [27]).

Theorem 3.1.8. *Given $a \in S(m_1,g)$, $b \in S(m_2,g)$ then*

$$a^{\mathrm{w}}(x,D)\, b^{\mathrm{w}}(x,D) = (a\sharp b)^{\mathrm{w}}(x,D),$$

where for any given $N \in \mathbb{Z}_+$

$$(a\sharp b)(X) = \sum_{j=0}^{N} \frac{1}{j!}\Big(\frac{i}{2}\sigma(D_X;D_Y)\Big)^j a(X)b(Y)\big|_{X=Y} + r_{N+1}(X), \tag{3.3}$$

with $r_{N+1} \in S(h^{N+1}m_1m_2, g)$. *In particular, for any given* $N \in \mathbb{Z}_+$,

$$\sigma(D_X;D_Y)^N a(X)b(Y)\big|_{X=Y} \in S(h^N m_1 m_2, g). \tag{3.4}$$

Here we write, by (2.2) with $X = (x,\xi)$ *and* $Y = (y,\eta)$,

$$\sigma(D_X;D_Y) := \sigma(D_x, D_\xi; D_y, D_\eta) := \langle D_\xi, D_y\rangle - \langle D_\eta, D_x\rangle.$$

This formula comes from considering in the first place the composition $a^{\mathrm{w}}b^{\mathrm{w}} = c^{\mathrm{w}}$ for symbols $a,b \in \mathscr{S}(\mathbb{R}^{2n})$ and, by using the relation

$$\iint f(x,y)e^{2ixy}dxdy = \frac{1}{4\pi}\iint \hat{f}(\xi,\eta)e^{-i\xi\eta/2}d\xi d\eta,\ f \in \mathscr{S}(\mathbb{R}^2),$$

through the formula

$$\begin{aligned} c(X) &= \pi^{-2n}\iint e^{-2i\sigma(X-Y,X-Z)}a(Y)b(Z)dYdZ \\ &= \pi^{-2n}\iint e^{-2i\sigma(T,Z)}a(X+T)b(X+Z)dTdZ \\ &= e^{i\sigma(D_X;D_Y)/2}(a(X)b(Y))\big|_{X=Y}, \end{aligned} \tag{3.5}$$

where (recall) $X = (x,\xi)$, $Y = (y,\eta)$, $Z = (z,\zeta)$. This is then extended to general symbols a and b of the kind considered in the theorem.

Remark 3.1.9. It follows from (3.3) that

$$a^{\mathrm{w}}(x,D)b^{\mathrm{w}}(x,D) = (ab)^{\mathrm{w}}(x,D) - \frac{i}{2}\{a,b\}^{\mathrm{w}}(x,D) + \ldots$$

Hence, for **scalar** symbols we have for the commutator

$$[a^{\mathrm{w}}(x,D), b^{\mathrm{w}}(x,D)] = \frac{1}{i}\{a,b\}^{\mathrm{w}}(x,D) + \ldots.$$

This is no longer true for **matrix-valued** symbols, simply because matrices do not commute. What one can say in this case is just that

$$[a^{\mathrm{w}}(x,D), b^{\mathrm{w}}(x,D)] = [a,b]^{\mathrm{w}}(x,D) - \frac{i}{2}\big(\{a,b\} - \{b,a\}\big)^{\mathrm{w}}(x,D) + \ldots,$$

where $[a,b] = ab - ba$ denotes the usual matrix-commutator. △

The next important result is the existence of a partition of unity associated with the metric g (see Hörmander [29] and Bony-Lerner [3]). Notice that no smoothness assumption on either the metric g or the g-admissible weights m was made. The existence of the partition of unity makes it possible to regularize the metric and the g-admissible weights.

Lemma 3.1.10. *Let g be an admissible metric, and let $r^2 < C_0^{-1}$. Then there exists a sequence of centers $\{X_\nu\}_{\nu\in\mathbb{Z}_+}$, a covering of $\mathbb{R}^{2n}$ made of g-balls $B^g_{\nu,r} = \{X; g_{X_\nu}(X - X_\nu) < r^2\}$ centered at X_ν and radius r, and a sequence of functions $\{\varphi_\nu\}$* **uniformly** *in $S(1,g)$, with $\operatorname{supp}\varphi_\nu \subset B^g_{\nu,r}$, such that $\sum_{\nu\in\mathbb{Z}_+}\varphi_\nu^2 = 1$. Moreover, for any given r_* such that $r^2 \le r_*^2 < C_0^{-1}$, there exists an integer N_{r_*} such that no more than N_{r_*} balls B^g_{ν,r_*} can intersect at each time, i.e. one has an a-priori finite number of overlappings of the* **dilates** *by r_*/r of the $B^g_{\nu,r}$; hence*

$$\forall E \subset \mathbb{N},\ \sharp E > N_{r_*} \Longrightarrow \bigcap_{\nu\in E} B^g_{\nu,r_*} = \emptyset.$$

In addition, with

$$g^\sigma_X(B - B') := \inf_{Y\in B,Y'\in B'} g^\sigma_X(Y - Y'),\ B, B' \subset \mathbb{R}^{2n},$$

and

$$\Delta_{\mu\nu}(r_*) := \max\left\{1, g^\sigma_{X_\mu}(B^g_{\mu,r_*} - B^g_{\nu,r_*}), g^\sigma_{X_\nu}(B^g_{\mu,r_*} - B^g_{\nu,r_*})\right\}^{1/2},$$

there exist constants $\tilde{N}$ and $\tilde{C}$ such that

$$\sup_\mu \sum_\nu \Delta_{\mu\nu}(r_*)^{-\tilde{N}} < \tilde{C}.$$

Moreover, for all $k \in \mathbb{Z}_+$ there exist $C > 0$ and $\ell \in \mathbb{Z}_+$ such that for any given $a \in S(m,g)$ and $b \in \mathrm{Conf}(1,g,X,r)$ one has

$$\|a\sharp b\|_{k,\mathrm{Conf}(1,g,X,r)} \le Cm(X)\|a\|_{\ell,S(m,g)}\|b\|_{\ell,\mathrm{Conf}(1,g,X,r)}. \tag{3.6}$$

Finally, for all $k, N \in \mathbb{Z}_+$ there exist $C = C_{k,N} > 0$ and $\ell \in \mathbb{Z}_+$ such that for every $\mu, \nu \in \mathbb{N}$ and every $a \in \mathrm{Conf}(1,g,X_\mu,r)$ and $b \in \mathrm{Conf}(1,g,X_\nu,r)$ one has

$$\begin{aligned}&\|a\sharp b\|_{k,\mathrm{Conf}(1,g,X_\mu,r)} + \|a\sharp b\|_{k,\mathrm{Conf}(1,g,X_\nu,r)}\\ &\qquad \le C\|a\|_{\ell,\mathrm{Conf}(1,g,X_\mu,r)}\|b\|_{\ell,\mathrm{Conf}(1,g,X_\nu,r)}\Delta_{\mu\nu}(r)^{-N}.\end{aligned} \tag{3.7}$$

One has also the following useful lemma, due to Bony and Lerner [3].

Lemma 3.1.11. *Let g be an admissible metric, and let m be a g-admissible weight. Let B_ν be a g-ball as in Lemma 3.1.10. Let $g_\nu = g_{X_\nu}$ and $m_\nu = m(X_\nu)$. Let $\{a_\nu\}_{\nu\in\mathbb{Z}_+}$ be a sequence of symbols with $a_\nu \in S(m_\nu, g_\nu)$, such that for any given integer $k \in \mathbb{Z}_+$*

$$\sup_{\nu\in\mathbb{Z}_+} \|a_\nu\|_{k,\mathrm{Conf}(m_\nu,g_\nu,X_\nu,r)} < +\infty.$$

Then $a := \sum_{\nu\in\mathbb{Z}_+} a_\nu$ belongs to $S(m,g)$. The sequence $\{a_\nu\}_{\nu\in\mathbb{Z}_+}$ is said to be **uniformly confined** *in $S(m,g)$. When $m=1$ we have from the Cotlar-Stein Lemma (see Hörmander [29], Lemma 18.6.5; see also Lemma 3.1.14 below) that $a^{\mathrm{w}} = \sum_\nu a_\nu^{\mathrm{w}}$ is a bounded operator in L^2.*

We next come to the action of the affine symplectic group. We have the following theorem (see Hörmander [29]).

Theorem 3.1.12 (Symplectic invariance). *For every affine symplectic transformation $\chi\colon \mathbb{R}^n\times\mathbb{R}^n \longrightarrow \mathbb{R}^n\times\mathbb{R}^n$ (i.e. $\chi^*\sigma=\sigma$, where $\chi^*\sigma$ is the pull-back of the 2-form σ by χ, defined by $(\chi^*\sigma)(v,w)=\sigma(\chi'v,\chi'w)$ where $\chi'\colon \mathbb{R}^{2n}\longrightarrow\mathbb{R}^{2n}$ is the tangent map associated with χ) there exists a unitary $U_\chi\colon L^2(\mathbb{R}^n)\longrightarrow L^2(\mathbb{R}^n)$,* **uniquely** *determined apart from a complex constant factor of modulus 1, such that for all $a\in S(m,g)$ one has*

$$U_\chi^{-1}a^{\mathrm{w}}(x,D)U_\chi = (a\circ\chi)^{\mathrm{w}}(x,D).$$

One has that U_χ is also an automorphism of $\mathscr{S}(\mathbb{R}^n)$ and $\mathscr{S}'(\mathbb{R}^n)$.

The proof of the theorem is obtained by considering the generators of the affine symplectic group of $\mathbb{R}^n\times\mathbb{R}^n$ and the associated **metaplectic operator** U_χ. One has that

- when $\chi\colon (x,\xi)\longmapsto(x+x_0,\xi)$, then $(U_\chi f)(x)=f(x-x_0)$;
- when $\chi\colon (x,\xi)\longmapsto(x,\xi+\xi_0)$, then $(U_\chi f)(x)=e^{i\langle x,\xi_0\rangle}f(x)$;
- when χ maps $(x_j,\xi_j)\longmapsto(\xi_j,-x_j)$ (for some j) keeping the other variables fixed, then U_χ is the normalized partial Fourier transform in the x_j variable $U_\chi f=(2\pi)^{-1/2}\mathscr{F}_{x_j\to\xi_j}f$;
- when $\chi\colon (x,\xi)\longmapsto(Tx,{}^tT^{-1}\xi)$, where $T\colon \mathbb{R}^n\longrightarrow\mathbb{R}^n$ is a linear isomorphism, then $(U_\chi f)(x)=|\det T|^{-1/2}f(T^{-1}x)$;
- when $\chi\colon (x,\xi)\longmapsto(x,\xi-Ax)$, for some $A={}^tA$, then $(U_\chi f)(x)=e^{-i\langle Ax,x\rangle/2}f(x)$.

We close this section by recalling the following fundamental result about the L^2-continuity.

Theorem 3.1.13. *Let $a\in S(m,g)$. Then $a^{\mathrm{w}}(x,D)\colon L^2(\mathbb{R}^n)\longrightarrow L^2(\mathbb{R}^n)$ is* **bounded** *iff m is* **bounded**. *Moreover, the operator $a^{\mathrm{w}}(x,D)$ is compact in $L^2(\mathbb{R}^n)$ iff $m\to 0$ at ∞. Finally, there exist constants $C>0$ and $k\in\mathbb{Z}_+$ (depending only on the dimension and the "structural" constants relative to g and m) such that*

$$\|a^{\mathrm{w}}\|_{L^2\to L^2}\le C\|a\|_{k;S(m,g)}.$$

When $a\in\mathscr{S}(\mathbb{R}^{2n})$ we have

$$\|a^{\mathrm{w}}\|_{L^2\to L^2}\le(2\pi)^{-2n}\|\hat{a}\|_{L^1}.$$

The proof of this theorem uses Lemma 3.1.10 and the very important Cotlar-Stein Lemma (see [29]), of which we give a proof for the sake of completeness.

Lemma 3.1.14. *Let H_1, H_2 be Hilbert spaces. Let $\{A_j\}_{j=1,\dots,+\infty}$ be linear bounded operators from H_1 to H_2 (i.e. they belong to $\mathscr{L}(H_1,H_2)$). Suppose that there exists a constant $M>0$ such that*

$$\sup_{j\in\mathbb{N}}\sum_{k=1}^{+\infty}\|A_j^*A_k\|^{1/2}\le M,\quad \sup_{j\in\mathbb{N}}\sum_{k=1}^{+\infty}\|A_jA_k^*\|^{1/2}\le M. \tag{3.8}$$

Then $Au:=\sum_{k=1}^{+\infty}(A_ku)$ exists for any given $u\in H_1$, with strong convergence in H_2, and for the norm of A we have $\|A\|\le M$.

Proof. We remark in the first place that if $T\in\mathscr{L}(H_1,H_2)$, then

$$\|T^*\|=\|T\|,\ \text{ and }\ \|T^*T\|=\|T\|^2, \tag{3.9}$$

so that, when $T=T^*$,

$$\|T^2\|=\|T\|^2.$$

In order to prove (3.9), one first notices that (with obvious notation)

$$\|T\|:=\sup_{\|f\|_{H_1}=\|g\|_{H_2}=1}|(Tf,g)_{H_2}|, \tag{3.10}$$

for on the one hand, for any given $\tilde f\in H_1$ and $\tilde g\in H_2$ with $\|\tilde f\|_{H_1}=\|\tilde g\|_{H_2}=1$, one has

$$|(T\tilde f,\tilde g)_{H_2}|\le \sup_{\|f\|_{H_1}=\|g\|_{H_2}=1}|(Tf,g)_{H_2}|,$$

whence

$$\|T\|\le \sup_{\|f\|_{H_1}=\|g\|_{H_2}=1}|(Tf,g)_{H_2}|,$$

because

$$\|T\tilde f\|_{H_2}=\sup_{\|g\|_{H_2}=1}|(T\tilde f,g)_{H_2}|\le \sup_{\|f\|_{H_1}=\|g\|_{H_2}=1}|(Tf,g)_{H_2}|,$$

and on the other

$$\sup_{\|f\|_{H_1}=\|g\|_{H_2}=1}|(Tf,g)_{H_2}|\le\|T\|.$$

Now, from (3.10) it follows that

$$\|T\|:=\sup_{\|f\|_{H_1}=\|g\|_{H_2}=1}|(Tf,g)_{H_2}|=\sup_{\|f\|_{H_1}=\|g\|_{H_2}=1}|(f,T^*g)_{H_1}|=\|T^*\|,$$

$$\|T^*T\|\le\|T^*\|\|T\|=\|T\|^2,$$

and

$$\|T\|^2 = \sup_{\|f\|_{H_1}=1} (Tf,Tf)_{H_2} = \sup_{\|f\|_{H_1}=1} |(T^*Tf,f)_{H_1}| \le \|T^*T\|$$

which proves (3.9).

The above remark yields $\|T\|^{2m} = \|(T^*T)^m\|$. Define now, for $N \in \mathbb{N}$, $\mathsf{A}_N := \sum_{k=1}^{N} A_k$. For the general term in $(\mathsf{A}_N^*\mathsf{A}_N)^m$ we have

$$\begin{aligned}&\|A_{j_1}^*A_{j_2}\dots A_{j_{2m-1}}^*A_{j_{2m}}\|\\ &\quad\le \min\{\|A_{j_1}^*A_{j_2}\|\dots\|A_{j_{2m-1}}^*A_{j_{2m}}\|, \|A_{j_1}^*\|\|A_{j_2}A_{j_3}^*\|\dots\|A_{j_{2m-2}}A_{j_{2m-1}}^*\|\|A_{j_{2m}}\|\}.\end{aligned}$$

Since $\|A_{j_1}^*\|, \|A_{j_{2m}}\| \le M$ by hypothesis, taking the geometric mean of the above bounds gives

$$\begin{aligned}\|\mathsf{A}_N\|^{2m} = \|(\mathsf{A}_N^*\mathsf{A}_N)^m\| &\le \sum_{j_1,j_2,\dots,j_{2m}=1}^{N} \|A_{j_1}^*A_{j_2}\dots A_{j_{2m-1}}^*A_{j_{2m}}\|\\ &\le M\sum_{j_1,j_2,\dots,j_{2m}=1}^{N} \|A_{j_1}^*A_{j_2}\|^{1/2}\|A_{j_2}A_{j_3}^*\|^{1/2}\dots\|A_{j_{2m-1}}^*A_{j_{2m}}\|^{1/2}\\ &\le M\sum_{j_1=1}^{N} M^{2m-1} = NM^{2m},\end{aligned}$$

where in the last inequality we have used (3.8). Thus

$$\|\mathsf{A}_N\| \le N^{1/(2m)}M \longrightarrow M \text{ as } m\to+\infty,\ \forall N\in\mathbb{N}. \tag{3.11}$$

We now show that the linear operator $A\colon H_1 \ni u \longmapsto Au := \sum_{k=1}^{+\infty}(A_ku) \in H_2$ is well-defined and bounded.

Suppose that $u = A_j^*v$ for some j and some $v\in H_2$. Then, since $\|A_kA_j^*\| = \|A_jA_k^*\| \le M\|A_jA_k^*\|^{1/2}$, we get that for $N_1+1\le N_2$

$$\|\mathsf{A}_{N_2}u - \mathsf{A}_{N_1}u\|_{H_2} = \|\sum_{k=N_1+1}^{N_2} A_kA_j^*v\|_{H_2} \le M\sum_{k=N_1+1}^{N_2} \|A_kA_j^*\|^{1/2}\|v\|_{H_2} \longrightarrow 0$$

as $N_1,N_2\to+\infty$, whence $\sum_{k=1}^{+\infty}(A_ku)$ exists in this case. It thus also follows that Au is also defined when $u\in\sum_{\text{finite}}\mathrm{Im}(A_j^*)$. Define now

$$W := \sum_{j=1}^{+\infty}\mathrm{Im}(A_j^*) := \{u\in H_1;\ \exists\{v_j\}_{j\in\mathbb{N}}\subset H_2, u = \lim_{N\to+\infty}\sum_{j=1}^{N}A_j^*v_j\}.$$

Define $u_N := \sum_{j=1}^{N} A_j^* v_j$. Then $\{u_N\}_N$ is a Cauchy-sequence in H_1. Moreover, from (3.11) one obtains that for any given $m \geq 1$,

$$\|\sum_{k=1}^{m} A_k(u_{N_1} - u_{N_2})\|_{H_2} \leq M\|u_{N_1} - u_{N_2}\|_{H_1}$$

uniformly in m, whence since Au_{N_1} and Au_{N_2} are defined (all N_1, N_2) one gets

$$\|Au_{N_1} - Au_{N_2}\|_{H_2} \leq M\|u_{N_1} - u_{N_2}\|_{H_1} \longrightarrow 0 \text{ as } N_1, N_2 \to +\infty.$$

Hence there exists

$$Au = \lim_{N\to+\infty} \sum_{k=1}^{+\infty} (A_k u_N), \quad \text{and} \quad \|\sum_{k=1}^{+\infty} A_k u\|_{H_2} \leq M\|u\|_{H_1}, \ \forall u \in W.$$

It follows that A is a bounded operator in W, whence it is also bounded in $\overline{W}$. Since

$$W^{\perp} = \overline{W}^{\perp} = \bigcap_{k\in\mathbb{N}} \mathrm{Ker}(A_k), \text{ one has } u \in W^{\perp} \implies Au = 0,$$

whence, as $H_1 = \overline{W} \oplus W^{\perp}$, we get $A \in \mathscr{L}(H_1, H_2)$, with $\|A\| \leq M$. □

In the case of matrix-valued and vector-valued symbols, Definitions 3.1.4 and 3.1.6, the composition formula (3.3) (keeping track, where necessary, of the order of the terms), Theorem 3.1.12 and the continuity theorem Theorem 3.1.13 all hold true. Upon denoting by M_N the set of $N \times N$ complex matrices, we shall write $S(m,g;\mathsf{M}_N) = S(m,g) \otimes \mathsf{M}_N$, and $S^{\mu}(g;\mathsf{M}_N) = S^{\mu}(g) \otimes \mathsf{M}_N$, for the matrix-valued analogue of the symbol spaces $S(m,g)$ and $S^{\mu}(g)$ considered above. In general, analogous notation will be used for the spaces $S(m,g;\mathsf{V}) = S(m,g) \otimes \mathsf{V}$ (where V is some real or complex finite-dimensional vector space) etc.

Remark 3.1.15. Note that for $a \in S(m,g)$ (or in general $a \in S(m,g;\mathsf{V})$), the **formal adjoint** (i.e. on $\mathscr{S}$ and $\mathscr{S}'$) of $a^{\mathrm{w}}(x,D)$ is

$$a^{\mathrm{w}}(x,D)^* = (a^*)^{\mathrm{w}}(x,D),$$

a^* being the complex conjugate $\bar{a}$ of a when a is scalar, or the linear form ${}^t\bar{a} \in \mathsf{V}^*$ when a is a vector in V, or the adjoint matrix $a^* = {}^t\bar{a}$ in case $\mathsf{V} = \mathsf{M}_N$. △

It is also useful to have the rule for passing from the Weyl-quantization to the ordinary quantization "to the right" (the so-called *left-quantization*: "taking derivatives before taking multiplications").

Theorem 3.1.16. *Let g be an admissible metric, and suppose that $g_{x,\xi}(y,\eta) = g_{x,\xi}(y,-\eta)$, for all (x,ξ), $(y,\eta) \in \mathbb{R}^{2n}$. Let m be a g-admissible weight. Then* $\exp\langle i\kappa D_x, D_\xi\rangle$ *is a weakly continuous isomorphism of $S(m,g)$ for every $\kappa \in \mathbb{R}$,*

$$e^{i\kappa\langle D_x,D_\xi\rangle}a(x,\xi)-\sum_{j=0}^{N}\frac{\langle i\kappa D_x,D_\xi\rangle^j}{j!}a(x,\xi)\in S(h^{N+1}m,g),$$

for all $N\in\mathbb{Z}_+$. *Hence, given* $a,b\in S(m,g)$ *we have*

$$a^{\mathrm{w}}(x,D)=b(x,D),$$

with

$$b(x,D)u(x)=(2\pi)^{-n}\iint e^{i\langle x-y,\xi\rangle}b(x,\xi)u(y)dyd\xi,\ u\in\mathscr{S},$$

where

$$b(x,\xi)=e^{i\langle D_x,D_\xi\rangle/2}a(x,\xi),\quad a(x,\xi)=e^{-i\langle D_x,D_\xi\rangle/2}b(x,\xi).\tag{3.12}$$

Equalities (3.12) follow in a way similar to that of the composition formula (3.3).

Remark 3.1.17. In Hörmander [29, p. 159], Theorem 3.1.16 is stated in a more general form, and takes care also of the relation of the Weyl-quantization to the so-called *right-quantization* ("taking derivatives after taking multiplications"). △

3.2 Global Metrics and Global Pseudodifferential Operators

We now consider the Weyl-Hörmander pseudodifferential calculus modelled after the harmonic oscillator.

Let us hence consider, for $X\in\mathbb{R}^{2n}$,

$$m(X):=(1+|X|^2)^{1/2},\ g_X=\frac{|dX|^2}{m(X)^2},\tag{3.13}$$

so that

$$g^\sigma_X=m(X)^2|dX|^2,\ h(X)=m(X)^{-2}.$$

The metric g is called the **global metric**. Let us check that g is indeed admissible, and that m and m^{-1} are g-admissible.

- *The metric g is slowly varying.* Let $c_0\in(0,1/2)$ to be picked, and suppose $g_X(Y-X)\le c_0$, that is, $|Y-X|^2\le c_0 m(X)^2$. Then, on the one hand

$$\begin{aligned}m(Y)^2&=1+|Y|^2\le 1+(|X|+|X-Y|)^2\\&\le 1+2|X|^2+2|X-Y|^2\le 2(c_0+1)m(X)^2,\end{aligned}$$

  and on the other

$$m(X)^2=1+|X|^2\le 1+2|Y|^2+2|X-Y|^2\le 2m(Y)^2+2c_0m(X)^2,$$

that is, in the end,

$$\frac{1-2c_0}{2}m(X)^2 \le m(Y)^2 \le 2(c_0+1)m(X)^2,$$

or, equivalently,

$$\frac{1-2c_0}{2}g_Y \le g_X \le 2(c_0+1)g_Y.$$

We may hence pick $c_0 = 1/4$ (say), and get that g is slowly varying with slowness constant $C_0^{-1} = 1/10$.

- *Uncertainty.* It is trivial:

$$g_X = \frac{|dX|^2}{m(X)^2} \le m(X)^2|dX|^2 = g_X^\sigma.$$

- *Temperateness.* We must find universal constants $C_1 > 0$ and $N_0 \in \mathbb{Z}_+$ such that

$$g_X(Z) \le C_1 g_Y(Z)\Big(1+g_X^\sigma(X-Y)\Big)^{N_0}, \quad \forall X,Y,Z \in \mathbb{R}^{2n},$$

that is to say,

$$\frac{|Z|^2}{m(X)^2} \le C_1\frac{|Z|^2}{m(Y)^2}\Big(1+m(X)^2|X-Y|^2\Big)^{N_0}.$$

We hence consider $m(Y)^2/m(X)^2$. Since

$$\begin{aligned}\frac{m(Y)^2}{m(X)^2} &\le \frac{1+2|X|^2+2|X-Y|^2}{m(X)^2} \le 2\frac{m(X)^2+|X-Y|^2}{m(X)^2}\\ &\le 2\Big(1+m(X)^2|X-Y|^2\Big),\end{aligned} \tag{3.14}$$

as $m(X) \ge 1$ for all X, the temperateness follows.
Hence g is admissible.

- To prove that m and m^{-1} are g-admissible, note that the first part of the above argument shows also that

$$g_X(Y-X) \le C_0^{-1} \Longrightarrow m(X) \approx m(Y),$$

that

$$\frac{m(Y)}{m(X)} \lesssim 1+g_X^\sigma(X-Y),$$

and that

$$\begin{aligned}\frac{m(X)^2}{m(Y)^2} &\le \frac{1+2|Y|^2+2|X-Y|^2}{m(Y)^2} \le 2\Big(1+\frac{|X-Y|^2}{m(Y)^2}\Big)\\ &\le 2\Big(1+m(X)^2|X-Y|^2\Big),\end{aligned} \tag{3.15}$$

for one has $m(Y)^{-1} \leq m(X)$ for all X, Y. Inequalities (3.14) and (3.15) prove that m and m^{-1} are g-admissible weights.

The weight m is called the **global weight**.

This shows that we may consider the class $S(m^\mu, g)$, $\mu \in \mathbb{R}$. Hence $a \in S(m^\mu, g)$ if for all $\alpha \in \mathbb{Z}_+^{2n}$ there exists $C_\alpha > 0$ such that

$$|\partial_X^\alpha a(X)| \leq C_\alpha m(X)^{\mu - |\alpha|}, \ \forall X \in \mathbb{R}^{2n}.$$

Definition 3.2.1. When $a \in S(m^\mu, g)$ we shall say that a has order μ.

Next, we introduce the basic notion of *elliptic* elements.

Definition 3.2.2. A symbol $a \in S(m^\mu, g)$ is said to be **globally elliptic** when there exist constants $c, C > 0$ such that

$$|X| \geq C \Longrightarrow |a(X)| \geq c\, m(X)^\mu.$$

Hence, we may say that not only does the symbol $a(X)$ grow at most like $m(X)^\mu$, but that for large X it is **equivalent** to the weight $m(X)^\mu$.

The next definition is introduced for keeping track of the important class of differential operators with polynomial coefficients, that are Weyl-quantizations of polynomials of the kind

$$\sum_{j=0}^{d} \sum_{|\alpha|+|\beta|=2d-2j} a_{\alpha\beta} x^\alpha \xi^\beta, \ a_{\alpha\beta} \in \mathbb{C}.$$

The grading $2d - 2j$, in place of the more general one $2d - j$, is *natural*, when considering spectral properties of Weyl-quantizations of *degree* 2 polynomials in $(x, \xi) \in \mathbb{R}^{2n}$.

Definition 3.2.3. We say that a symbol $a \in S(m^\mu, g)$ is **classical**, and write $a \in S_{\mathrm{cl}}(m^\mu, g)$, if there exists a sequence $\{a_{\mu-2j}\}_{j \geq 0} \subset C^\infty(\mathbb{R}^{2n} \setminus \{0\})$ such that for all $j \geq 0$

$$a_{\mu-2j}(tX) = t^{\mu-2j} a_{\mu-2j}(X), \ \forall t > 0, \ \forall X \neq 0$$

(that is, the $a_{\mu-2j}$ are **positively homogeneous** of degree $\mu - 2j$), and for any given $N \in \mathbb{Z}_+$

$$a(X) - \chi(X) \sum_{j=0}^{N} a_{\mu-2j}(X) \in S(m^{\mu-2(N+1)}, g),$$

χ being some excision function, that is a function $0 \leq \chi \leq 1$ that is supported away from 0, for instance $\chi \equiv 0$ for $|X| \leq 1/2$, and $\chi \equiv 1$ for $|X| \geq 1$. We shall write

$$a \sim \sum_{j \geq 0} a_{\mu-2j}.$$

Remark 3.2.4. More generally, one may consider also **semi-regular classical** symbols $a \in S(m^\mu, g)$, that is symbols for which there exists a sequence $\{a_{\mu-j}\}_{j\geq 0} \subset C^\infty(\mathbb{R}^{2n} \setminus \{0\})$ such that for all $j \geq 0$

$$a_{\mu-j}(tX) = t^{\mu-j} a_{\mu-j}(X), \ \forall t > 0, \ \forall X \neq 0,$$

and

$$a \sim \sum_{j\geq 0} a_{\mu-j}.$$

We shall write $a \in S_{\mathrm{srcl}}(m^\mu, g)$. However, our interest will rest mainly on **classical** symbols. △

Remark 3.2.5. One may also define classical symbols by requiring the sequence $\{a_{\mu-2j}\}_{j\geq 0}$ to be smooth on the whole $\mathbb{R}^{2n}$, but with the positive homogeneity property holding for all $t > 1$ and $|X| > 1$ only. This does not make any difference in the theory (it merely avoids the use of excision functions) and it is just a matter of taste.

△

Note hence that, with $d \in \mathbb{Z}_+$,

$$\sum_{j=0}^{d} \sum_{|\alpha|+|\beta|=2d-2j} a_{\alpha\beta} x^\alpha \xi^\beta \in S_{\mathrm{cl}}(m^{2d}, g),$$

whereas

$$\sum_{j=0}^{d} \sum_{|\alpha|+|\beta|=2d-j} a_{\alpha\beta} x^\alpha \xi^\beta \in S_{\mathrm{srcl}}(m^{2d}, g).$$

Definition 3.2.6. Let $\mu \in \mathbb{Z}_+$. A classical symbol $a \in S_{\mathrm{cl}}(m^\mu, g)$ is a **global polynomial differential** (GPD for short) symbol of order μ if

$$a = \sum_{j=0}^{[\mu/2]} a_{\mu-2j}$$

(with $[\mu/2]$ denoting, as usual, the integer part of $\mu/2$), where the entries of the $a_{\mu-2j}$ are **homogeneous polynomials** in $X \in \mathbb{R}^{2n}$ of degree $\mu - 2j$.

A **global polynomial differential operator** (GPDO for short) of order μ is the Weyl-quantization of a GPD symbol of order μ.

To recognize globally elliptic *classical* symbols, it is very useful to have the following lemma.

Lemma 3.2.7. *Let $a \in S_{\mathrm{cl}}(m^\mu, g)$. Then*

$$\min_{|X|=1} |a_\mu(X)| > 0, \tag{3.16}$$

iff a is globally elliptic.

Proof. Let χ be an excision function as in Definition 3.2.3. Then

$$a(X) = \chi(X)a_\mu(X) + O(m(X)^{\mu-2}).$$

We may hence take $|X| \geq 1$, so that in the above identity $\chi = 1$. Now, for $|X| \geq 1$ we also have $m(X) \approx |X|$, whence, using the homogeneity of a_μ we get

$$a(X) = m(X)^\mu \Big(\frac{|X|^\mu}{m(X)^\mu} a_\mu(X/|X|) + O(m(X)^{-2})\Big), \quad |X| \geq 1.$$

Hence

$$a(X) \approx m(X)^\mu, \quad \forall X \text{ with } |X| \gtrsim 1,$$

iff (3.16) holds. □

Lemma 3.2.7 then induces the following definition.

Definition 3.2.8. We say that $a \in S_{\mathrm{cl}}(m^\mu, g)$ is globally elliptic when (3.16) holds.

In the vector-valued case, we will use the following definition of globally elliptic symbols.

Definition 3.2.9. Let V be a finite-dimensional vector space (complex or real). We denote by $S(m^\mu, g; \mathsf{V})$, resp. $S_{\mathrm{cl}}(m^\mu, g; \mathsf{V})$, the class of vector-valued symbols, resp. vector-valued classical symbols. In other words, we have $S(m^\mu, g; \mathsf{V}) = S(m^\mu, g) \otimes \mathsf{V}$, and likewise for $S_{\mathrm{cl}}(m^\mu, g; \mathsf{V})$. In particular, when $\mathsf{V} = \mathsf{M}_N$, having that $a \in S(m^\mu, g; \mathsf{M}_N)$, resp. $S_{\mathrm{cl}}(m^\mu, g; \mathsf{M}_N)$, means that **each entry** a_{jk} of the matrix a belongs to $S(m^\mu, g)$, resp. $S_{\mathrm{cl}}(m^\mu, g)$, and having an $N \times N$ **global polynomial differential system** of order μ means having a symbol $a \in S_{\mathrm{cl}}(m^\mu, g; \mathsf{M}_N)$ such that each entry is a GPD symbol of order μ (see Definition 3.2.6).

We shall say that a symbol $a \in S(m^\mu, g; \mathsf{M}_N)$ is **globally elliptic** if there exist constants $c, C > 0$ such that

$$|\det a(X)| \geq c\, m(X)^{\mu N}, \quad \text{whenever } |X| \geq C.$$

Equivalently, when $a \in S(m^\mu, g; \mathscr{L}(\mathsf{V}, \mathsf{W}))$, where W is another finite-dimensional vector space, we require (for some norms $|\cdot|_{\mathsf{V}}$ in V and $|\cdot|_{\mathsf{W}}$ in W)

$$|a(X)v|_{\mathsf{W}} \approx m(X)^\mu |v|_{\mathsf{V}}, \quad \forall v \in \mathsf{V}, \forall X \text{ with } |X| \gtrsim 1,$$

and when $a = a^* \in S(m^\mu, g; \mathsf{M}_N)$ is positive-definite or negative-definite as an Hermitian matrix, we may equivalently require

$$|\langle a(X)v, v\rangle_{\mathbb{C}^N}| \approx m(X)^\mu |v|^2_{\mathbb{C}^N}, \quad \forall v \in \mathbb{C}^N, \ \forall X \text{ with } |X| \gtrsim 1.$$

When $a \in S_{\mathrm{cl}}(m^\mu, g; \mathsf{M}_N)$ is classical, we shall then require

$$\min_{|X|=1} |\det a_\mu(X)| > 0,$$

or, equivalently,

$$|a_\mu(X)v|_{\mathbb{C}^N} \approx |v|_{\mathbb{C}^N}, \ \forall v \in \mathbb{C}^N, \forall X \text{ with } |X| = 1,$$

and when $a_\mu = a_\mu^*$ is positive/negative-definite as an Hermitian matrix,

$$|\langle a_\mu(X)v, v\rangle_{\mathbb{C}^N}| \approx |v|^2_{\mathbb{C}^N}, \ \forall v \in \mathbb{C}^N, \ \forall X \text{ with } |X| = 1.$$

Analogous requirements for classical symbols are possible in the case one has $\mathsf{V} = \mathscr{L}(\mathbb{C}^{N_1}, \mathbb{C}^{N_2})$ etc. (of course det can be used only where meaningful).

We shall say that a symbol $a = a^* \in S(m^\mu, g; \mathsf{M}_N)$ is **globally positive elliptic** if

$$\langle a(X)v, v\rangle_{\mathbb{C}^N} \approx m(X)^\mu |v|^2_{\mathbb{C}^N}, \ \forall v \in \mathbb{C}^N, \forall X \text{ with } |X| \gtrsim 1.$$

When $a = a^* \in S_{\mathrm{cl}}(m^\mu, g; \mathsf{M}_N)$, we shall say that a is **globally positive elliptic** if

$$\langle a_\mu(X)v, v\rangle_{\mathbb{C}^N} \approx |v|^2_{\mathbb{C}^N}, \ \forall v \in \mathbb{C}^N, \ \forall X \text{ with } |X| = 1. \tag{3.17}$$

We shall say that a matrix-valued GPDO of order μ is **elliptic** (resp. **positive elliptic**) if its principal symbol a_μ is globally elliptic (resp. positive elliptic).

Remark 3.2.10. Recall that

$$a(X)^{-1} = \frac{1}{\det a(X)}\, {}^{\mathrm{co}}a(X),$$

where ${}^{\mathrm{co}}a \in S(m^{(N-1)\mu}, g; \mathsf{M}_N)$ is the cofactor matrix (i.e. the transpose of the the matrix whose ij-entry is the algebraic cofactor of the ij-entry of a) associated with $a(X)$. One then sees that, equivalently:

- $a \in S(m^\mu, g; \mathsf{M}_N)$ is globally elliptic if $a(X)^{-1}$ exists for all $X \in \mathbb{R}^{2n}$ with $|X| \gtrsim 1$ and $\chi a^{-1} \in S(m^{-\mu}, g; \mathsf{M}_N)$, where χ is some excision function;
- $a \in S_{\mathrm{cl}}(m^\mu, g; \mathsf{M}_N)$ is globally elliptic if $a_\mu(X)^{-1}$ exists for all $X \in \mathbb{R}^{2n}$ with $|X| = 1$. △

Definition 3.2.11. A **non-commutative harmonic oscillator** (NCHO for short) is the Weyl-quantization of any given 2nd-order $N \times N$ GPD system $a \in S_{\mathrm{cl}}(m^2, g; \mathsf{M}_N)$. Hence $a = a_2 + a_0$, where a_2 is a matrix whose entries are homogeneous quadratic forms in $X = (x, \xi) \in \mathbb{R}^{2n}$, and a_0 is a constant matrix.

We shall be particularly interested in the following two-parameter family of NCHOs. Let $\alpha, \beta \in \mathbb{C}$, $J = \begin{bmatrix} 0 & -1 \\ 1 & 0 \end{bmatrix}$, and

$$Q_{(\alpha,\beta)}(x, \xi) = \begin{bmatrix} \alpha & 0 \\ 0 & \beta \end{bmatrix} \frac{x^2 + \xi^2}{2} + iJx\xi, \ x, \xi \in \mathbb{R}. \tag{3.18}$$

Hence, $Q_{(\alpha,\beta)} \in S_{\mathrm{cl}}(m^2, g; \mathsf{M}_2)$ and the NCHO $Q^{\mathrm{w}}_{(\alpha,\beta)}(x,D)$ is then the system of GPDOs of order 2

$$Q^{\mathrm{w}}_{(\alpha,\beta)}(x,D) = \begin{bmatrix} \alpha & 0 \\ 0 & \beta \end{bmatrix} \Big(-\frac{\partial_x^2}{2} + \frac{x^2}{2}\Big) + J\Big(x\partial_x + \frac{1}{2}\Big). \tag{3.19}$$

A NCHO is **elliptic** (resp. **positive elliptic**) when it is elliptic (resp. positive elliptic) as a GPDO.

Remark 3.2.12. The NCHO $Q_{(\alpha,\beta)}(x,\xi)$, for $\alpha, \beta > 0$ and $\alpha\beta > 1$ is **positive elliptic**. (We leave it to the reader to check this claim.) △

It is important to make the following observation on **formally self-adjoint positive elliptic** $N \times N$ systems of GPDOs.

Lemma 3.2.13. *Let $a = a^* \in S_{\mathrm{cl}}(m^\mu, g; \mathsf{M}_N)$ be a globally positive elliptic* GPD *system. Then $\mu \in \mathbb{Z}_+$ is* **even**. *Hence $a = a_\mu + a_{\mu-2} + \ldots + a_0$, where a_0 is an $N \times N$* **constant** *Hermitian matrix, and all the $a_{\mu-2j}$, $0 \le j \le (\mu-2)/2$, are $N \times N$ Hermitian matrices.*

Proof. By (3.17), taking $v = e_1$ (the first vector of the canonical basis of $\mathbb{C}^N$), we have for the 11-entry of a_μ

$$a_{\mu,11}(X) \approx |X|^\mu, \ \ \forall X \in \mathbb{R}^{2n}.$$

Hence $a_{\mu,11}$ is a *nonnegative homogeneous polynomial* in $X \in \mathbb{R}^{2n}$ of degree μ, which is *positive* when $X \neq 0$. Hence μ must be an *even* (nonnegative) integer. □

It is also important to define the *smoothing* elements.

Definition 3.2.14. A symbol

$$a \in \bigcap_{\mu \in \mathbb{R}} S(m^\mu, g) =: S(m^{-\infty}, g) = \mathscr{S}(\mathbb{R}^n \times \mathbb{R}^n)$$

is said to be a **smoothing** symbol. In fact, in this case

$$a^{\mathrm{w}}(x,D) \colon \mathscr{S}'(\mathbb{R}^n) \longrightarrow \mathscr{S}(\mathbb{R}^n)$$

is **continuous**, and for its Schwartz kernel $\mathsf{K}_{a^{\mathrm{w}}}$ one has $\mathsf{K}_{a^{\mathrm{w}}} \in \mathscr{S}(\mathbb{R}^n \times \mathbb{R}^n)$. Analogously in the matrix-valued case.

In general, any continuous map $R \colon \mathscr{S}'(\mathbb{R}^n; \mathbb{C}^N) \longrightarrow \mathscr{S}(\mathbb{R}^n; \mathbb{C}^N)$ is called **a smoothing operator**. Equivalently, for the Schwartz kernel K_R of a smoothing operator R we have $\mathsf{K}_R \in \mathscr{S}(\mathbb{R}^n \times \mathbb{R}^n; \mathsf{M}_N)$.

We next have the following useful result (of basic importance when constructing a *parametrix* of an elliptic operator in these global classes).

Proposition 3.2.15. *Let $\mu_j \searrow -\infty$, $\mu_j > \mu_{j+1}$, $j \in \mathbb{N}$, be a monotone strictly decreasing sequence of real numbers. Let $a_j \in S(m^{\mu_j}, g)$, $j \in \mathbb{N}$. Then there exists $a \in S(m^{\mu_1}, g)$ such that*

$$a \sim \sum_{j\geq 1} a_j,$$

that is, for all $r \in \mathbb{N}$ we have

$$a - \sum_{j=1}^{r} a_j \in S(m^{\mu_{r+1}}, g).$$

If another a' has the same property, then $a - a' \in \mathscr{S}(\mathbb{R}^{2n})$.

Proof. Let χ be an excision function, with $0 \leq \chi \leq 1$, such that $\chi(X) = 0$ if $|X| \leq 1/2$ and $\chi(X) = 1$ if $|X| \geq 1$. In the first place we show that we can choose a monotone strictly increasing sequence of positive $R_j \to +\infty$, increasing so quickly as $j \to +\infty$ that for any given $j \geq 2$ and for all $\alpha \in \mathbb{Z}^{2n}_{+}$ with $|\alpha| \leq j$,

$$\left|\partial_X^\alpha \Big(\chi(X/R_j)a_j(X)\Big)\right| \leq 2^{-j} m(X)^{\mu_j+1-|\alpha|}. \tag{3.20}$$

To see this, note that

$$|\partial_X^\alpha(\chi(X/R))| \leq C_\alpha m(X)^{-|\alpha|}, \quad \text{for } R \geq 1. \tag{3.21}$$

In fact,

$$\partial_X^\alpha(\chi(X/R)) = R^{-|\alpha|}(\partial_X^\alpha \chi)(X/R),$$

and

$$|\alpha| \geq 1,\ X \in \operatorname{supp}(\partial_X^\alpha \chi)(\cdot/R) \Longrightarrow R/2 \leq |X| \leq R,$$

from which (3.21) follows, and shows that $\chi(\cdot/R) \in S(1, g)$ **uniformly** in $R \geq 1$. Then for all $\alpha \in \mathbb{Z}^{2n}_{+}$

$$\left|\partial_X^\alpha \Big(\chi(X/R)a_j(X)\Big)\right| \leq C_{j,\alpha} m(X)^{\mu_j-|\alpha|}, \tag{3.22}$$

if $R \geq 1$, whence

$$\chi(\cdot/R)a_j \in S(m^{\mu_j}, g), \quad j \in \mathbb{N}, \quad R \geq 1.$$

Of course, this is seen also by noting that $\chi(X/R)a_j(X) = a_j(X)$ for X large, for

$$\operatorname{supp}\Big(\chi(\cdot/R)a_j\Big) \subset \{X \in \mathbb{R}^{2n};\ |X| \geq R/2\}, \quad \forall j \in \mathbb{N}. \tag{3.23}$$

Now, given any $j \geq 1$, if $R \geq 1$ **and** $|\alpha| \leq j$ we have

$$\left|\partial_X^\alpha \Big(\chi(X/R)a_j(X)\Big)\right| \leq \max_{|\alpha|\leq j}\{C_{j,\alpha}\} m(X)^{\mu_j-|\alpha|} =: C_j m(X)^{\mu_j-|\alpha|}. \tag{3.24}$$

On the other hand,

$$m(X)^{\mu_j-|\alpha|} \le \varepsilon\, m(X)^{\mu_j+1-|\alpha|},$$

when X is such that

$$m(X) = (1+|X|^2)^{1/2} \ge \frac{1}{\varepsilon}.$$

So, to satisfy (3.20), it suffices to choose $\varepsilon = 1/(2^j C_j)$, and take

$$(1+R_j^2/4)^{1/2} \ge 2^j C_j, \quad j \ge 2.$$

That is, it suffices to take

$$R_j \ge 2^{j+1} C_j, \quad j \ge 2.$$

Hence we may choose

$$R_1 = 1, \quad R_j = 2^{j+1}(C_j+1) + R_{j-1}, \; j \ge 2.$$

Now, (3.23) and $R_j \nearrow +\infty$ yield that the sum

$$a(X) := \sum_{j\ge 1} \chi(X/R_j) a_j(X)$$

is locally finite, hence $a \in C^\infty$. On the other hand, given any $r \in \mathbb{N}$ and any $\alpha \in \mathbb{Z}_+^{2n}$, we may find $N \in \mathbb{N}$ so large that $|\alpha| \le N+1$ and $\mu_{N+1} + 1 \le \mu_r$. Hence

$$\begin{aligned}\Big|\partial_X^\alpha \Big(a(X) - \sum_{j=1}^{N} \chi(X/R_j)a_j(X)\Big)\Big| &= \Big|\partial_X^\alpha \Big(\sum_{j=N+1}^{\infty} \chi(X/R_j)a_j(X)\Big)\Big| \\ &\le \sum_{j=N+1}^{\infty} \frac{1}{2^j} m(X)^{\mu_j+1-|\alpha|} \\ &\le 2^{-N} m(X)^{\mu_r-|\alpha|} \qquad (3.25)\end{aligned}$$

(recall that $\mu_j > \mu_{j+1}$). Therefore, by choosing $r = 1$ we obtain that for any given $\alpha \in \mathbb{Z}_+^{2n}$ (with N large depending on α as above)

$$\partial_X^\alpha a(X) = \partial_X^\alpha \Big(a(X) - \sum_{j=1}^{N} \chi(X/R_j)a_j(X)\Big) + \partial_X^\alpha \Big(\sum_{j=1}^{N} \chi(X/R_j)a_j(X)\Big),$$

whence, by (3.22) and (3.25),

$$|\partial_X^\alpha a(X)| \le \frac{1}{2^N} m(X)^{\mu_1-|\alpha|} + \sum_{j=1}^{N} C_{j,\alpha} m(X)^{\mu_j-|\alpha|} \le C_\alpha m(X)^{\mu_1-|\alpha|},$$

that is $a \in S(m^{\mu_1}, g)$.

On the other hand, we also obtain from (3.25) that for any given $r \in \mathbb{N}$ and any given $\alpha \in \mathbb{Z}_+^{2n}$ with $N \geq r+1$ so large depending on α and r that $|\alpha| \leq N+1$ and $\mu_{N+1} + 1 \leq \mu_{r+1}$,

$$\begin{aligned}\Big|\partial_X^\alpha\Big(a(X) - \sum_{j=1}^{r} a_j(X)\Big)\Big| &\leq \Big|\partial_X^\alpha\Big(a(X) - \sum_{j=1}^{N} \chi(X/R_j)a_j(X)\Big)\Big| \\ &\quad + \sum_{j=1}^{r}\Big|\partial_X^\alpha\Big((1-\chi(X/R_j))a_j(X)\Big)\Big| \\ &\quad + \sum_{j=r+1}^{N}\Big|\partial_X^\alpha\Big(\chi(X/R_j)a_j(X)\Big)\Big| \\ &\leq C_{\alpha,r} m(X)^{\mu_{r+1}-|\alpha|},\end{aligned}$$

for we have

$$(1-\chi(\cdot/R_j))a_j \in S(m^{-\infty}, g), \ \forall j \in \mathbb{N},$$

and

$$\chi(\cdot/R_j)a_j \in S(m^{\mu_{r+1}}, g), \ \forall j \geq r+1.$$

This shows that $a \sim \sum_{j\geq 1} a_j$.

Finally, if $a' \in S(m^{\mu_1}, g)$ has this last property, then for all $r \in \mathbb{N}$,

$$a - a' = \Big(a - \sum_{j=1}^{r} a_j\Big) - \Big(a' - \sum_{j=1}^{r} a_j\Big) \in S(m^{\mu_{r+1}}, g),$$

that is $a - a' \in S(m^{-\infty}, g) = \mathscr{S}(\mathbb{R}^{2n})$, which concludes the proof. □

The following proposition (that we state without proof, for which we address the reader to Helffer [17] or Shubin [67]) is also useful.

Proposition 3.2.16. *Let $a_j \in S(m^{\mu_j}, g)$, $j \in \mathbb{N}$, where μ_j is monotone strictly decreasing to $-\infty$ as $j \to +\infty$. Let $a \in C^\infty(\mathbb{R}^{2n})$ be such that for any given $\alpha \in \mathbb{Z}_+^{2n}$ there are constants ν_α and C_α such that*

$$|\partial_X^\alpha a(X)| \leq C_\alpha m(X)^{\nu_\alpha}.$$

Finally, let there exist ℓ_j and C_j with ℓ_j monotone strictly decreasing to $-\infty$ as $j \to +\infty$, and suppose the following estimate holds

$$|a(X) - \sum_{j=1}^{r-1} a_j(X)| \leq C_r m(X)^{\ell_r},$$

for any given $r \in \mathbb{N}$. Then $a \sim \sum_{j\geq 1} a_j$.

In these global classes we have that for any given

$$\begin{aligned} Au(x) &:= a^{\mathrm{w}}(x,D)u(x) \\ &= \lim_{\varepsilon\to 0+}(2\pi)^{-n}\iint e^{i\langle x-y,\xi\rangle}a\big(\frac{x+y}{2},\xi\big)\chi(\varepsilon x,\varepsilon y,\varepsilon\xi)u(y)dyd\xi, \end{aligned}$$

the limit being **independent of** $\chi\in\mathscr{S}(\mathbb{R}^{3n})$ such that $\chi(0)=1$. In this case we have, using the relations

$$(1+|x-y|^2)^{-M}(1-\Delta_\xi)^M e^{i\langle x-y,\xi\rangle}=e^{i\langle x-y,\xi\rangle},$$

$$(1+|\xi|^2)^{-N}(1-\Delta_y)^N e^{i\langle x-y,\xi\rangle}=e^{i\langle x-y,\xi\rangle},$$

and integrating by parts, that

$$\begin{aligned} Au(x) = (2\pi)^{-n}\iint e^{i\langle x-y,\xi\rangle}&(1+|x-y|^2)^{-M} \\ &\times(1-\Delta_\xi)^M(1-\Delta_y)^N\Big(\frac{1}{(1+|\xi|^2)^N}a\big(\frac{x+y}{2},\xi\big)u(y)\Big)dyd\xi. \end{aligned}$$

In fact, for large M and N the integral becomes absolutely convergent, for if $a\in S(m^\mu,g)$ then (see e.g. Shubin [67]), for $b(x,y,\xi)=a((x+y)/2,\xi)$ we have

$$|\partial_x^\alpha\partial_y^\beta\partial_\xi^\gamma b(x,y,\xi)|\le C_{\alpha\beta\gamma}(1+|x|+|y|+|\xi|)^{\mu-|\alpha|-|\beta|-|\gamma|}(1+|x-y|)^{|\mu|+|\alpha|+|\beta|+|\gamma|},$$

for all $(x,y,\xi)\in\mathbb{R}^{3n}$, which allows us to conclude the convergence. In this case the Schwartz kernel $\mathsf{K}_A\in\mathscr{S}'(\mathbb{R}^n\times\mathbb{R}^n)$ is given by

$$\langle\mathsf{K}_A|\psi\rangle_{\mathscr{S}',\mathscr{S}}=\lim_{\varepsilon\to 0+}(2\pi)^{-n}\iiint e^{i\langle x-y,\xi\rangle}a\big(\frac{x+y}{2},\xi\big)\chi(\varepsilon x,\varepsilon y,\varepsilon\xi)\psi(x,y)dydxd\xi,$$

for all $\psi\in\mathscr{S}(\mathbb{R}^{2n})$.

Also in the case of these global classes we may pass from the Weyl-quantization to the ordinary quantization (left-quantization). Theorem 3.1.16 is thus stated in the following form.

Theorem 3.2.17. *Given* $a\in S(m^\mu,g)$ *we have*

$$a^{\mathrm{w}}(x,D)=b(x,D),$$

where

$$b(x,\xi)=e^{i\langle D_x,D_\xi\rangle/2}a(x,\xi),\quad a(x,\xi)=e^{-i\langle D_x,D_\xi\rangle/2}b(x,\xi).\tag{3.26}$$

We next summarize some properties of $a^{\mathrm{w}}(x,D)$.

Theorem 3.2.18.

1. *Given any $a \in S(m^\mu, g)$ we have:*
   - $a^{\mathrm{w}}(x,D)\colon \mathscr{S}(\mathbb{R}^n) \longrightarrow \mathscr{S}(\mathbb{R}^n)$ *and* $a^{\mathrm{w}}(x,D)\colon \mathscr{S}'(\mathbb{R}^n) \longrightarrow \mathscr{S}'(\mathbb{R}^n)$ *is* **continuous**;
   - *When a is real-valued, then $a^{\mathrm{w}}(x,D)$ is* **formally self-adjoint**.

   The same holds true for $a \in S(m^\mu, g; \mathsf{M}_N)$.
2. *If $a \in S(1,g)$ then $a^{\mathrm{w}}(x,D)\colon L^2(\mathbb{R}^n) \longrightarrow L^2(\mathbb{R}^n)$ is* **continuous**. *The same holds true for $a \in S(1,g;\mathsf{M}_N)$.*
3. *If $a \in S(m^{-s}, g)$, with $s > 0$, then $a^{\mathrm{w}}(x,D)\colon L^2(\mathbb{R}^n) \longrightarrow L^2(\mathbb{R}^n)$ is* **compact**. *The same holds true for $a \in S(m^{-s}, g; \mathsf{M}_N)$, $s > 0$.*

We next define general global pseudodifferential operators.

Definition 3.2.19.

1. We say that $A \in \mathrm{OP}S(m^\mu, g)$ (i.e., that A is a **global pseudodifferential operator of order** μ) if there exists $a \in S(m^\mu, g)$ such that
$$A = a^{\mathrm{w}}(x,D) + R, \ \text{ with } \mathsf{K}_R \in \mathscr{S}(\mathbb{R}^n \times \mathbb{R}^n).$$
2. We say that $A \in \mathrm{OP}S_{\mathrm{cl}}(m^\mu, g)$ if $a \in S_{\mathrm{cl}}(m^\mu, g)$.
3. We say that $A \in \mathrm{OP}S(m^\mu, g)$ is **elliptic** when a is globally elliptic.
4. If $A \in \mathrm{OP}S_{\mathrm{cl}}(m^\mu, g)$, $a \sim \sum_{j \geq 0} a_{\mu - 2j}$, we call a_μ **the principal symbol of** A and $a_{\mu-2}$ **the subprincipal symbol of** A.

Analogously for $N \times N$ systems.

Example 3.2.20. For example, for the "translated" harmonic oscillator $p_{0,n}(x,\xi) + C$ (C a real constant), which is a classical symbol of order 2 (and also a GPD symbol of order 2), we have that $p_{0,n}(x,\xi)$ is the principal symbol and the constant C is the subprincipal one.

As another example, consider the symbol of the NCHO
$$Q_{(\alpha,\beta)}(x,\xi) = \begin{bmatrix} \alpha & 0 \\ 0 & \beta \end{bmatrix} \frac{x^2 + \xi^2}{2} + \begin{bmatrix} 0 & -i \\ i & 0 \end{bmatrix} x\xi.$$

The principal part is $Q_{(\alpha,\beta)}(x,\xi)$ itself, and there is no subprincipal part here. Observe hence that in the corresponding Weyl-quantization
$$Q^{\mathrm{w}}_{(\alpha,\beta)}(x,D) = \begin{bmatrix} \alpha & 0 \\ 0 & \beta \end{bmatrix} \frac{-\partial_x^2 + x^2}{2} + \begin{bmatrix} 0 & -1 \\ 1 & 0 \end{bmatrix} (x\partial_x + \frac{1}{2}),$$

the part $\frac{1}{2}\begin{bmatrix} 0 & -1 \\ 1 & 0 \end{bmatrix}$ is **not** the subprincipal term. △

Theorem 3.2.21. *Let $a \in S_{\mathrm{cl}}(m^{\mu_1}, g)$ and $b \in S_{\mathrm{cl}}(m^{\mu_2}, g)$. Then*

$$a^{\mathrm{w}}(x,D)b^{\mathrm{w}}(x,D) \in \mathrm{OP}S_{\mathrm{cl}}(m^{\mu_1+\mu_2}, g),$$

and for the symbol composition $a\sharp b$ we have (using the fact that the derivative of a function which is homogeneous of degree μ is a function homogeneous of degree $\mu - 1$)

$$a\sharp b \sim \sum_{j\geq 0} (a\sharp b)_{\mu_1+\mu_2-2j}, \tag{3.27}$$

so that the principal symbol of the composition is

$$(a\sharp b)_{\mu_1+\mu_2} = a_{\mu_1} b_{\mu_2}, \tag{3.28}$$

and the subprincipal symbol is

$$(a\sharp b)_{\mu_1+\mu_2-2} = a_{\mu_1} b_{\mu_2-2} + a_{\mu_1-2} b_{\mu_2} - \frac{i}{2}\{a_{\mu_1}, b_{\mu_2}\}. \tag{3.29}$$

Exercise 3.2.22. Write down the composition formula for semiregular classical symbols. △

We now come to a central construction in the theory of elliptic global pseudodifferential operators: the existence of a (two-sided) parametrix.

Theorem 3.2.23. *Let $A \in \mathrm{OP}S(m^{\mu}, g)$ be elliptic. Then there exists $B \in \mathrm{OP}S(m^{-\mu}, g)$ such that*

$$BA = I + R, \quad AB = I + R', \tag{3.30}$$

where R, R' are **smoothing operators**.

Furthermore, if $A \in \mathrm{OP}S_{\mathrm{cl}}(m^{\mu}, g)$ then $B \in \mathrm{OP}S_{\mathrm{cl}}(m^{-\mu}, g)$.

One calls B a **two-sided parametrix**.

Proof. We shall give the proof when A is "classical".

In the first place, it suffices to see that $BA = I + R$ and $AB' = I + \tilde{R}$, for some B and B'. In fact, we then have

$$BAB' = B(I + \tilde{R}) = (I + R)B',$$

that is

$$B = B' + \underbrace{(RB' - B\tilde{R})}_{\text{smoothing}}.$$

We hence prove that we may find B as in the statement, such that $BA = I + R$ (i.e. B is a *left-parametrix*). The construction of a *right-parametrix* is completely analogous.

Let χ be an excision function. Let

$$b_{-\mu} = \frac{\chi}{a_{\mu}} \in S_{\mathrm{cl}}(m^{-\mu}, g).$$

Then

$$B_1 = (b_{-\mu})^{\mathrm{w}}(x,D) \in \mathrm{OP}S_{\mathrm{cl}}(m^{-\mu},g),$$

and

$$B_1 A = I + R_1,\ R_1 \in \mathrm{OP}S_{\mathrm{cl}}(m^{-1},g),$$

and actually $R_1 \in \mathrm{OP}S_{\mathrm{cl}}(m^{-2},g)$, because we are working with classical symbols. We hence "Neumann-invert" $I+R_1$ as follows: for any given $N \in \mathbb{Z}_+$ we have

$$\Big(\sum_{j=0}^{N}(-1)^j R_1^j\Big) B_1 A = I - (-1)^{N+1} R_1^{N+1}.$$

If $r_1^{(j)}$ is the symbol of R_1^j (which is classical), that is $r_1^{(0)} = 1$ and $r_1^{(j)} = \underbrace{r_1 \sharp \ldots \sharp r_1}_{j \text{ times}}$, then by Proposition 3.2.15 we may find a symbol $s \in S_{\mathrm{cl}}(1,g)$ such that

$$s \sim \sum_{j=0}^{\infty}(-1)^j r_1^{(j)}.$$

Hence $S = s^{\mathrm{w}}(x,D)$ satisfies

$$SB_1A = I + R, \quad \text{with } R \text{ smoothing}.$$

We thus set $B := SB_1$. This concludes the proof of the theorem. □

Remark 3.2.24.

1. Theorem 3.2.21 also holds for matrix-valued operators.
2. Theorem 3.2.23 also holds for matrix-valued elliptic global pseudodifferential operators.
3. An equivalent proof may be given by using the composition formula (3.3) and by considering, at each degree of homogeneity, the equation

$$b \sharp a = 1 + r,$$

and finally by using an excision function to "re-sum" the homogeneous terms. Hence, for instance, we have the equations

$$b_{-\mu}a_\mu = 1,\ \ b_{-\mu}a_{\mu-2} + b_{-\mu-2}a_\mu - \frac{i}{2}\{b_{-\mu}, a_\mu\} = 0,\ \ldots,$$

so that we take

$$b_{-\mu} = a_\mu^{-1},\ \ b_{-\mu-2} = \Big(\frac{i}{2}\{a_\mu^{-1}, a_\mu\} - a_\mu^{-1}a_{\mu-2}\Big)a_\mu^{-1},\ \ldots.$$

Of course, in the matrix-valued case we must be careful in keeping the right order of the factors. △

We are now ready to define the natural spaces on which global operators act.

Let $\Lambda^2 = 1 + |x|^2 + |D|^2$ (where $|D|^2$ is a pseudodifferential notation for $-\Delta_x$) be the harmonic oscillator in n-dimensions, translated by 1. Let $s \in \mathbb{R}$. By virtue of our knowledge of the spectrum of the harmonic oscillator (see Theorem 2.2.1 and its n-dimensional extension Theorem 2.2.2) we may define the $s/2$-power $\Lambda^s = (1+|x|^2+|D|^2)^{s/2}$ (see also Section 5.3 of Chapter 5 below, where a few facts of the functional calculus are recalled). Using the proof of Theorem 5.5.1 below (see Helffer [17, pp. 46–52]) in the special case of Λ^2, one has that $\Lambda^s = \ell_s^{\mathrm{w}}(x,D) \in \mathrm{OP}S_{\mathrm{cl}}(m^s, g)$ is globally elliptic, with principal symbol $(|x|^2+|\xi|^2)^{s/2}$. Hence for all $s, s' \in \mathbb{R}$, one has that on $\mathscr{S}(\mathbb{R}^n)$ (and $\mathscr{S}'(\mathbb{R}^n)$)

$$\Lambda^s \Lambda^{s'} = \Lambda^{s+s'}, \ \Lambda^{-s} = (\Lambda^s)^{-1}, \text{ and } \Lambda^s = (\Lambda^s)^*$$

also **as global pseudodifferential operators**.

Definition 3.2.25. Let $s \in \mathbb{R}$. Define the spaces

$$B^s(\mathbb{R}^n) = \{u \in \mathscr{S}'(\mathbb{R}^n);\ \Lambda^s u \in L^2(\mathbb{R}^n)\},$$

endowed with the norm and inner-product respectively given by

$$\|u\|_{B^s} = \|\Lambda^s u\|_0,\ (u,v)_s := (\Lambda^s u, \Lambda^s v)_0,\ u,v \in B^s(\mathbb{R}^n).$$

We define the $\mathbb{C}^N$-valued case as follows:

$$B^s(\mathbb{R}^n;\mathbb{C}^N) := B^s(\mathbb{R}^n) \otimes \mathbb{C}^N,$$

endowed with the natural norms and inner-products induced by the norm and inner-product of B^s and $\mathbb{C}^N$.

Here are some important properties of the B^s, that are derived by the continuity properties and the existence of a parametrix of the operators Λ^s (see Helffer [17]).

Proposition 3.2.26.

1. *For each $s \in \mathbb{R}$, $B^s(\mathbb{R}^n)$ is a Hilbert space, and $\mathscr{S}(\mathbb{R}^n) \subset B^s(\mathbb{R}^n)$ is* **densely** *embedded.*
2. *For $s \in \mathbb{Z}_+$,*

$$B^s(\mathbb{R}^n) = \{u \in L^2(\mathbb{R}^n);\ \sum_{|\alpha|+|\beta|\le s} \|x^\alpha \partial_x^\beta u\|_0^2 < \infty\},$$

and

$$\sum_{|\alpha|+|\beta|\le s} \|x^\alpha \partial_x^\beta u\|_0^2 \approx \|u\|_{B^s}^2,$$

that is, the two norms are equivalent.

3. *For any given $s \in \mathbb{R}$, the L^2 inner-product defined for $u, v \in \mathscr{S}$ extends to a sesquilinear form which is continuous on $B^s(\mathbb{R}^n) \times B^{-s}(\mathbb{R}^n)$, and one has that $B^{-s}(\mathbb{R}^n)$ is identified with $(B^s(\mathbb{R}^n))^*$ (the dual of $B^s(\mathbb{R}^n)$).*
4. *For any given $s' < s$ we have that $B^s(\mathbb{R}^n) \subset B^{s'}(\mathbb{R}^n)$ with* **compact and dense** *range.*
5. *We have*

$$\mathscr{S}(\mathbb{R}^n) = \bigcap_{s \in \mathbb{R}} B^s(\mathbb{R}^n), \quad \mathscr{S}'(\mathbb{R}^n) = \bigcup_{s \in \mathbb{R}} B^s(\mathbb{R}^n).$$

Hence, given any $p, q, \in \mathbb{Z}_+$, if

$$|u|_{p,q} = \sup_{|\alpha| \leq q} \sup_{x \in \mathbb{R}^n} (1+|x|)^p |\partial_x^\alpha u(x)|$$

denotes a seminorm in $\mathscr{S}$, there is $s_0 = s_0(p,q,n)$ such that for all $s \geq s_0$ there exists $C_s > 0$ for which we have

$$|u|_{p,q} \leq C_s \|u\|_{B^s}, \quad \forall u \in \mathscr{S}(\mathbb{R}^n). \tag{3.31}$$

6. *For any given $s \in \mathbb{R}$, any given $A \in \mathrm{OPS}(m^\mu, g)$ is* **bounded** *as an operator*

$$A \colon B^s(\mathbb{R}^n) \longrightarrow B^{s-\mu}(\mathbb{R}^n).$$

7. *The spaces $B^s(\mathbb{R}^n)$ do not depend on Λ^s, that is, for any given globally elliptic $A \in \mathrm{OPS}_{\mathrm{cl}}(m^s, g)$ one has*

$$B^s(\mathbb{R}^n) = \{u \in \mathscr{S}'(\mathbb{R}^n);\ Au \in L^2(\mathbb{R}^n)\}.$$

8. *Properties 1–7 above hold true also in the vector-valued and matrix-valued cases.*

It is worth noting that Property 5 of Proposition 3.2.26 does not hold for the more usual Sobolev spaces $H^s(\mathbb{R}^n)$, for in this case we have, upon setting

$$H^{+\infty}(\mathbb{R}^n) := \bigcap_{s \in \mathbb{R}} H^s(\mathbb{R}^n), \ H^{-\infty}(\mathbb{R}^n) := \bigcup_{s \in \mathbb{R}} H^s(\mathbb{R}^n),$$

that

$$\mathscr{S}(\mathbb{R}^n) \subsetneq H^{+\infty}(\mathbb{R}^n) \subsetneq H^{-\infty}(\mathbb{R}^n) \subsetneq \mathscr{S}'(\mathbb{R}^n).$$

Notice also that

$$\mathscr{S}(\mathbb{R}^n) = \bigcap_{k \in \mathbb{Z}} B^k(\mathbb{R}^n).$$

3.3 Spectral Properties of Globally Elliptic Pseudodifferential Operators

We now want to study how the spectrum of an elliptic global psedudodifferential operator depends on the spaces B^s.

Remark that *all* the results stated in this section *hold true also for matrix-valued operators.*

We have the following very important lemma, which shows that the kernel of an elliptic global pseudodifferential operator does not depend on the spaces B^s in which the operator is realized (in fact, it consists exclusively of Schwartz functions).

Lemma 3.3.1. *Let $A \in \mathrm{OPS}(m^\mu, g)$ be elliptic. Then*

$$\mathrm{Ker}(A\colon B^{s+\mu} \to B^s) = \mathrm{Ker}(A\colon \mathscr{S} \to \mathscr{S}) = \mathrm{Ker}(A\colon \mathscr{S}' \to \mathscr{S}'), \ \forall s \in \mathbb{R}.$$

Proof. Take $\psi \in \mathscr{S}'$ such that $A\psi = 0$. Then, by the existence of a parametrix Q, $0 = QA\psi = \psi + R\psi$, i.e. $\psi = -R\psi \in \mathscr{S}$, since R is smoothing. Hence

$$\mathrm{Ker}(A\colon \mathscr{S}' \to \mathscr{S}') \subset \mathrm{Ker}(A\colon \mathscr{S} \to \mathscr{S}) \subset \mathrm{Ker}(A\colon \mathscr{S}' \to \mathscr{S}').$$

Since it is trivial that

$$\mathrm{Ker}(A\colon \mathscr{S} \to \mathscr{S}) \subset \mathrm{Ker}(A\colon B^{s+\mu} \to B^s) \subset \mathrm{Ker}(A\colon \mathscr{S}' \to \mathscr{S}'),$$

we obtain the claim. □

Next we show that the kernel of an elliptic global pseudodifferential operator has a finite dimension.

Lemma 3.3.2. *Let $A \in \mathrm{OPS}(m^\mu, g)$ be elliptic. Then*

$$\dim \mathrm{Ker} A < +\infty.$$

Proof. Let Q be a two-sided parametrix of A, with $QA = I + R$, R smoothing. Since

$$\mathrm{Ker}(A\colon B^\mu \to L^2) \subset \mathrm{Ker}(QA\colon B^\mu \to B^\mu) = \mathrm{Ker}(I + R\colon B^\mu \to B^\mu),$$

and since

$$R\colon B^\mu \longrightarrow \mathscr{S} \hookrightarrow B^{\mu+1} \hookrightarrow\!\!\hookrightarrow B^\mu,$$

where $\hookrightarrow\!\!\hookrightarrow$ denotes compact embedding, we get that $R\colon B^\mu \longrightarrow B^\mu$ is compact, whence the claim. □

The range of an elliptic global pseudodifferential operator enjoys some nice properties as well. They are stated in the following results.

Lemma 3.3.3. *Let* $A \in \mathrm{OPS}(m^\mu, g)$ *be elliptic. For all* $s \in \mathbb{R}$ *one has:*

(i) $\mathrm{Im}(A\colon B^{\mu+s} \to B^s)$ *is* **closed**, *with* **finite** *codimension;*
(ii) $\mathrm{codim}\,\mathrm{Im}(A\colon B^{\mu+s} \to B^s) = \dim \mathrm{Ker} A^*$ *(hence* **independent** *of s).*

Proof. Let Q be a two-sided parametrix of A, with $AQ = I + R'$, R' smoothing. One has

$$\mathrm{Im}(A\colon B^{\mu+s} \to B^s) \supset \mathrm{Im}(AQ\colon B^s \to B^s) = \mathrm{Im}(I + R'\colon B^s \to B^s),$$

and since $R'\colon B^s \to B^s$ is **compact**, $\mathrm{Im}(I + R'\colon B^s \to B^s)$ is closed, with finite codimension. This proves (i).

We prove (ii). By the Closed Range Theorem we have

$$\dim(B^s / A(B^{\mu+s})) = \dim \mathrm{Ker}(A^+\colon B^s \to B^{\mu+s}),$$

where the operator A^+ is defined by

$$(Au, g)_s = (u, A^+ g)_{\mu+s}, \quad u \in B^{\mu+s},\ g \in B^s.$$

(Recall that $(Au, g)_s = (\Lambda^s A u, \Lambda^s g)_0$.) Taking sequences $\{u_j\}_j, \{g_j\}_j \subset \mathscr{S}$ with $u_j \overset{B^{\mu+s}}{\longrightarrow} u$ and $g_j \overset{B^s}{\longrightarrow} g$, as $j \to +\infty$, yields

$$\begin{aligned}(\Lambda^s A u, \Lambda^s g)_0 &= \lim_{j\to+\infty} (\Lambda^s A u_j, \Lambda^s g_j)_0 = \lim_{j\to+\infty} (u_j, A^* \Lambda^{2s} g_j)_0 \\ &= \lim_{j\to+\infty} (\Lambda^{\mu+s} u_j, \Lambda^{-(\mu+s)} A^* \Lambda^{2s} g_j)_0 = (\Lambda^{\mu+s} u, \Lambda^{-(\mu+s)} A^* \Lambda^{2s} g)_0.\end{aligned}$$

We thus have

$$(\Lambda^{\mu+s} u, \Lambda^{-(\mu+s)} A^* \Lambda^{2s} g)_0 = (u, A^+ g)_{\mu+s} = (\Lambda^{\mu+s} u, \Lambda^{\mu+s} A^+ g)_0.$$

Then, by comparison, $\Lambda^{\mu+s} A^+ g = \Lambda^{-(\mu+s)} A^* \Lambda^{2s} g$, whence

$$A^+ g = \Lambda^{-2(\mu+s)} A^* \Lambda^{2s} g, \quad \forall g \in B^s,$$

and, finally,

$$\dim \mathrm{Ker}(A^+\colon B^s \to B^{\mu+s}) = \dim \mathrm{Ker} A^*,$$

which concludes the proof. □

Lemma 3.3.4. *Let* $A \in \mathrm{OPS}(m^\mu, g)$ *be elliptic. One has:*

(i) $\mathrm{Im}(A\colon \mathscr{S} \to \mathscr{S})$ *is* **closed**, *with* **finite** *codimension equal to* $\dim \mathrm{Ker} A^*$;
(ii) $\mathrm{Im}(A\colon \mathscr{S}' \to \mathscr{S}')$ *is* **closed**, *with* **finite** *codimension equal to* $\dim \mathrm{Ker} A^*$.

Proof. To see (i), take $\{u_j\}_j \subset \mathscr{S}$ such that $Au_j \xrightarrow{\mathscr{S}} g \in \mathscr{S}$, as $j \to +\infty$. Then, in particular, $Au_j \xrightarrow{L^2} g$, whence $g \in \mathrm{Im}(A\colon B^\mu \longrightarrow L^2)$. Therefore there exists $u \in B^\mu$ such that $Au = g$ and, by the existence of a parametrix, we have that $u \in \mathscr{S}$, since $g \in \mathscr{S}$. This proves that $\mathrm{Im}(A\colon \mathscr{S} \to \mathscr{S})$ is closed. We now claim that

$$\mathrm{Im}(A\colon \mathscr{S} \to \mathscr{S}) = \{u \in \mathscr{S};\ (u,g)_0 = 0,\ \forall g \in \mathrm{Ker}A^*\} =: V.$$

(Recall that $\mathrm{Ker}A^* \subset \mathscr{S}$.) To prove the claim, observe that the inclusion $\subset$ is trivial. To prove $\supset$, suppose in the first place that $\mu = 0$. Then, since $L^2 = \mathrm{Im}A \oplus \mathrm{Ker}A^*$ ($A\colon L^2 \to L^2$ is now continuous), we have that if $u \in \mathscr{S}$, then $u = Au' + g$, with $g \in \mathscr{S}$ and $u' \in L^2$, whence $Au' \in \mathscr{S}$. Thus, by the existence of a parametrix, $u' \in \mathscr{S}$. When $u \in V$, then $u = Au'$ and $u \in A(\mathscr{S})$, which proves the claim for $\mu = 0$. When $\mu \neq 0$, the assertion follows by observing that

$$\mathrm{Im}(A\colon \mathscr{S} \to \mathscr{S}) = \mathrm{Im}(A\Lambda^{-\mu}\colon \mathscr{S} \to \mathscr{S}), \quad \text{and that} \quad \mathrm{Ker}A^* = \mathrm{Ker}\,(\Lambda^{-\mu}A^*).$$

This reduces matters to the case $\mu = 0$ and proves (i).

To prove (ii), we put, for $\psi \in \mathscr{S}'$ and $\varphi \in \mathscr{S}$,

$$(\psi,\varphi)_{\mathscr{S}',\mathscr{S}} := \langle \psi | \bar{\varphi} \rangle_{\mathscr{S}',\mathscr{S}},$$

so that

$$(A\psi,\varphi)_{\mathscr{S}',\mathscr{S}} = (\psi, A^*\varphi)_{\mathscr{S}',\mathscr{S}},\ \forall \psi \in \mathscr{S}', \forall \varphi \in \mathscr{S}.$$

We claim that

$$\mathrm{Im}(A\colon \mathscr{S}' \to \mathscr{S}') = \{\psi \in \mathscr{S}';\ (\psi,g)_{\mathscr{S}',\mathscr{S}} = 0,\ \forall g \in \mathrm{Ker}A^*\} =: V'.$$

The inclusion $\subset$ is trivial. To prove $\supset$, take $\psi \in V'$. Since $\psi \in B^{s_0}$ for some $s_0 \in \mathbb{R}$, we get (recalling that the dual of B^s is B^{-s})

$$(\psi,g)_{\mathscr{S}',\mathscr{S}} = (\psi,g)_{B^{s_0},B^{-s_0}}.$$

Suppose at first that $\mu = 0$. Then

$$A\colon B^{s_0} \to B^{s_0},\ A^*\colon B^{-s_0} \to B^{-s_0},$$

and, by construction,

$$(A\psi,g)_{B^{s_0},B^{-s_0}} = (\psi, A^*g)_{B^{s_0},B^{-s_0}}.$$

Since A^* has closed range, the Closed Range Theorem yields that if

$$(\psi,g)_{B^{s_0},B^{-s_0}} = 0,\ \forall g \in \mathrm{Ker}A^*,$$

then $\psi \in \mathrm{Im}(A\colon B^{s_0} \to B^{s_0})$, whence $\psi \in \mathrm{Im}(A\colon \mathscr{S}' \to \mathscr{S}')$. This proves the claim and also that $\mathrm{Im}(A\colon \mathscr{S}' \to \mathscr{S}')$ is closed with codimension equal to $\dim \mathrm{Ker} A^*$, when $\mu = 0$. When $\mu \neq 0$, we reduce matters to the case $\mu = 0$, by observing that

$$\mathrm{Im}(A\colon \mathscr{S}' \to \mathscr{S}') = \mathrm{Im}(A\Lambda^{-\mu}\colon \mathscr{S}' \to \mathscr{S}'), \quad \text{and that} \quad \mathrm{Ker} A^* = \mathrm{Ker}(\Lambda^{-\mu} A^*).$$

This concludes the proof. □

We next have the following useful proposition.

Proposition 3.3.5. *Let $A \in \mathrm{OP}S(m^\mu, g)$ be elliptic. The following assertions are equivalent:*

(i) $\mathrm{Ker} A = \{0\}$;
(ii) $A^*\colon \mathscr{S} \to \mathscr{S}$ *is onto;*
(iii) $A^*\colon \mathscr{S}' \to \mathscr{S}'$ *is onto;*
(iv) $A^*\colon B^{\mu+s} \to B^s$ *is onto, for all* $s \in \mathbb{R}$*;*
(v) $A^*\colon B^{\mu+s} \to B^s$ *is onto, for some* $s \in \mathbb{R}$*.*

Proof. We start by observing that (iv)⇒(iii), and that, in turn, by the existence of a parametrix, (iii)⇒(ii). Of course, (iv) implies (v). We now prove that (i) implies any of the (ii) to (v). Since

$$\dim(B^s / A^*(B^{\mu+s})) = \dim \mathrm{Ker} A^{**} = \dim \mathrm{Ker} A = 0,$$

we immediately get that $A^*\colon B^{\mu+s} \to B^s$ is onto, for any given $s \in \mathbb{R}$ (since we already know that $\mathrm{Im}(A^*\colon B^{\mu+s} \to B^s)$ is closed). It follows that for any given $u \in \mathscr{S} \subset L^2$ there exists $g \in B^\mu$ such that $A^* g = u$, whence, as $Q^* A^* = I + R'$ with R' smoothing,

$$\mathscr{S} \ni Q^* u = Q^* A^* g = g + \underbrace{R' g}_{\in \mathscr{S}} \Longrightarrow g \in \mathscr{S}.$$

Thus $A^*(\mathscr{S}) = \mathscr{S}$.

That $A^*(\mathscr{S}') = \mathscr{S}'$ is trivial. In fact, $u \in \mathscr{S}'$ implies $u \in B^{s_0}$ for some $s_0 \in \mathbb{R}$, whence the existence of $g \in B^{\mu+s_0} \subset \mathscr{S}'$ (as we already know that (i)⇒(iv)) such that $A^* g = u$. This concludes the proof that (i)⇒(ii), (iii), (iv) and (v).
We finally prove that (v)⇒(i). If $A^*\colon B^{\mu+s} \to B^s$ is onto for some s, it follows that

$$\dim \mathrm{Ker} A = \mathrm{codim}\, A^*(B^{\mu+s}) = 0.$$

This concludes the proof. □

Corollary 3.3.6. *Let $A \in \mathrm{OP}S(m^\mu, g)$ be elliptic. If $A = A^*$ then the injectivity of A is equivalent to the surjectivity of A, which is in turn equivalent to the invertibility of A (on any of the spaces $\mathscr{S}$, $\mathscr{S}'$, or B^s).*

Recall that (see Helffer [17], or Hörmander [29], or Shubin [67]) a linear continuous operator $T\colon B_1 \to B_2$, B_1 and B_2 Banach spaces, is a **Fredholm** operator if

- $\dim \operatorname{Ker} T < +\infty$,
- $\dim \operatorname{Coker} T < +\infty$, where $\operatorname{Coker} T = B_2/T(B_1)$,

so that one may define the **index** of T as

$$\operatorname{ind} T = \dim \operatorname{Ker} T - \dim \operatorname{Coker} T.$$

Recall that in this situation $T(B_1)$ is then closed.

From the preceding results one has the following Fredholm property of a globally elliptic psedudodifferential operator (see [17] or [67]).

Theorem 3.3.7. *Let $A \in \mathrm{OP}S(m^\mu, g)$ be elliptic. Then:*

- *$A\colon B^{\mu+s} \to B^s$ is a Fredholm operator, for any given $s \in \mathbb{R}$;*
- *$\operatorname{ind} A$ is* **independent of** *s and is expressed by the formula*

$$\operatorname{ind} A = \dim \operatorname{Ker} A - \dim \operatorname{Ker} A^*.$$

In particular, if $A = A^*$, that is, if A is formally self-adjoint, then $\operatorname{ind} A = 0$.

Here is another important consequence of global ellipticity, and of the previous results.

Theorem 3.3.8. *Let $A \in \mathrm{OP}S_{\mathrm{cl}}(m^\mu, g)$ be elliptic and suppose there exists $s_0 \in \mathbb{R}$ such that $A\big|_{B^{s_0}} : B^{s_0}(\mathbb{R}^n) \longrightarrow B^{s_0-\mu}(\mathbb{R}^n)$ is an* **isomorphism** *(that is, it is invertible with linear and continuous inverse). Then*

$$A^{-1}\colon B^s(\mathbb{R}^n) \longrightarrow B^{s+\mu}(\mathbb{R}^n) \text{ is continuous } \forall s \in \mathbb{R},$$

and $(A\big|_{B^s})^{-1}$ is the restriction to $B^{s-\mu}(\mathbb{R}^n)$ of an operator in $\mathrm{OP}S_{\mathrm{cl}}(m^{-\mu}, g)$. The same holds true in the case of $N \times N$ systems.

Proof. We give a proof in the scalar case. From Lemma 3.3.1 and Lemma 3.3.3 we have that $A\big|_{B^s} : B^s(\mathbb{R}^n) \longrightarrow B^{s-\mu}(\mathbb{R}^n)$ is an **isomorphism** for all $s \in \mathbb{R}$. By the global ellipticity, we know that there is $Q \in \mathrm{OP}S_{\mathrm{cl}}(m^{-\mu}, g)$ such that

$$QA = I + R,\ AQ = I + R',\ R, R' \text{ smoothing}.$$

By hypothesis, upon setting $A_0 := A\big|_{B^{s_0}}$, we have

$$A_0 A_0^{-1} = I_{B^{s_0-\mu}},\ A_0^{-1} A_0 = I_{B^{s_0}}.$$

Then

$$Q\big|_{B^{s_0-\mu}} = A_0^{-1} A_0 Q\big|_{B^{s_0-\mu}} = A_0^{-1}(I_{B^{s_0-\mu}} + R'\big|_{B^{s_0-\mu}}),$$

whence it follows that

$$A_0^{-1} = Q\big|_{B^{s_0-\mu}} - A_0^{-1} R'\big|_{B^{s_0-\mu}}. \tag{3.32}$$

On the other hand, we also have

$$Q\big|_{B^{s_0-\mu}} = Q\big|_{B^{s_0-\mu}} A_0 A_0^{-1} = (I_{B^{s_0}} + R\big|_{B^{s_0}}) A_0^{-1} = A_0^{-1} + R\big|_{B^{s_0}} A_0^{-1},$$

whence it follows, using (3.32), that

$$A_0^{-1} = Q\big|_{B^{s_0-\mu}} - R\big|_{B^{s_0}} A_0^{-1} = Q\big|_{B^{s_0-\mu}} - R\big|_{B^{s_0}} Q\big|_{B^{s_0-\mu}} + R\big|_{B^{s_0}} A_0^{-1} R'\big|_{B^{s_0-\mu}}. \quad (3.33)$$

But RQ is **smoothing**, as well as $RA_0^{-1}R'$, for

$$\mathscr{S}'(\mathbb{R}^n) \xrightarrow{R'} \mathscr{S}(\mathbb{R}^n) \hookrightarrow B^{s_0-\mu}(\mathbb{R}^n) \xrightarrow{A_0^{-1}} B^{s_0}(\mathbb{R}^n) \xrightarrow{R} \mathscr{S}(\mathbb{R}^n)$$

is continuous. Hence from (3.33) it follows that $A^{-1} \in \mathrm{OP}S_{\mathrm{cl}}(m^{-\mu}, g)$ and this concludes the proof. □

Let us now suppose that $\mu > 0$ (the case $\mu = 0$ is trivial). Consider

$$P_A\colon D(P_A) \subset L^2(\mathbb{R}^n) \longrightarrow L^2(\mathbb{R}^n), \ \ D(P_A) = \{u \in L^2;\ P_A u = Au \in L^2\},$$

where Au is understood in the sense of distributions. The unbounded operator P_A is called the **maximal operator** (associated with A). When A is an elliptic global pseudodifferential operator, it is straightforward to see that P_A is a closed operator, that is (see also Chapter 2)

$$\left.\begin{array}{l} D(P_A) \ni u_j \xrightarrow{L^2} u, \ \text{as } j \to +\infty \\[2ex] P_A u_j \xrightarrow{L^2} v, \ \text{as } j \to +\infty \end{array}\right\} \Longrightarrow u \in D(P_A) \ \text{and} \ P_A u = v.$$

In fact, from the hypothesis we have $Au_j \xrightarrow{\mathscr{S}'} Au$, so that $Au = v \in L^2$ whence $u \in D(P_A)$, which proves that P_A is a closed operator. Moreover,we have the following lemma (see also point 7 of Proposition 3.2.26).

Lemma 3.3.9. *One has*

$$D(P_A) = B^{\mu}.$$

Proof. To see this, just note that we obviously have $B^{\mu} \subset D(P_A)$. If now $u \in L^2$ with $Au \in L^2$, by the existence of a parametrix we get

$$B^{\mu} \ni QAu = u + Ru \Longrightarrow u \in B^{\mu},$$

which gives $D(P_A) \subset B^{\mu}$. □

This immediately gives back the fact that $D(H)$ (see (2.4)), where H is the harmonic oscillator $p_{0,n}^{\mathrm{w}}(x,D) = (|D|^2 + |x|^2)/2$, is exactly the Hilbert space $B^2(\mathbb{R}^n)$. Hence, in the case of the NCHO $Q_{(\alpha,\beta)}^{\mathrm{w}}(x,D)$, for $\alpha, \beta > 0$ and $\alpha\beta > 1$, we have

$$D(Q^{\mathrm{w}}_{(\alpha,\beta)}(x,D)) = B^2(\mathbb{R};\mathbb{C}^2).$$

The next proposition gives that when $A = A^*$, i.e. when A is *formally self-adjoint*, then $P_A = P_A^*$, i.e. P_A is *self-adjoint.*

Proposition 3.3.10. *One has $P_A^* = P_{A^*}$. Hence if A is formally self-adjoint then P_A is self-adjoint.*

Proof. Let $g \in D(P_A^*)$, that is $g \in L^2$ and there exists $h \in L^2$ such that

$$(Au,g)_0 = (u,h)_0, \ \ \forall u \in D(P_A) = B^\mu. \tag{3.34}$$

Hence (3.34) holds in particular for all $u \in \mathscr{S}$. Now, from

$$(Au,g)_0 = \langle \bar{g}|Au\rangle_{\mathscr{S}',\mathscr{S}} = \langle {}^tA\bar{g}|u\rangle_{\mathscr{S}',\mathscr{S}} = \langle \overline{A^*g}|u\rangle_{\mathscr{S}',\mathscr{S}} = \langle \bar{h}|u\rangle_{\mathscr{S}',\mathscr{S}}, \ \ \forall u \in \mathscr{S},$$

we obtain $A^*g \overset{\mathscr{S}'}{=} h \in L^2$. The existence of a parametrix then yields $g \in B^\mu$. Thus $D(P_A^*) \subset B^\mu$, and $P_A^* g = A^*g$ when $g \in D(P_A^*)$.

On the other hand, $B^\mu \subset D(P_A^*)$. In fact, if $g \in B^\mu$ take $u \in B^\mu$ and sequences $\{u_j\}_j, \{g_j\}_j \subset \mathscr{S}$ such that

$$u_j \xrightarrow{B^\mu} u, \ \ g_j \xrightarrow{B^\mu} g, \ \text{as } j \to +\infty.$$

Since $A^* \in \mathrm{OPS}(m^\mu, g)$, we thus have that $A^*g_j \xrightarrow{L^2} A^*g$, as $j \to +\infty$, and

$$(Au,g)_0 = \lim_{j\to+\infty} (Au_j, g_j)_0 = \lim_{j\to+\infty} (u_j, A^*g_j)_0 = (u, A^*g)_0,$$

whence $g \in D(P_A^*)$, and this concludes the proof. □

One defines the **minimal operator** P'_A by

$$P'_A\colon D(P'_A) = \mathscr{S}(\mathbb{R}^n) \subset L^2(\mathbb{R}^n) \longrightarrow L^2(\mathbb{R}^n), \ \ P'_A u = Au, \ \forall u \in D(P'_A).$$

By the same method used in the case of the maximal operator, one sees (exercise for the reader!) that P'_A is *closable*, i.e. that

$$\left.\begin{array}{l} D(P'_A) \ni u_j \xrightarrow{L^2} 0, \ \text{as } j \to +\infty \\[2ex] P'_A u_j \xrightarrow{L^2} v, \ \text{as } j \to +\infty \end{array}\right\} \Longrightarrow v = 0.$$

Moreover, we have the following proposition.

Proposition 3.3.11. *One has* $(P'_A)^* = P_{A^*}$

Proof. As before, $D((P'_A)^*) \subset B^\mu$, and $(P'_A)^* g = A^* g$, for all $g \in D((P'_A)^*)$. Take now $g \in B^\mu$ and $u \in \mathscr{S}$, and consider a sequence $\{g_j\}_j \subset \mathscr{S}$ such that $g_j \xrightarrow{B^\mu} g$, as $j \to +\infty$. Then, as before,

$$(Au,g)_0 = \lim_{j\to+\infty} (Au,g_j)_0 = \lim_{j\to+\infty} (u,A^*g_j)_0 = (u,A^*g)_0,$$

whence $g \in D((P'_A)^*)$ and $(P'_A)^* g = A^* g$. □

Corollary 3.3.12. $A = A^*$ *implies* $P_A = P_A^* = (P'_A)^*$.

We may now give the following theorem about the spectrum of a formally self-adjoint positive elliptic global pseudodifferential operator (or system) of order $\mu \geq 1$.

Theorem 3.3.13. *Let* $A \in \mathrm{OP}S_{\mathrm{cl}}(m^\mu, g)$, $\mu \geq 1$, *be elliptic, with real symbol and positive principal symbol* a_μ. *Then* $A\colon D(A) = B^\mu \subset L^2 \longrightarrow L^2$, *thought of as an unbounded operator, is self-adjoint, with a discrete spectrum, made of an increasing to* $+\infty$ *and bounded from below sequence* $\{\lambda_j\}_{j\in\mathbb{N}}$ *of eigenvalues with finite multiplicities,*

$$-\infty < \lambda_1 \leq \lambda_2 \leq \ldots \longrightarrow +\infty,$$

with repetitions according to the multiplicity. The corresponding eigenfunctions belong to $\mathscr{S}$ *and, possibly after an orthonormalization procedure, form a basis of* L^2.

Proof. The fact that A is a self-adjoint unbounded operator with a discrete spectrum and Schwartz eigenfunctions follows from the arguments above (B^μ is compactly embedded in L^2, and if $\lambda \in \mathrm{Spec}(A)$ then $\mathrm{Ker}(A-\lambda) \subset \mathscr{S}$ with finite dimension). We have to prove that the spectrum is bounded from below. Since $a_\mu > 0$, using the calculus of global operators we may find $Q \in \mathrm{OP}S_{\mathrm{cl}}(m^{\mu/2}, g)$ such that

$$A = Q^*Q + R, \;\; R \in \mathrm{OP}S_{\mathrm{cl}}(m^{\mu-2}, g). \tag{3.35}$$

Then

$$(Au,u) = \|Qu\|_0^2 + (Ru,u), \;\; \forall u \in \mathscr{S}.$$

Since Q is elliptic, with principal symbol $q_{\mu/2} = a_\mu^{1/2}$, we have, by the existence of a parametrix M for Q, that for some $c > 0$

$$\|Qu\|_0^2 + \|u\|_0^2 \geq c\|u\|_{B^{\mu/2}}^2, \;\; \forall u \in \mathscr{S}. \tag{3.36}$$

In fact, from $MQ = I + \tilde{R}$ ($\tilde{R}$ smoothing) it follows $u = MQu - \tilde{R}u$, so that

$$\|u\|_{B^{\mu/2}}^2 = \|\Lambda^{\mu/2}u\|_0^2 = \|\Lambda^{\mu/2}MQu - \Lambda^{\mu/2}\tilde{R}u\|_0^2 \lesssim \|Qu\|_0^2 + \|u\|_0^2, \;\; \forall u \in \mathscr{S},$$

since, $M\colon L^2 \longrightarrow B^{\mu/2}$ being continuous, the operator $\Lambda^{\mu/2}M$ has order zero and $\Lambda^{\mu/2}\tilde{R}$ is smoothing, and hence they are both $L^2 \to L^2$ bounded. This proves (3.36).

Now, as the inverse $(\Lambda^{\mu/2})^{-1}$ of $\Lambda^{\mu/2}$ belongs to $S(m^{-\mu/2},g)$, we have

$$|(Ru,u)| = (((\Lambda^{\frac{\mu-1}{2}})^{-1})^* R(\Lambda^{\frac{\mu-1}{2}})^{-1}\Lambda^{\frac{\mu-1}{2}}u, \Lambda^{\frac{\mu-1}{2}}u)| \le C\|u\|^2_{B^{(\mu-1)/2}},\ \forall u \in \mathscr{S}, \tag{3.37}$$

for $((\Lambda^{(\mu-1)/2})^{-1})^* R(\Lambda^{(\mu-1)/2})^{-1}$ has order -1 and hence, in particular, is $L^2 \to L^2$ bounded.

On the other hand, for any given $\varepsilon > 0$ (to be picked later) there exists $C_\varepsilon > 0$ such that (note that $\mu - 1 \ge 0$)

$$\|u\|^2_{B^{(\mu-1)/2}} \le \varepsilon \|u\|^2_{B^{\mu/2}} + C_\varepsilon \|u\|^2_0,\ \forall u \in \mathscr{S}. \tag{3.38}$$

From (3.36), (3.37) and (3.38) we finally get

$$(Au,u) \ge (c - \varepsilon C)\|u\|^2_{B^{\mu/2}} + O(\|u\|^2_0) \ge -C_0\|u\|^2_0,\ \forall u \in \mathscr{S},$$

by choosing $0 < \varepsilon < c/C$. □

Exercise 3.3.14. Let $\mu \ge 1$. Prove inequality (3.38). △

Corollary 3.3.15. *In the hypotheses of Theorem 3.3.13 we have that* Spec(A) **does not depend on** $s \in \mathbb{R}$. *More precisely, if we think of A as an unbounded operator*

$$A_s\colon D(A_s) = B^{\mu+s} \subset B^s \longrightarrow B^s,\ A_s u = Au,\ u \in D(A_s),$$

then it follows that $A_s^* = A_s = A^* = A$ *and, by Lemma 3.3.1,*

$$\mathrm{Spec}(A_s) = \mathrm{Spec}(A_0) =: \mathrm{Spec}(A).$$

Remark 3.3.16. Theorem 3.3.13 and Corollary 3.3.15 hold true also in the matrix-valued case. The only delicate point is to obtain (3.35) in the case of systems, by constructing the expansion of the symbol of Q. One can actually find $Q = Q^* \in \mathrm{OP}S_{\mathrm{cl}}(m^{\mu/2},g;\mathrm{M}_N)$ such that

$$A = Q^2 + R,\ R \in \mathrm{OP}S(m^{-\infty},g;\mathrm{M}_N),$$

for at the level of the principal symbol one has to find a matrix $q_{\mu/2} = q^*_{\mu/2} > 0$ such that

$$q^2_{\mu/2} = a_\mu,\ X \ne 0, \tag{3.39}$$

and at the level of lower degrees of homogeneity one has to solve

$$j-\text{th term of } (q^*_{\mu/2} \sharp q_{\mu/2}) = a_{\mu-2j},\ X \ne 0,\ \forall j \in \mathbb{Z}_+.$$

In particular, the subprincipal part is the solution of

$$q_{\mu/2}q_{\mu/2-2}+q_{\mu/2-2}q_{\mu/2}=a_{\mu-2}+\frac{i}{2}\{q_{\mu/2},q_{\mu/2}\},\ X\neq 0, \tag{3.40}$$

with $q^*_{\mu/2-2}=q_{\mu/2-2}$. Lemma 3.3.17 below takes care of (3.39) and Lemma 3.3.19 below takes care of (3.40). △

Lemma 3.3.17. *Let* $0<a_\mu=a^*_\mu\in C^\infty(\mathbb{R}^{2n}\setminus\{0\};\mathsf{M}_N)$ *be positively homogenous of degree* μ, *and let* $0<c_1<c_2$ *be such that*

$$c_1|v|^2_{\mathbb{C}^N}\leq\langle a_\mu(\omega)v,v\rangle_{\mathbb{C}^N}\leq c_2|v|^2,\ \ \forall v\in\mathbb{C}^N,\ \forall\omega\in\mathbb{S}^{2n-1}.$$

Then there exists a unique $0<q_{\mu/2}=q^*_{\mu/2}\in C^\infty(\mathbb{R}^{2n}\setminus\{0\};\mathsf{M}_N)$ *positively homogeneous of degree* $\mu/2$ *such that* $q_{\mu/2}(X)^*q_{\mu/2}(X)=a_\mu(X)$ *for all* $X\neq 0$.

Proof. We construct $q_{\mu/2}$ on $\mathbb{S}^{2n-1}$, for then it suffices to put

$$q_{\mu/2}(X)=|X|^{\mu/2}q_{\mu/2}(\frac{X}{|X|}),\ \ \forall X\neq 0.$$

From the hypothesis we have that $\operatorname{Spec}(a_\mu(\omega))\subset[c_1,c_2]\subset\mathbb{R}_+$, for all $\omega\in\mathbb{S}^{2n-1}$. Let hence $\gamma\subset\mathbb{C}$ be a closed, counter-clockwise oriented curve contained in $\operatorname{Re}\zeta>0$ enclosing $[c_1,c_2]$ such that $\bar\gamma=\gamma$, and define

$$q_{\mu/2}(\omega)=\frac{1}{2\pi i}\int_\gamma\zeta^{1/2}(\zeta-a_\mu(\omega))^{-1}d\zeta,$$

where $\zeta^{1/2}$ is the branch of the square-root that is positive when ζ is positive. It is then clear that $q_{\mu/2}$ is smooth and that $q^*_{\mu/2}=q_{\mu/2}$. Let γ' be another curve (closed, counter-clockwise oriented) contained in $\operatorname{Re}\zeta>0$ which encloses γ, with $\gamma\cap\gamma'=\emptyset$. It follows from the Cauchy formula that

$$q_{\mu/2}(\omega)=\frac{1}{2\pi i}\int_{\gamma'}\zeta^{1/2}(\zeta-a_\mu(\omega))^{-1}d\zeta,$$

so that, using the resolvent identity

$$(\zeta'-\zeta)(\zeta-a_\mu(\omega))^{-1}(\zeta'-a_\mu(\omega))^{-1}=(\zeta-a_\mu(\omega))^{-1}-(\zeta'-a_\mu(\omega))^{-1},$$

we get

$$\begin{aligned}q_{\mu/2}(\omega)^2&=\frac{1}{(2\pi i)^2}\int_\gamma\int_{\gamma'}\zeta^{1/2}(\zeta')^{1/2}(\zeta-a_\mu(\omega))^{-1}(\zeta'-a_\mu(\omega))^{-1}d\zeta d\zeta'\\&=\frac{1}{2\pi i}\int_\gamma\Big(\frac{1}{2\pi i}\int_{\gamma'}\frac{(\zeta')^{1/2}}{\zeta'-\zeta}d\zeta'\Big)\zeta^{1/2}(\zeta-a_\mu(\omega))^{-1}d\zeta\end{aligned}$$

$$-\frac{1}{2\pi i}\int_{\gamma'}\Big(\frac{1}{2\pi i}\int_{\gamma}\frac{\zeta^{1/2}}{\zeta'-\zeta}d\zeta\Big)(\zeta')^{1/2}(\zeta'-a_\mu(\omega))^{-1}d\zeta'$$
$$=\frac{1}{2\pi i}\int_{\gamma}\zeta(\zeta-a_\mu(\omega))^{-1}d\zeta=a_\mu(\omega),$$

for

$$\frac{1}{2\pi i}\int_{\gamma}\frac{\zeta^{1/2}}{\zeta'-\zeta}d\zeta=0,$$

since $\zeta'\in\gamma'$ is "outside" γ. This concludes the proof. □

Remark 3.3.18. When working with a symbol $0<a=a^*\in S(m^\mu,g;\mathsf{M}_M)$ such that for $0<c_1<c_2$

$$c_1|v|^2_{\mathbb{C}^N}\le\frac{\langle a(X)v,v\rangle_{\mathbb{C}^N}}{m(X)^\mu}\le c_2|v|^2_{\mathbb{C}^N},\ \forall v\in\mathbb{C}^N,\ \forall X\in\mathbb{R}^{2n}\ |X|\ge c>0,$$

one has the analogue of Lemma 3.3.17, by working with $\dfrac{a(X)}{m(X)^\mu}$, $|X|\ge c$. △

Lemma 3.3.19. *The equation*

$$q_{\mu/2}q_{\mu/2-2}+q_{\mu/2-2}q_{\mu/2}=a_{\mu-2}+\frac{i}{2}\{q_{\mu/2},q_{\mu/2}\},\ X\neq0,\tag{3.41}$$

has a unique smooth solution $q_{\mu/2-2}=q^*_{\mu/2-2}$ defined for $X\neq0$, positively homogeneous of degree $\mu/2-2$.

Proof. Call $c_{\mu-2}$ the right-hand side of (3.41). Notice that $c^*_{\mu-2}=c_{\mu-2}$. Again, we work on $\mathbb{S}^{2n-1}$ and then extend the result by homogeneity of degree $\mu/2-2$ as before. Since by Lemma 3.3.17

$$\mathrm{Spec}(q_{\mu/2}(\omega))\subset[c'_1,c'_2]\subset\mathbb{R}_+,$$

we have

$$\inf_{\substack{\zeta\in\mathrm{Spec}(q_{\mu/2}(\omega))\\ \zeta'\in\mathrm{Spec}(-q_{\mu/2}(\omega))}}|\zeta-\zeta'|\ge c_0>0,\ \forall\omega\in\mathbb{S}^{2n-1}.$$

We therefore readily see that, with $\gamma\subset\{\mathrm{Re}\,\zeta>0\}$ a closed, counter-clockwise oriented curve encircling $[c'_1,c'_2]$ with $\gamma=\bar{\gamma}$,

$$q_{\mu/2-2}(\omega)=\frac{1}{2\pi i}\int_{\gamma}(\zeta-q_{\mu/2}(\omega))^{-1}c_{\mu-2}(\omega)(\zeta+q_{\mu/2}(\omega))^{-1}d\zeta.$$

Hence $q_{\mu/2-2}$ is smooth, with $q_{\mu/2-2}(\omega)^*=q_{\mu/2-2}(\omega)$ for all $\omega\in\mathbb{S}^{2n-1}$, and defining $q_{\mu/2-2}(X)=|X|^{\mu/2-2}q_{\mu/2-2}(X/|X|)$ for $X\neq0$ proves the lemma. □

Remark 3.3.20. Given an operator A as in Theorem 3.3.13, it is useful to introduce a calculus with parameters tailored to A, for understanding, for example, the resolvent $(A-\lambda)^{-1}$. For $X=(x,\xi)\in\mathbb{R}^{2n}$, $\lambda\in\mathbb{C}$, consider (for reasons of homogeneity, as suggested by the example $|x|^2+|\xi|^2-\lambda$) the weight m_λ and the metric g_λ defined by

$$m_\lambda(X):=(1+|X|^2+|\lambda|^{2/\mu})^{1/2},$$

$$g_{\lambda,X}:=\frac{|dX|^2}{m_\lambda(X)^2}.$$

Then one sees that the metric g_λ is admissible (with structural constants **independent of** λ) and that $m_\lambda(X)^{\pm 1}$ is a g-admissible function for all $\lambda\in\mathbb{C}$ (**uniformly** in λ). One may then work with the class of symbol $S(m_\lambda^\mu, g_\lambda;\lambda\in\Omega)$ of functions $a\in C^\infty(\mathbb{R}^{2n}\times\Omega)$ satisfying the following estimates: *For any given* $\alpha\in\mathbb{Z}_+^{2n}$ *there is* $C_\alpha>0$ *such that*

$$|\partial_X^\alpha a(X,\lambda)|\le C_\alpha m_\lambda(X)^{\mu-|\alpha|},\ \forall X\in\mathbb{R}^{2n},\ \forall\lambda\in\Omega.$$

In this framework, the **classical** symbols are defined to be those symbols $a\in S(m_\lambda^\mu, g_\lambda;\lambda\in\Omega)$, Ω now a cone of $\mathbb{C}$, for which there exists $\{a_{\mu-2j}\}_{j\ge 0}\subset C^\infty(\mathbb{R}^{2n}\times\Omega\setminus\{(0,0)\})$ such that

$$a(X,\lambda)\sim\sum_{j\ge 0}a_{\mu-2j}(X,\lambda),$$

where

$$a_{\mu-2j}(tX,t^\mu\lambda)=t^{\mu-2j}a_{\mu-2j}(X,\lambda),\ \forall t>0,\ |X|+|\lambda|\neq 0.$$

In this case the excision function χ to be used is a function constructed by taking $\omega\in C^\infty(\mathbb{R}_t)$, with $0\le\omega\le 1$, $\omega\equiv 0$ for $|t|\le 1/2$ and $\omega\equiv 1$ for $|t|\ge 1$, and then by putting $\chi(X,\lambda):=\omega(|X|^2+|\lambda|^{2/\mu})$. (See Helffer [17] and Shubin [67] for more on this.) △

Remark 3.3.21. It is interesting to note that, when $\mu=2$, in Theorem 3.3.13 above it is not necessary to assume that $a_2>0$ for deducing the semi-boundedness from below of the operator A. It suffices that $a_2\ge 0$. One has in fact the following result of Hörmander's, his celebrated Sharp-Gårding inequality, which is valid even for infinite-size systems (see Hörmander [29, Theorem 18.6.14]).

Theorem 3.3.22 (Hörmander's Sharp-Gårding inequality). *Let g be admissible (hence with uncertainty Planck's function $h\le 1$ everywhere). Suppose that $0\le a=a^*\in S(h^{-1},g;\mathscr{L}(\mathsf{H},\mathsf{H}))$, where H is a Hilbert space. Then there exists $C>0$ such that*

$$(a^{\mathrm{w}}(x,D)u,u)\ge -C\|u\|_0^2,\ \forall u\in\mathscr{S}(\mathbb{R}^n;\mathsf{H}).$$

Since, in the case of the global metric, saying that $a\in S(h^{-1},g)$ is equivalent to saying that $a\in S(m^2,g)$, we conclude from this theorem that any given "second order" (in the sense of the growth measured by m) global system, even infinite-dimensional, is semi-bounded from below in L^2. The "fourth order" case (that is, the case $S(h^{-2},g) = S(m^4,g)$) for nonnegative global symbols is completely different, and is the celebrated Fefferman-Phong inequality. It is known that in general this inequality does not hold for systems (see the references to Brummelhuis, Hörmander and Parmeggiani in [53]). However, we have recently proved that it is indeed valid for large classes of systems of PDEs (see Parmeggiani [53, 54, 56, 57]). △

3.4 Notes

For the Weyl-Hörmander calculus, the reader is addressed to the original paper by Hörmander [27] and Hörmander's book [29].

For the global calculus, the reader is addressed to Helffer's book [17] and to Shubin's book [67].

The L^2-continuity theorem was first proved in the class $S^m_{1/2,1/2}$ by Calderón and Vaillancourt and then generalized by many people (Beals and Fefferman, Cotlar and Stein, Unterberger and others; see the references in Hörmander [27] or [29]) to more general pseudodifferential calculi.

Chapter 4
The Spectral Counting Function $N(\lambda)$ and the Behavior of the Eigenvalues: Part 1

To understand the eigenvalues of an elliptic global operator, a very useful, basic and general tool is the *Minimax Principle*. After recalling it, we shall use the Minimax Principle to study the first properties of the spectral counting function, and of the behavior of the large eigenvalues, of an elliptic global operator. Remark that everything we say in this section holds also for matrix-valued operators.

4.1 The Minimax Principle

In the first place, it is very useful to recall the **Minimax Principle** in the abstract setting.

Theorem 4.1.1 (Minimax Principle). *Let $A\colon D(A)\subset H \longrightarrow H$ be an unbounded, densely defined and closed self-adjoint operator in the (infinite-dimensional) Hilbert space H, semi-bounded from below (that is, $A \geq -CI$ on $D(A)$ for some constant $C>0$), and with compact resolvent. Then, with* Spec(A) *given by the sequence*

$$-\infty < \lambda_1 \leq \lambda_2 \leq \ldots \to +\infty,$$

where the eigenvalues are **repeated** *according to multiplicity, for the j-th eigenvalue one has*

$$\lambda_j = \sup_{u_1,\ldots,u_{j-1}\,\text{lin. ind.}} \left[\inf_{\substack{u\in\text{Span}\{u_1,\ldots,u_{j-1}\}^{\perp}\cap D(A)\\ \|u\|=1}} (Au,u) \right], \tag{4.1}$$

or, in other words,

$$\lambda_j = \sup_{\substack{\text{codim}\,V=j-1\\ V\,\text{closed}}} \left[\inf_{\substack{u\in V\cap D(A)\\ \|u\|=1}} (Au,u) \right].$$

Proof. (We follow Dimassi-Sjöstrand's book [7].) The first step is to see that $D(A)\cap E$ is always dense in E, where $E = \text{Span}\{u_1,\ldots,u_{j-1}\}^{\perp}$ and where

A. Parmeggiani, *Spectral Theory of Non-Commutative Harmonic Oscillators: An Introduction*, Lecture Notes in Mathematics 1992,
DOI 10.1007/978-3-642-11922-4_4, © Springer-Verlag Berlin Heidelberg 2010

$u_1,\dots,u_{j-1}$ are linearly independent. Let $v_1,\dots,v_{j-1} \in D(A)$ be close to $u_1,\dots,u_{j-1}$ in norm. Then the space $F := \mathrm{Span}\{v_1,\dots,v_{j-1}\} \subset D(A)$ has dimension $j-1$ and is transversal to E, whence it follows that there exists a unique bounded projection $\pi_E\colon H \longrightarrow H$ with $\mathrm{Im}\,\pi_E = E$, $\mathrm{Ker}\,\pi_E = F$. Then

$$D(A) \ni u = \underbrace{u'}_{\in F} + \underbrace{u''}_{\in E} = \underbrace{u'}_{\in D(A)} + \pi_E(u) \Longrightarrow \pi_E(u) = u - u' \in D(A).$$

Thus $\pi_E\big|_{D(A)}\colon D(A) \longrightarrow D(A)$, and we also have that $\pi_E(D(A)) = D(A) \cap E$ is dense in E.

The next step is to prove that

$$\Lambda_j(u_1,\dots,u_{j-1}) := \inf_{\substack{u\in E\cap D(A)\\ \|u\|=1}} (Au,u) \le \lambda_j. \tag{4.2}$$

Let in fact $e_1,\dots,e_j \in D(A)$ be an *orthonormal* (ON) system of eigenfunctions of A corresponding to the eigenvalues $\lambda_1 \le \dots \le \lambda_j$. Then $\mathrm{Span}\{e_1,\dots,e_j\} \cap E \ne \{0\}$ (notice that the first space has dimension j and the second one has codimension $j-1$). We then let $u := \sum_1^j c_k e_k$ be a normalized vector in the intersection, so that $\sum_1^j |c_k|^2 = 1$. It is clear that $u \in D(A) \cap E$, and that

$$(Au,u) = \sum_{k=1}^{j} \lambda_k |c_k|^2 \le \lambda_j \sum_{k=1}^{j} |c_k|^2 = \lambda_j,$$

from which we deduce (4.2). We finally prove (4.1). We have just seen that the right-hand side is $\le \lambda_j$. But if we take $u_k = e_k$, $1 \le k \le j-1$, and consider the sequence $\{e_j\}_{j\ge 1} \subset D(A)$ of the ON eigenvectors of A, we may then take $u \in \mathrm{Span}\{e_1,\dots,e_{j-1}\}^\perp \cap D(A)$ of the form $u = \sum_{k\ge j} c_k e_k$ with

$$\sum_{k\ge j} |\lambda_k c_k|^2 < +\infty, \text{ and } \sum_{k\ge j} |c_k|^2 = 1,$$

so that

$$(Au,u) = \sum_{k\ge j} \lambda_j |c_k|^2 \ge \lambda_j \sum_{k\ge j} |c_k|^2 = \lambda_j,$$

which yields

$$\inf_{\substack{u\in \mathrm{Span}\{e_1,\dots,e_{j-1}\}^\perp \cap D(A)\\ \|u\|=1}} (Au,u) = \lambda_j.$$

This concludes the proof. □

Corollary 4.1.2. *Formula (4.1) is equivalent to*

$$\lambda_j = \sup_{u_1,\ldots,u_{j-1}} \left[\inf_{\substack{u \in \mathrm{Span}\{u_1,\ldots,u_{j-1}\}^\perp \cap D(A) \\ \|u\|=1}} (Au,u) \right], \tag{4.3}$$

where it is important to notice that the $u_1,\ldots,u_{j-1}$ may be taken **not necessarily** *linearly independent.*

Proof. Denote by λ_j' the right-hand side of (4.3). Then it is clear that $\lambda_j \le \lambda_j'$. Suppose that $\lambda_j' > \lambda_j$, that is, λ_j' is obtained by using vectors $u_1,\ldots,u_{j-1}$ which are **not** linearly independent. Hence for any given $\varepsilon > 0$ there exist $u_1^\varepsilon,\ldots,u_{j-1}^\varepsilon$ with $\dim \mathrm{Span}\{u_1^\varepsilon,\ldots,u_{j-1}^\varepsilon\} \le j-2$, such that (with the notation of (4.2), where $E = \mathrm{Span}\{u_1^\varepsilon,\ldots,u_{j-1}^\varepsilon\}^\perp$)

$$\lambda_j' - \varepsilon < \Lambda_j(u_1^\varepsilon,\ldots,u_{j-1}^\varepsilon) \le \lambda_{j-1} \le \lambda_j,$$

by Theorem 4.1.1, whence a contradiction. □

Remark 4.1.3. Formula (4.3) also follows by observing that the suprema of the $\Lambda_j(u_1,\ldots,u_{j-1})$ in (4.1) and (4.3) are equal, because $j-1$ vectors can be always ε-approximated by $j-1$ **linearly independent** vectors (in other words, systems of $j-1$ vectors that are linearly independent are always dense in systems of $j-1$ vectors). △

To prove this claim, it suffices to show that given j vectors $u_1,\ldots,u_j$ and given any $\varepsilon > 0$ we may always find vectors $v_1,\ldots,v_j$ that are linearly independent and such that $\|v_k - u_k\| < \varepsilon$, $1 \le k \le j$. To see this we first consider $\mathrm{Span}\{u_1,\ldots,u_j\} =: V = \mathrm{Span}\{u_{i_1},\ldots,u_{i_\ell}\}$, where $I := \{i_1,\ldots,i_\ell\} \subset \{1,\ldots,j\} =: J$ and $u_{i_1},\ldots,u_{i_\ell}$ are linearly independent. We order the elements in $J \setminus I = I'$ (that we suppose non-empty, for otherwise there is nothing to prove) in increasing order, so that $I' = \{i_1',\ldots,i_{j-\ell}'\}$, with $i_1' < \ldots < i_{j-\ell}'$. Recall that $u_k \in V$ for all $k \in I'$. Consider then $V^\perp$, so that $H = V \oplus V^\perp$, with orthogonal sum. Let $\{e_k\}_{k\ge1} \subset V^\perp$ be an ON basis of $V^\perp$. Define

$$v_{i_k'} = u_{i_k'} + \frac{\varepsilon}{j} e_k, \quad k = 1,\ldots,j-\ell,$$

and

$$v_{i_k} = u_{i_k}, \quad 1 \le k \le \ell.$$

It is then clear that the $v_1,\ldots,v_j$ are now linearly independent and that

$$\sum_{i=1}^{j} \|u_i - v_i\| < \varepsilon.$$

This gives the claim of the remark. □

We conclude this section with the following important application (called "the Rayleigh-Ritz technique") of Corollary 4.1.2 (see Reed-Simon [61, p. 82]).

Theorem 4.1.4. *Let A be as in Theorem 4.1.1. Let $V \subset D(A)$ be a j-dimensional subspace, and let $\Pi_V : H \longrightarrow H$ be the orthogonal projection of H onto V. Write $A_V := \Pi_V A \Pi_V$, and denote by $\hat{\lambda}_1 \leq \ldots \leq \hat{\lambda}_j$ the eigenvalues of the $j \times j$ matrix $A_V\big|_V$. Then*

$$\lambda_k \leq \hat{\lambda}_k, \ 1 \leq k \leq j.$$

Proof. By the Minimax Principle (in the form (4.3)), $A_V\big|_V$ has eigenvalues given by

$$\begin{aligned}
\hat{\lambda}_k &= \sup_{u_1,\ldots,u_{k-1}\in V} \left[\inf_{\substack{u\in V\cap \mathrm{Span}\{u_1,\ldots,u_{k-1}\}^{\perp} \\ \|u\|=1}} (Au,u) \right] \\
&= \sup_{u_1,\ldots,u_{k-1}\in H} \left[\inf_{\substack{u\in V\cap \mathrm{Span}\{\Pi_V(u_1),\ldots,\Pi_V(u_{k-1})\}^{\perp} \\ \|u\|=1}} (Au,u) \right] \\
&\qquad \text{(using the fact that for } u\in V \text{ one has } (u,\Pi_V(u')) = (u,u')) \\
&= \sup_{u_1,\ldots,u_{k-1}\in H} \left[\inf_{\substack{u\in V\cap \mathrm{Span}\{u_1,\ldots,u_{k-1}\}^{\perp} \\ \|u\|=1}} (Au,u) \right] \\
&\geq \sup_{u_1,\ldots,u_{k-1}\in H} \left[\inf_{\substack{u\in D(A)\cap \mathrm{Span}\{u_1,\ldots,u_{k-1}\}^{\perp} \\ \|u\|=1}} (Au,u) \right] = \lambda_k,
\end{aligned}$$

which concludes the proof. □

4.2 The Spectral Counting Function

Fix an elliptic $A \in \mathrm{OP}S_{\mathrm{cl}}(m^{\mu}, g)$ (possibly matrix-valued), $\mu \geq 1$, with real-valued symbol and principal symbol $a_{\mu} > 0$. Then we know that $A = A^*$. It is no restriction, by Theorem 3.3.13 and Remark 3.3.16, to assume that $A > 0$, for it suffices to consider $A + c$ for a suitable constant $c \in \mathbb{R}$. Then the spectrum $\mathrm{Spec}(A)$ of A is made by a sequence of eigenvalues

$$0 < \lambda_1 \leq \lambda_2 \leq \ldots \longrightarrow +\infty,$$

where the eigenvalues are repeated according to their multiplicities.

Remark 4.2.1. Notice that if $A \in \mathrm{OP}S_{\mathrm{cl}}(m^{\mu}, g)$, $\mu > 0$, and c is a constant, then $A + c \in \mathrm{OP}S(m^{\mu}, g)$, and it is **classical** only when $\mu \in 2\mathbb{N}$. However, since c times the identity operator commutes with everything and is continuous on any desired space and, in case $A = A^*$ has a discrete spectrum, adding c amounts to shifting each eigenvalue of A by the same constant c, this won't be causing any problems throughout these notes, and we shall keep thinking of $A + c$ as a classical operator.

$\triangle$

Let $\lambda \in [0, +\infty) =: \overline{\mathbb{R}}_+$ (recall also that $\mathbb{R}_+ = (0, +\infty)$), and define the **spectral counting functions** (associated with A)

$$N(\lambda) = \sharp\{j \in \mathbb{N};\ \lambda_j < \lambda\}, \text{ and } N_0(\lambda) = \sharp\{j \in \mathbb{N};\ \lambda_j \leq \lambda\}. \tag{4.4}$$

When comparing two elliptic global pseudodifferential operators A and B, we shall at times write N_A and N_B to denote the respective counting functions $N(\lambda)$. We shall study in this chapter the first basic properties of the function N. The function N_0 is also introduced (some authors prefer using it in place of N). As far as the asymptotic properties of N and N_0 are concerned, there is no difference, and it is just a matter of taste.

We next list a few immediate properties of the spectral counting functions:

- *$N(\lambda)$ and $N_0(\lambda)$ are non-decreasing functions, with limit $+\infty$ at $+\infty$;*
- *One has*
$$N(\lambda_j) < N_0(\lambda_j),\ \ \forall \lambda_j \in \mathrm{Spec}(A),$$
and
$$N(\lambda) = N_0(\lambda),\ \ \forall \lambda \notin \mathrm{Spec}(A);$$
- *$N(\lambda_j) \leq j$, for all $\lambda_j \in \mathrm{Spec}(A)$.*
Notice that this might not be true for $N_0(\lambda)$. One may just think of a case in which $\lambda_j = \lambda_{j+1} = \ldots = \lambda_{j+k_j}$, for some j and k_j, so as to have $N_0(\lambda_j) = j + k_j$.

The next proposition (see Helffer [17]) shows how a-priori inequalities are used to compare the behavior of eigenvalues related to different elliptic global pseudodifferential operators (this also holds for elliptic operators on a compact boundaryless manifold).

This will then be used (see Theorem 4.3.4, Theorem 4.4.1, and Corollary 4.4.3 below) to obtain a first information on the behavior of the eigenvalues $\lambda_j(A)$ as $j \to +\infty$ of an elliptic global pseudodifferential operator, by using the harmonic oscillator as a reference operator B.

Proposition 4.2.2. *Suppose that $A, B \colon D \subset L^2(\mathbb{R}^n; \mathbb{C}^N) \longrightarrow L^2(\mathbb{R}^n; \mathbb{C}^N)$ be positive, self-adjoint operators with compact resolvent, with same domain D such that $\mathscr{S}(\mathbb{R}^n; \mathbb{C}^N) \subset D$. Suppose there exist $C_1, C_2 > 0$ with*

$$\|Au\|^2 \leq C_1\left(\|Bu\|^2 + \|u\|^2\right),\ \ \forall u \in \mathscr{S}(\mathbb{R}^n; \mathbb{C}^N),$$

$$\|Bu\|^2 \leq C_2\left(\|Au\|^2 + \|u\|^2\right),\ \ \forall u \in \mathscr{S}(\mathbb{R}^n; \mathbb{C}^N).$$

Then

$$\lambda_j(A) \approx \lambda_j(B), \text{ as } j \to +\infty,$$

that is, one can find $c_1, c_2 > 0$ *and* $j_0 \in \mathbb{N}$ *such that*

$$c_1\lambda_j(A) \le \lambda_j(B) \le c_2\lambda_j(A), \ \forall j \ge j_0.$$

Hence, by possibly changing the constants, we may always suppose that

$$c_1\lambda_j(A) \le \lambda_j(B) \le c_2\lambda_j(A), \ \forall j \in \mathbb{N}.$$

Proof. One rewrites the above inequalities as

$$(A^2u,u) \le C_1\big((B^2+I)u,u\big), \ (B^2u,u) \le C_2\big((A^2+I)u,u\big), \ \forall u \in \mathscr{S}(\mathbb{R}^n;\mathbb{C}^N),$$

whence from the Minimax Principle (see (4.1))

$$\lambda_j(A)^2 \le C_1(\lambda_j(B)^2+1), \text{ and } \lambda_j(B)^2 \le C_2(\lambda_j(A)^2+1),$$

for all $j \in \mathbb{N}$. Since $\lambda_j(A), \lambda_j(B) \to +\infty$ as $j \to +\infty$, there exists $j_0 \in \mathbb{N}$ such that $\lambda_j(A), \lambda_j(B) \ge 1$ for all $j \ge j_0$, whence the desired conclusion. □

To see how this preliminary information influences the behavior of the function $N(\lambda)$ we show the following corollary.

Corollary 4.2.3. *Let* A,B *be as in Proposition 4.2.2. Suppose there exist* $C_1,C_2>0$ *such that*

$$C_1\lambda_j(A) \le \lambda_j(B) \le C_2\lambda_j(A), \ \forall j \in \mathbb{N}. \tag{4.5}$$

Then, upon writing N_A *(resp.* N_B*) for the counting-fuction associated with* A *(resp.* B*), we have*

$$N_A(C_2^{-1}\lambda) \le N_B(\lambda) \le N_A(C_1^{-1}\lambda), \ \forall \lambda \in [0,+\infty). \tag{4.6}$$

Proof. Let $\tau \ge 0$ and let

$$J_A(\tau) := \{j \in \mathbb{N};\ \lambda_j(A) < \tau\}, \ J_B(\tau) := \{j \in \mathbb{N};\ \lambda_j(B) < \tau\}.$$

Then from (4.5) we have that

$$J_A(\tau) \subset J_B(C_2\tau) \subset J_A(\frac{C_2}{C_1}\tau),$$

whence

$$N_A(\tau) \le N_B(C_2\tau) \le N_A(\frac{C_2}{C_1}\tau),$$

for all $\tau \ge 0$. Setting $\tau = \lambda/C_2$ gives (4.6). □

Remark 4.2.4. From Corollary 4.2.3 we have that if we know that (say) $N_A(\lambda) = c_A\lambda^m(1+o(1))$ as $\lambda \to +\infty$, then we may conclude only that $N_B(\lambda) \approx \lambda^m$ as $\lambda \to +\infty$, that is, there are constants $C_1, C_2, \lambda_0 > 0$ such that

$$C_1\lambda^m \le N_B(\lambda) \le C_2\lambda^m, \ \forall \lambda \ge \lambda_0.$$

△

We now prepare the ground for exploiting Proposition 4.2.2 in the case in which A is an elliptic global pseudodifferential system (with positive-definite principal symbol) in $\mathbb{R}^n$ and $B = p_0^{\mathrm{w}}(x,D)I$, where I is the 2×2, or $N \times N$, identity matrix, and where $p_0(x,\xi) = (|x|^2+|\xi|^2)/2$ (for simplicity, we shall drop the dependence on n in the notation for the harmonic oscillator). In particular, the results of the next section apply to the case in which A is our elliptic NCHO $Q^{\mathrm{w}}_{(\alpha,\beta)}(x,D)$ (for $\alpha,\beta > 0$, $\alpha\beta > 1$).

In order to deduce the growth-rate of the eigenvalues $\lambda_j(A)$ of A, we have to understand $N_B(\lambda)$ for B and deduce the growth of the eigenvalues $\lambda_j(B)$ as j^{power}. This is done in the next section.

4.3 The Spectral Counting Function of the (Scalar) Harmonic Oscillator

We have the following well-known lemmas.

Lemma 4.3.1. *Let $p(x,\xi) = |x|^2+|\xi|^2$. For the spectral counting function $N(\lambda)$ associated with $p^{\mathrm{w}}(x,D)$ we have*

$$N(\lambda) \sim \Big((2\pi)^{-n} \iint_{p(x,\xi)\le 1} dx d\xi\Big)\lambda^n, \ \lambda \to +\infty.$$

Proof. It is well-known that the eigenvalues of $p^{\mathrm{w}}(x,D)$ are

$$\lambda = \sum_{j=1}^n (2\alpha_j+1) = n+2\sum_{j=1}^n \alpha_j, \ \alpha = (\alpha_1,\ldots,\alpha_n) \in \mathbb{Z}_+^n.$$

Hence, if $\mu = \alpha_1+\ldots+\alpha_n (= |\alpha|)$, one has $\lambda = n+2\mu$. Now,

$$\begin{aligned}\sharp\{\alpha\in\mathbb{Z}_+^n;\ |\alpha|<\mu+1\} &= \sharp\{\alpha\in\mathbb{Z}_+^n;\ |\alpha|\le\mu\} \\ &= \sum_{j=0}^{\mu}\sharp\{\alpha\in\mathbb{Z}_+^n; |\alpha|=j\} = \sum_{j=0}^{\mu}\binom{j+n-1}{n-1} = \binom{\mu+n}{n} \\ &= \frac{(\mu+n)(\mu+n-1)\ldots(\mu+1)}{n!} \sim \frac{\mu^n}{n!}, \end{aligned} \tag{4.7}$$

as $\mu \to +\infty$, so that, for $\lambda = n+2+2\mu$,

$$N(n+2+2\mu) \sim \frac{\mu^n}{n!}, \text{ as } \mu \to +\infty.$$

On the other hand,

$$(2\pi)^{-n} \iint_{p(x,\xi)\leq\lambda} dxd\xi = \lambda^n (2\pi)^{-n} \iint_{|x|^2+|\xi|^2\leq 1} dxd\xi = \frac{\lambda^n}{2^n n!},$$

and when $\lambda = n+2+2\mu$,

$$\frac{\lambda^n}{2^n n!} \sim \frac{\mu^n}{n!}, \text{ as } \mu \to +\infty,$$

which proves the claim. □

For the sake of completeness we prove formula (4.7), that is, we show that

$$\binom{j+n}{n} = \sharp\{\alpha \in \mathbb{Z}_+^n;\ |\alpha| \leq j\}. \tag{4.8}$$

Proof (of (4.8)). We proceed by induction on the dimension n. The case $n = 1$ is obviously true. Now, we have

$$\sharp\{\alpha \in \mathbb{Z}_+^n;\ |\alpha| \leq j\} = \sum_{k=0}^{j} \sharp\{\alpha \in \mathbb{Z}_+^n;\ |\alpha| = k\},$$

and, writing $\alpha = (\alpha', \alpha_n) \in \mathbb{Z}_+^{n-1} \times \mathbb{Z}_+$,

$$\begin{aligned}\sharp\{\alpha \in \mathbb{Z}_+^n;\ |\alpha| = k\} &= \sum_{\alpha_n=0}^{k} \sharp\{\alpha' \in \mathbb{Z}_+^{n-1};\ |\alpha'| = k-\alpha_n\} \\ &= \sharp\{\alpha' \in \mathbb{Z}_+^{n-1};\ |\alpha'| \leq k\} = \binom{k+n-1}{n-1},\end{aligned}$$

by the induction hypothesis. We must thus prove that

$$\sum_{k=0}^{j} \binom{k+n-1}{n-1} = \binom{j+n}{n}.$$

But this follows from the formula

$$\binom{n}{k} = \binom{n-1}{k} + \binom{n-1}{k-1},$$

for (with the convention that $\binom{n-1}{n}=0$)

$$\sum_{k=0}^{j}\binom{k+n-1}{n-1}=\sum_{k=0}^{j}\left(\binom{k+n}{n}-\binom{k+n-1}{n}\right)=\binom{j+n}{n}-\binom{n-1}{n},$$

which proves the claim. □

We may hence prove the following lemma, which gives the behavior of the eigenvalues of $p_0^{\mathrm{w}}(x,D)$ as $j\to+\infty$. Note that having the asymptotic behavior of $N(\lambda)$ given by Lemma 4.3.1 would suffice in case the eigenvalues of $p_0^{\mathrm{w}}(x,D)$ were all simple, for $N(\lambda_{j+1})=j$ then, but this is not the case for $p_0^{\mathrm{w}}(x,D)$ when $n\geq 2$. Note that a harmonic oscillator with simple eigenvalues is given, for example, by the Weyl-quantization of

$$\sum_{j=1}^{n}\mu_j(x_j^2+\xi_j^2),\ \textit{where } \mu_1,\dots,\mu_n>0 \textit{ are } \mathbb{Q}\textit{-independent.}$$

Lemma 4.3.2. *For the eigenvalues λ_j (repeated according to the multiplicity) of $p_0^{\mathrm{w}}(x,D)$ we have*

$$\lambda_j\approx j^{1/n},\ \textit{as}\ j\to+\infty. \tag{4.9}$$

Proof. The case $n=1$ is trivial, for the eigenvalues of $p_0^{\mathrm{w}}(x,D)$ are then all simple and, by Lemma 4.3.1, we have

$$j=N(\lambda_{j+1})\approx\lambda_j,\ \text{as}\ j\to+\infty,$$

where $\lambda_j=j-1/2$, $j\geq 1$. Therefore we consider the case $n>1$. Let m_{k+1} be the multiplicity of the eigenvalues $\lambda_j=|\alpha|+n/2$, when $\alpha\in\mathbb{Z}_+^n$ and $|\alpha|=k$. Then

$$m_{k+1}=\binom{k+n-1}{n-1},$$

and, for $k\geq 0$,

$$\binom{k+n}{n}=\sum_{\ell=0}^{k}m_{\ell+1}=\sum_{\ell=0}^{k}\binom{\ell+n-1}{n-1}=:M_{k+1}.$$

We now consider the intervals of integers $I_k:=\mathbb{N}\cap(M_k,M_{k+1}]$, $k\geq 1$, and write

$$\mathbb{N}=\{M_1\}\cup\bigcup_{k\geq 1}I_k,\quad M_1=m_1=1,$$

the union being disjoint. Since $\lambda_1 \leq \lambda_2 \leq \ldots$ are the eigenvalues of $p_0^{\mathrm{w}}(x,D)$, we have that

for every $j > 1$ there exists $k = k_j \in \mathbb{N}$ such that $j \in I_k$ and $\lambda_j = \lambda_{M_{k+1}}$,

and that

$$N(\lambda_{M_{k+1}}) = M_k \approx (\lambda_{M_{k+1}})^n, \text{ as } k \to +\infty,$$

by Lemma 4.3.1. As

$$M_k = \sum_{\ell=1}^{k} m_\ell = \sum_{\ell=0}^{k-1} m_{\ell+1} = \sum_{\ell=0}^{k-1} \binom{\ell+n-1}{n-1} = \binom{k-1+n}{n},$$

we have that

$$M_k \sim \frac{(k-1)^n}{n!} \sim \frac{k^n}{n!}, \quad k \to +\infty.$$

Since $M_k < j \leq M_{k+1}$, we get that $j \to +\infty$ iff $k \to +\infty$ and

$$(k-1)^n \lesssim j \lesssim k^n, \text{ as } j \to +\infty,$$

whence $j \approx k^n$ and finally

$$\lambda_j = \lambda_{M_{k+1}} \approx k \approx j^{1/n}, \text{ as } j \to +\infty.$$

This concludes the proof of the lemma. □

Remark 4.3.3. The above argument applies also in the case of $p_0^{\mathrm{w}}(x,D)I_N$. In this case we have an increase of multiplicity due to the presence of the $N \times N$ identity matrix I_N. △

Theorem 4.3.4. *Let $\mu > 0$, and let $P_0^{\mu/2}$ be the $\mu/2$-functional power of $P_0 = p_0^{\mathrm{w}}(x,D)$. By Theorem 5.5.1 below, $P_0^{\mu/2} \in \mathrm{OPS}_{\mathrm{cl}}(m^\mu, g)$, that is $P_0^{\mu/2}$ is an elliptic classical global pseudodifferential operator of order μ. Denoting by $\{\lambda_j(P_0^{\mu/2})\}_{j\geq 1}$ the non-decreasing sequence of the eigenvalues of $P_0^{\mu/2}$, repeated according to multiplicity, we have*

$$\lambda_j(P_0^{\mu/2}) \approx j^{\mu/2n}, \quad as \ j \to +\infty.$$

It is important to note the power of j: it is the order of the operator divided by twice the dimension.

Proof. Since

$$\lambda_j(P_0^{\mu/2})^{2/\mu} = \lambda_j, \quad j \in \mathbb{N},$$

the proof follows immediately from (4.9). □

Remark 4.3.5. Of course, when $\mu \in 2\mathbb{N}$ Theorem 4.3.4 holds without making use of Theorem 5.5.1. △

4.4 Consequences on the Spectral Counting Function of an Elliptic Global ψdo

Using Proposition 4.2.2 we may now obtain the following behavior of the large eigenvalues of an elliptic global pseudodifferential operator, and in particular of our NCHO $Q^{\mathrm{w}}_{(\alpha,\beta)}(x,D)$ when $\alpha,\beta>0$ and $\alpha\beta>1$.

The first result we state is the following.

Theorem 4.4.1. *Let* $0<A=A^*\in \mathrm{OPS}_{\mathrm{cl}}(m^{\mu},g;\mathsf{M}_N)$, $\mu>0$, *be an elliptic, self-adjoint and positive global pseudodifferential operator. Then, upon denoting by* $\lambda_j(A)$, $j\geq 1$, *the sequence of the repeated eigenvalues of A we have*

$$\lambda_j(A)\approx j^{\mu/2n},\quad as\ j\to+\infty.$$

Proof. We start by remarking that, by virtue of the results of Section 3.3, A has a discrete spectrum, made of eigenvalues with finite multiplicities which, due to the positivity of A, diverge to $+\infty$.

Now, by Theorem 5.5.1 below, the $\mu/2$-power $P_0^{\mu/2}$ of $P_0=p_0^{\mathrm{w}}(x,D)$ is an elliptic classical global pseudodifferential operator of order μ. Put, for short, $B=P_0^{\mu/2}I_N$, where I_N is the $N\times N$ identity matrix. Let therefore Q_A and Q_B be two-sided parametrices (of order $-\mu$) for A and B, respectively, so that

$$Q_AA=I+R_A,\quad Q_BB=I+R_B,$$

where R_A and R_B are smoothing operators. Then, for all $u\in\mathscr{S}(\mathbb{R}^n;\mathbb{C}^N)$,

$$Au=A(Q_BB-R_B)u=(AQ_B)Bu-AR_Bu,$$

$$Bu=B(Q_AA-R_A)u=(BQ_A)Au-BR_Au,$$

so that, by the L^2-boundedness of AQ_B, AR_B, BQ_A and of BR_A, we get

$$\|Au\|_0^2\lesssim\|Bu\|_0^2+\|u\|_0^2,\quad\forall u\in\mathscr{S}(\mathbb{R}^n;\mathbb{C}^N),$$

and, analogously,

$$\|Bu\|_0^2\lesssim\|Au\|_0^2+\|u\|_0^2,\quad\forall u\in\mathscr{S}(\mathbb{R}^n;\mathbb{C}^N).$$

The result therefore follows from Proposition 4.2.2 and Theorem 4.3.4. □

Remark 4.4.2. Of course, when $\mu\in 2\mathbb{N}$ Theorem 4.4.1 holds without making use of Theorem 5.5.1. △

In particular, for our elliptic NCHO $Q^{\mathrm{w}}_{(\alpha,\beta)}(x,D)$, $\alpha,\beta>0$ and $\alpha\beta>1$, we therefore have the following result.

Corollary 4.4.3. *Let $A = Q^{\mathrm{w}}_{(\alpha,\beta)}(x,D)$, with $\alpha,\beta > 0$ and $\alpha\beta > 1$. Hence A is a positive elliptic system of* GPDOs *of order two, to which we may apply our results. Then*

$$\lambda_j(A) \approx j, \text{ as } j \to +\infty.$$

Recall that, given a self-adjoint operator $A > 0$ with a discrete spectrum $\{\lambda_j\}_{j\geq 1}$, the **spectral zeta function** associated with A is by definition the series

$$\zeta_A(s) = \sum_{j\geq 1} \frac{1}{\lambda_j^s}, \quad s \in \mathbb{C} \text{ with } \mathrm{Re}\, s \text{ sufficiently large.} \tag{4.10}$$

From Theorem 4.4.1 and Corollary 4.4.3 we get the following immediate information about a first region of convergence of the spectral zeta functions ζ_A and ζ_Q associated with an elliptic A and an elliptic $Q^{\mathrm{w}}_{(\alpha,\beta)}(x,D)$, respectively. We have in fact the following corollary.

Corollary 4.4.4. *The spectral zeta function ζ_A is holomorphic for* $\mathrm{Re}\, s > 2n/\mu$. *In particular, for $\alpha,\beta > 0$ and $\alpha\beta > 1$ the spectral zeta function ζ_Q is holomorphic for* $\mathrm{Re}\, s > 1$.

Proof. The proof follows immediately from the behaviors

$$\lambda_j(A) \approx j^{\mu/2n}, \text{ and } \lambda_j(Q) \approx j, \text{ as } j \to +\infty,$$

and the fact that the harmonic series $\sum_{j\geq 1} j^{-s}$ converges for $\mathrm{Re}\, s > 1$. □

Remark 4.4.5. It is well-known that for the spectral zeta function of the harmonic oscillator $H = p_0^{\mathrm{w}}(x,D)$ $(n = 1)$ one has (exercise for the reader)

$$\zeta_H(s) = \sum_{j\geq 0} \frac{1}{(j+1/2)^s} = (2^s - 1)\zeta(s), \tag{4.11}$$

where $\zeta(s)$ is the Riemann zeta function. △

Chapter 5
The Heat-Semigroup, Functional Calculus and Kernels

We shall review in this chapter some elementary properties of the heat-semigroup associated with a globally elliptic operator, along with its functional calculus and some properties of the Schwartz kernels involved that give rise to the definition of the "trace" of the operator.

5.1 Elementary Properties of the Heat-Semigroup

Let $a \in S_{\mathrm{cl}}(m^{\mu}, g; \mathsf{M}_N)$ be globally elliptic with $\mu \in \mathbb{N}$ and principal symbol $a_{\mu} = a_{\mu}^* > 0$ (as an Hermitian matrix), and suppose that $A^* = A = a^{\mathrm{w}}(x,D)$ be positive on $\mathscr{S}(\mathbb{R}^n;\mathbb{C}^N)$, that is there exists $c > 0$ such that

$$(Au,u) \geq c\|u\|_0^2,\ \forall u \in \mathscr{S}(\mathbb{R}^n;\mathbb{C}^N).$$

Let $s \in \mathbb{R}$ and let also

$$A_s \colon D(A_s) = B^{s+\mu}(\mathbb{R}^n;\mathbb{C}^N) \subset B^s(\mathbb{R}^n;\mathbb{C}^N) \longrightarrow B^s(\mathbb{R}^n;\mathbb{C}^N),\ A_s := A\big|_{B^s},$$

be a realization of A as an unbounded operator in $B^s(\mathbb{R}^n;\mathbb{C}^N)$. Then we know that $A_s = A_s^*$, $\mathrm{Spec}(A_s) = \mathrm{Spec}(A_0) =: \mathrm{Spec}(A)$, that the eigenfunctions of A_s are the eigenfunctions of A, and that they belong to $\mathscr{S}$, by the elliptic regularity (i.e., the existence of the parametrix). We have that the resolvent

$$R_s(\lambda) = (A_s - \lambda)^{-1} \colon B^s(\mathbb{R}^n;\mathbb{C}^N) \longrightarrow B^s(\mathbb{R}^n;\mathbb{C}^N)$$

is defined, bounded and holomorphic for all $\lambda \in \mathbb{C} \setminus \mathrm{Spec}(A)$. It is important to note that

$$s' \leq s \Longrightarrow R_{s'}(\lambda)\big|_{B^s} = R_s(\lambda),\ \forall \lambda \in \mathbb{C} \setminus \mathrm{Spec}(A), \tag{5.1}$$

for if $\lambda \notin \mathrm{Spec}(A)$, then $(A_{s'} - \lambda)u = f \in B^s$ is solvable in $B^{s'+\mu}$ and, by the elliptic regularity, for the solution u we have $u \in B^{s+\mu} \subset B^{s'+\mu}$.

A. Parmeggiani, *Spectral Theory of Non-Commutative Harmonic Oscillators: An Introduction*, Lecture Notes in Mathematics 1992,
DOI 10.1007/978-3-642-11922-4_5,

By virtue of (5.1), since $\bigcap_{s\in\mathbb{R}} B^s = \mathscr{S}$ and $\bigcup_{s\in\mathbb{R}} B^s = \mathscr{S}'$, we may hence define, as a continuous map,

$$R(\lambda)\colon \mathscr{S}'(\mathbb{R}^n;\mathbb{C}^N) \longrightarrow \mathscr{S}'(\mathbb{R}^n;\mathbb{C}^N),\ R(\lambda)\big|_{B^s} := R_s(\lambda),\ \forall\lambda\in\mathbb{C}\setminus\mathrm{Spec}(A). \quad (5.2)$$

Note that $R(\lambda)\colon \mathscr{S}(\mathbb{R}^n;\mathbb{C}^N) \longrightarrow \mathscr{S}(\mathbb{R}^n;\mathbb{C}^N)$ is also continuous. We may then simply write $R(\lambda)$ for the **resolvent** of A, regardless the domain used to realize A as an unbounded operator. It is hence well-known that for any given $s\in\mathbb{R}$ there is $C_s>0$ such that

$$\|R(\lambda)\|_{B^s\to B^s} \le \frac{C_s}{\mathrm{dist}(\lambda,\mathrm{Spec}(A))}.$$

Let $\delta_A := \min\mathrm{Spec}(A)$. We may therefore choose $\delta'\in(0,\delta_A)$ such that

$$D_{\delta'} := \{\lambda\in\mathbb{C};\ \mathrm{Re}\,\lambda\le\delta'\}\cup\{\lambda=\rho e^{i\phi};\ \rho\ge\frac{\delta'}{2},\ \phi\in[\frac{\pi}{4},\frac{7}{4}\pi]\}\subset\mathbb{C}\setminus\mathrm{Spec}(A).$$

Hence there exists $C_s'>0$ such that

$$\lambda\in D_{\delta'} \Longrightarrow \|R(\lambda)\|_{B^s\to B^s} \le \frac{C_s'}{1+|\lambda|}.$$

Let $\partial D_{\delta'}$ be oriented in such a way that $D_{\delta'}$ is kept to the right-hand side. Then (see Kato's book [35]) we may define

$$\begin{cases} e^{-tA} = \dfrac{1}{2\pi i}\displaystyle\int_{\partial D_{\delta'}} e^{-\lambda t}R(\lambda)d\lambda,\ t>0,\\ e^{-tA}\big|_{t=0} = I. \end{cases}$$

Hence, for any fixed $s\in\mathbb{R}$, $\{e^{-tA}\}_{t\ge0}$ is a strongly continuous semigroup in B^s with generator $-A_s$. Furthermore, for every $\delta\in(0,\delta_A)$, we have

$$\sup_{t\ge1} e^{\delta t}\|e^{-tA}\|_{B^s\to B^s} < +\infty.$$

Using the same arguments as in Chazarain-Piriou [6], one can prove the following lemma. (Recall that $\overline{\mathbb{R}}_+ = [0,+\infty)$.)

Lemma 5.1.1. *Let $f\in B^s(\mathbb{R}^n;\mathbb{C}^N)$. Put $F(t)=e^{-tA}f$. Then for every $p,j\in\mathbb{Z}_+$,*

$$t^p(\frac{d}{dt})^jF\in C^0(\overline{\mathbb{R}}_+;B^{s+(p-j)\mu}(\mathbb{R}^n;\mathbb{C}^N)),$$

and

$$\sup_{t\ge0}\|t^p(\frac{d}{dt})^jF\|_{B^{s+(p-j)\mu}} \le C_{p,j}\|f\|_{B^s}.$$

From the lemma we may therefore think of e^{-tA} as a map

$$e^{-tA}\colon \mathscr{S}(\mathbb{R}^n;\mathbb{C}^N) \longrightarrow \mathscr{S}(\overline{\mathbb{R}}_+;\mathscr{S}(\mathbb{R}^n;\mathbb{C}^N)),$$

and as a map

$$e^{-tA}\colon \mathscr{S}'(\mathbb{R}^n;\mathbb{C}^N) \longrightarrow \mathscr{S}(\overline{\mathbb{R}}_+;\mathscr{S}'(\mathbb{R}^n;\mathbb{C}^N)).$$

We also have the following fact about the heat-equation associated with $-A$.

Lemma 5.1.2. *For any given $f \in \mathscr{S}'(\mathbb{R}^n;\mathbb{C}^N)$ and $g \in \mathscr{S}(\overline{\mathbb{R}}_+;\mathscr{S}'(\mathbb{R}^n;\mathbb{C}^N))$, there exists a unique $u(t) = u \in \mathscr{S}(\overline{\mathbb{R}}_+;\mathscr{S}'(\mathbb{R}^n;\mathbb{C}^N))$ such that*

$$\begin{cases} \dfrac{du}{dt} + Au(t) = g(t), \ t > 0, \\ u(0) = f, \end{cases}$$

with

$$u(t) = e^{-tA}f + \int_0^t e^{-(t-t')A}g(t')dt'.$$

Moreover,

$$f \in \mathscr{S}(\mathbb{R}^n;\mathbb{C}^N),\ g \in \mathscr{S}(\overline{\mathbb{R}}_+;\mathscr{S}(\mathbb{R}^n;\mathbb{C}^N)) \Longrightarrow u \in \mathscr{S}(\overline{\mathbb{R}}_+;\mathscr{S}(\mathbb{R}^n;\mathbb{C}^N)).$$

5.2 Direct Definition of $\mathrm{Tr}\, e^{-tA}$

In this section we give a direct proof that the kernel $e^{-tA}(x,y)$, $t > 0$, of e^{-tA} (the **heat-kernel** of A) is a rapidly decreasing function of (x,y), and that

$$\mathrm{Tr}\, e^{-tA} = \sum_{j\geq 1} e^{-t\lambda_j},\ t > 0.$$

Let $\{\varphi_j\} \subset \mathscr{S}(\mathbb{R}^n;\mathbb{C}^N)$ be an ON systems of $L^2(\mathbb{R}^n;\mathbb{C}^N)$ made of eigenfunctions of A.

To start with, note that if $u = \sum_{j=1}^{\infty} u_j\varphi_j \in L^2$, i.e. $\sum_{j=1}^{\infty} |u_j|^2 < +\infty$, then

$$e^{-tA}u = \sum_{j=1}^{\infty} e^{-t\lambda_j}(u,\varphi_j)\varphi_j,$$

so that, $e^{-tA}\colon \mathscr{S} \longrightarrow \mathscr{S} \hookrightarrow \mathscr{S}'$ being continuous (and recalling that $(v^* \otimes w)\zeta = v^*(\zeta)w$), we have by the Schwartz-kernel Theorem that

$$e^{-tA}(x,y) = \sum_{j\geq 1} e^{-t\lambda_j}\varphi_j(y)^* \otimes \varphi_j(x) \in \mathscr{S}',\ t > 0.$$

If we proceed formally, we have

$$\mathsf{Tr}\, e^{-tA} = \int_{\mathbb{R}^n} \mathsf{Tr}\, e^{-tA}(x,x)dx = \sum_{j\geq 1} e^{-t\lambda_j} \underbrace{\int_{\mathbb{R}^n} |\varphi_j(x)|^2_{\mathbb{C}^N} dx}_{=1} = \sum_{j\geq 1} e^{-t\lambda_j},$$

where Tr denotes the matrix-trace.

We next show that the formal procedure is actually correct. Since $A\varphi_j = \lambda_j \varphi_j$, we get $A^r \varphi_j = \lambda_j^r \varphi_j$, for all $r \in \mathbb{N}$, and

$$\|\varphi_j\|^2_{B^{\mu r}} \lesssim \|\varphi_j\|_0^2 + \|A^r \varphi_j\|_0^2 = (1+\lambda_j^{2r})\|\varphi_j\|_0^2. \tag{5.3}$$

(The reader may prove (5.3) as an exercise by using a parametrix of A, or else look at Lemma 5.3.1 below.) From (5.3) and (3.31) we thus get that given any $\mathscr{S}$-seminorm $|\cdot|_{p,q}$, $p,q \in \mathbb{Z}_+$, there exists r so large that

$$|\varphi_j|_{p,q} \leq C_{pq}(1+\lambda_j^{2r})\|\varphi_j\|_0^2 = C_{pq}(1+\lambda_j^{2r}). \tag{5.4}$$

We now notice that

$$(t\lambda_j)^k e^{-t\lambda_j} \leq C e^{-t\lambda_j/2}, \ t>0, \tag{5.5}$$

where $C=(2k/e)^k$. Indeed, it is easy to see that

$$\tau^a e^{-\tau} \leq \left(\frac{a}{e}\right)^a, \ \forall \tau, a > 0, \tag{5.6}$$

so that

$$\tau^a e^{-\tau} = \tau^a e^{-\tau/2} e^{-\tau/2} = \left(\frac{\tau}{2}\right)^a e^{-\tau/2} 2^a e^{-\tau/2} \leq \left(\frac{a}{e}\right)^a 2^a e^{-\tau/2} = \left(\frac{2a}{e}\right)^a e^{-\tau/2}.$$

Hence we obtain from (5.4) and (5.5), recalling that $\lambda_j \nearrow +\infty$ as $j \to +\infty$,

$$|e^{-tA}(\cdot,\cdot)|_{p,q} \leq C'_{pq} t^{-k} \sum_{j\geq 1} e^{-t\lambda_j/2} < +\infty, \ \forall t > 0,$$

for we already know that $\lambda_j \approx j^{\mu/2n}$ as $j \to +\infty$. Hence,

$$t \mapsto e^{-tA}(\cdot,\cdot) \in C^\infty(\mathbb{R}_+; \mathscr{S}(\mathbb{R}^n \times \mathbb{R}^n; \mathsf{M}_N)),$$

and

$$\mathsf{Tr}\, e^{-tA} = \sum_{j\geq 1} e^{-t\lambda_j}, \ t > 0,$$

as we wanted.

The next step, to be carried out in Chapter 6, will be to study the singularity as $t \to 0+$ of the trace of the heat-kernel. We shall accomplish this by constructing a parametrix approximation of e^{-tA} (the parametrix we are referring to is a parametrix for $d/dt + A$), that will give the sought information. After that, we shall relate the singularity of $\operatorname{Tr} e^{-tA}$ as $t \to 0+$ to the counting function $N(\lambda)$ through the Karamata theorem.

But before doing that, we continue with a section about the abstract functional calculus of an elliptic global system, a section about kernels, and finally close the chapter with a section about $f(A)$ as a global pseudodifferential operator.

5.3 Abstract Functional Calculus

We recall in this section how to obtain an abstract functional calculus for an elliptic $0 < A = A^* \in \mathrm{OPS}_{\mathrm{cl}}(m^{\mu}, g; \mathrm{M}_N)$, $\mu > 0$, with a discrete spectrum, $\operatorname{Spec}(A)$, made of a sequence $0 < \lambda_1 \le \lambda_2 \le \ldots \to +\infty$ of eigenvalues repeated according to multiplicity, and with eigenfunctions $\{\varphi_j\}_{j\in\mathbb{N}} \subset \mathscr{S}(\mathbb{R}^n;\mathbb{C}^N)$ that form an orthonormal basis of $L^2(\mathbb{R}^n;\mathbb{C}^N)$. Let $f\colon \operatorname{Spec}(A) \longrightarrow \mathbb{C}$. Define the unbounded operator $f(A)\colon D(f(A)) \subset L^2(\mathbb{R}^n;\mathbb{C}^N) \longrightarrow L^2(\mathbb{R}^n;\mathbb{C}^N)$ by

$$D(f(A)) := \Big\{u = \sum_{j\ge 1} u_j\varphi_j \in L^2(\mathbb{R}^n;\mathbb{C}^N);\ \sum_{j\ge 1} |f(\lambda_j)|^2|u_j|^2 < +\infty\Big\},$$

$$D(f(A)) \ni u = \sum_{j\ge 1} u_j\varphi_j \longmapsto f(A)u := \sum_{j\ge 1} f(\lambda_j)u_j\varphi_j.$$

Then $f(A)$ is a closed operator with dense domain, since the $\varphi_j \in D(f(A))$. If f is real-valued, then $f(A) = f(A)^*$.

Now we want to have conditions on f which ensure that $f(A)\colon \mathscr{S}(\mathbb{R}^n;\mathbb{C}^N) \longrightarrow \mathscr{S}(\mathbb{R}^n;\mathbb{C}^N)$ be continuous. The first step is to understand $D(A^p)$, for $p \in \mathbb{N}$.

Lemma 5.3.1. *Define, for $p \in \mathbb{N}$, the unbounded operator $A^p\colon D(A^p) \subset L^2 \longrightarrow L^2$, by $D(A^p) = \{u \in D(A^{p-1});\ A^{p-1}u \in D(A)\}$, and $A^p u = A(A^{p-1}u)$. Remark that the domain is dense in L^2 since $\mathscr{S} \subset D(A^p)$. Using the eigenfunctions $\{\varphi_j\}_{j\in\mathbb{N}}$, we then have*

$$D(A^p) = \Big\{u = \sum_{j\ge 1} u_j\varphi_j \in L^2;\ \sum_{j\ge 1} \lambda_j^{2p}|u_j|^2 < +\infty\Big\} = B^{p\mu}(\mathbb{R}^n;\mathbb{C}^N),$$

and

$$Au = \sum_{j\ge 1} \lambda_j u_j \varphi_j.$$

Proof. Note in the first place that $\varphi_j \in \mathscr{S}(\mathbb{R}^n;\mathbb{C}^N) \subset B^s(\mathbb{R}^n;\mathbb{C}^N)$, for all $s \in \mathbb{R}$. If $u \in D(A^p)$ we want then to prove that $\Lambda^{p\mu}u \in L^2$, where, recall, $\Lambda^{p\mu}u$ is taken in the sense of distributions. Using a parametrix Q of A^p such that $QA^p = I + R$ gives

$$u \in D(A^p) \Longrightarrow \Lambda^{p\mu} u = \Lambda^{p\mu}(QA^p - R)u = \underbrace{(\Lambda^{p\mu} Q)}_{\text{order } 0} \underbrace{A^p u}_{\in L^2} - \underbrace{\Lambda^{p\mu} R u}_{\in L^2} \in L^2.$$

And conversely, supposing $u \in B^{p\mu}$, and using a parametrix E of $\Lambda^{p\mu}$ such that $E\Lambda^{p\mu} = I + \tilde{R}$, yields

$$\mathscr{S}' \ni A^p u = A^p(E\Lambda^{p\mu} - \tilde{R})u = \underbrace{(A^p E)}_{\text{order } 0} \underbrace{\Lambda^{p\mu} u}_{\in L^2} - \underbrace{A^p \tilde{R} u}_{\in L^2} \in L^2,$$

and this concludes the proof. □

Remark 5.3.2. Note hence that for $p \in \mathbb{N}$,

$$\|u\|^2_{B^{p\mu}} \approx \|u\|^2_{D(A^p)} := \|u\|_0^2 + \|A^p u\|_0^2 \approx \|A^p u\|_0^2, \tag{5.7}$$

for, by our assumption, $\mathrm{Spec}(A) \subset (0, +\infty)$. △

When f is slowly increasing on the spectrum of A one has that $f(A)$ is continuous from $\mathscr{S}$ into itself. One has in fact the following lemma.

Lemma 5.3.3. *Suppose f is slowly increasing on* $\mathrm{Spec}(A)$, *i.e. there exists $C > 0$ and $p \in \mathbb{Z}_+$ such that*

$$|f(\lambda_j)| \leq C(1 + \lambda_j)^p, \ \forall j \in \mathbb{N}. \tag{5.8}$$

Then $f(A)\colon \mathscr{S}(\mathbb{R}^n;\mathbb{C}^N) \longrightarrow \mathscr{S}(\mathbb{R}^n;\mathbb{C}^N)$ is continuous.

Proof. For all $r \in \mathbb{Z}_+$ we have

$$|(1+\lambda_j)^r f(\lambda_j) u_j| \leq C(1+\lambda_j)^{r+p}|u_j|, \ \forall j \in \mathbb{N}. \tag{5.9}$$

By Lemma 5.3.1 we have $D(A^p) = B^{p\mu}$, whence (5.9) implies the continuity of

$$f(A)\big|_{B^{(r+p)\mu}} : D(A^{r+p}) = B^{(r+p)\mu} \longrightarrow D(A^r) = B^{r\mu}, \ \forall r \in \mathbb{Z}_+,$$

the spaces being endowed with the respective Hilbert-space structures (and making use of (5.7)). This proves the lemma. □

Remark 5.3.4. One has the following immediate consequences.

1. If f is bounded on $\mathrm{Spec}(A)$ (e.g. $p = 0$ in (5.8)) then $f(A)\colon L^2 \longrightarrow L^2$ is continuous.
2. Since $A > 0$, we may consider $f(A)$ when $f(\lambda) = \lambda^s$, $\lambda > 0$, $s \in \mathbb{R}$.
3. If $f \in C_0^\infty(\mathbb{R})$ then $f(A)$ is smoothing and compact (and actually of finite rank).
4. If $f(\lambda) = e^{it\lambda}$, $t \in \mathbb{R}$, then

$$e^{itA}\colon L^2 \overset{\text{unitary}}{\longrightarrow} L^2, \ (e^{itA})^* = e^{-itA}.$$

5. If $f \in \mathscr{S}([0,+\infty))$, then $f(A)$ is smoothing. This is in particular the case when $f(\lambda) = e^{-t\lambda}$, for $t, \lambda > 0$. △

It is useful to have also the following proposition about the Schrödinger group.

Proposition 5.3.5. *The function*

$$t \longmapsto e^{-itA} \in C^\infty(\mathbb{R}_t; \mathscr{L}(\mathscr{S}, \mathscr{S})).$$

Moreover, with $D_t = -i\partial_t$, it solves the Cauchy problem

$$\begin{cases} (D_t + A)e^{-itA} = 0, \text{ in } C^\infty(\mathbb{R}_t; \mathscr{L}(\mathscr{S}, \mathscr{S})), \\ e^{-itA}\big|_{t=0} = I. \end{cases}$$

Proof. One directly sees, from the very definition, that for all $m, p \in \mathbb{N}$

$$t \longmapsto e^{-itA} \in C^{m-1}(\mathbb{R}_t; \mathscr{L}(D(A^{m+p}), D(A^p))),$$

and that the equation is satisfied. □

5.4 Kernels

We now come to the study of the Schwartz kernel of $f(A)$. Let hence f be slowly increasing on $\mathrm{Spec}(A)$. Then, by Lemma 5.3.3 $f(A)\colon \mathscr{S} \longrightarrow \mathscr{S}$ is continuous so that, using the embedding $\mathscr{S} \hookrightarrow \mathscr{S}'$, $f(A)\colon \mathscr{S} \longrightarrow \mathscr{S}'$ is also continuous. Hence, by the Schwartz-kernel theorem, $f(A)$ has a distribution kernel: for any given $u, v \in \mathscr{S}$ we have

$$(f(A)u, v) = \langle \mathsf{K} | \bar{v} \otimes u \rangle_{\mathscr{S}', \mathscr{S}}.$$

We have the following proposition.

Proposition 5.4.1. *If f is slowly increasing on* $\mathrm{Spec}(A)$, *then the distribution kernel* K *of $f(A)$ belongs to $\mathscr{S}'(\mathbb{R}^n \times \mathbb{R}^n; \mathsf{M}_N)$, and one has*

$$\mathsf{K} = \mathscr{S}'\text{-}\lim_{m \to +\infty} \mathsf{K}_m, \text{ where } \mathsf{K}_m(x,y) = \sum_{j=1}^{m} f(\lambda_j)\varphi_j(y)^* \otimes \varphi_j(x),$$

where $\varphi_j(y)^* \otimes \varphi_j(x)v = \langle v, \varphi_j(y)\rangle_{\mathbb{C}^N} \varphi_j(x)$, for all $v \in \mathbb{C}^N$ (that is, $\varphi_j(y)^* \otimes \varphi_j(x) = \varphi_j(x)^t \overline{\varphi_j(y)}$, colum-times-row).

Proof. Let $k_0 \in \mathbb{N}$ be such that

$$\frac{f(\lambda_j)}{(1+\lambda_j)^{2k_0}} \longrightarrow 0, \text{ as } j \to +\infty. \tag{5.10}$$

Let P_m be the L^2-orthogonal projection onto $\mathrm{Span}\{\varphi_j\}_{1\le j\le m}$. Then $R_m := f(A)P_m$ is the operator whose kernel is K_m. In addition, R_m can be extended as an operator belonging to $\mathscr{L}(D(A^{k_0}), D(A^{k_0})^*)$, where, recall, $D(A^{k_0})^*$ is the dual space of $D(A^{k_0})$. By (5.10), $\{R_m\}_{m\in\mathbb{N}}$ is a Cauchy sequence in $\mathscr{L}(D(A^{k_0}), D(A^{k_0})^*)$. Hence there exists

$$\lim_{m\to+\infty} R_m = R \in \mathscr{L}(D(A^{k_0}), D(A^{k_0})^*).$$

But for all $j \ge 1$

$$R\varphi_j = \lim_{m\to+\infty} R_m\varphi_j = \lim_{m\to+\infty} f(A)\varphi_j = f(A)\varphi_j.$$

Since $\mathscr{S} \hookrightarrow D(A^{k_0})$, and $D(A^{k_0})^* \hookrightarrow \mathscr{S}'$, we have

$$R \in \mathscr{L}(\mathscr{S}, \mathscr{S}'),$$

with

$$R\varphi_j = f(A)\varphi_j, \ \forall j \ge 1.$$

Hence $R = f(A)$ in $\mathscr{L}(\mathscr{S}, \mathscr{S}')$, so that $f(A) = \lim_{m\to+\infty} R_m$. □

Proposition 5.4.2. *If f is rapidly decreasing on* $\mathrm{Spec}(A)$ *then*

$$\mathsf{K}_m \xrightarrow{\mathscr{S}} \mathsf{K}, \ \textit{as } m \to +\infty.$$

Proof. From Proposition 5.4.1 with $k_0 = 0$ we obtain that $\mathsf{K}_m \xrightarrow{L^2} \mathsf{K}$ as $m \to +\infty$. Let A_x be the operator A acting on x-functions (or distributions). Then, with

$$(A_y^* \otimes A_x)(v^* \otimes u) = (A_y v)^* \otimes (A_x u),$$

we have, for $r \in \mathbb{N}$,

$$((A_y^*)^r \otimes A_x^r)\mathsf{K}_m(x,y) = \sum_{j=1}^{m} \lambda_j^{2r} f(\lambda_j)\varphi_j(y)^* \otimes \varphi_j(x).$$

Hence, on the one hand, by Proposition 5.4.1,

$$((A_y^*)^r \otimes A_x^r)\mathsf{K}_m \longrightarrow ((A_y^*)^r \otimes A_x^r)\mathsf{K}, \text{ as } m \to +\infty,$$

in the topology of $\mathscr{S}'(\mathbb{R}^n \times \mathbb{R}^n; \mathsf{M}_N)$, and on the other hand $\lambda_j^{2r} f(\lambda_j)$ is rapidly decreasing for $j \to +\infty$, whence it also follows that

$$((A_y^*)^r \otimes A_x^r)\mathsf{K}_m \longrightarrow ((A_y^*)^r \otimes A_x^r)\mathsf{K}, \text{ as } m \to +\infty,$$

in the topology of $L^2(\mathbb{R}^n \times \mathbb{R}^n; \mathsf{M}_N)$, for all $r \in \mathbb{Z}_+$. Since A_y^* and A_x are elliptic we obtain, by a natural vector-valued regularity theorem for A_x or A_y^* starting from $L^2_{x,y}$, that

$$\mathsf{K}_m \longrightarrow \mathsf{K}, \text{ as } m \to +\infty,$$

in the topology of $B_x^r \hat{\otimes} B_y^r$, for all $r \in \mathbb{Z}_+$. Since $B_x^r \hat{\otimes} B_y^r \subset B_{x,y}^r$ and $\bigcap_{r \in \mathbb{Z}_+} B_{x,y}^r = \mathscr{S}_{x,y}$, the claim follows. □

Corollary 5.4.3. *If f is rapidly decreasing on* Spec(A) *then*

$$\operatorname{Tr} f(A) := \sum_{j \geq 1} f(\lambda_j) = \int_{\mathbb{R}^n} \mathsf{Tr}\big(\mathsf{K}(x,x)\big)dx,$$

where, recall, Tr *denotes the matrix-trace.*

To make this section self-contained, we prove the following "rough" information on the behavior of the eigenvalues of A.

Proposition 5.4.4. *There exists $k_0 \in \mathbb{N}$ such that*

$$\sum_{j=1}^{\infty} \frac{1}{\lambda_j^{2k_0}} < +\infty.$$

Proof. For simplicity we consider the scalar case. We know that the operator $A^{-k} \in \mathrm{OP}S_{\mathrm{cl}}(m^{-k\mu}, g)$, $k \in \mathbb{N}$. Using Theorem 3.1.16 (or Theorem 3.2.17), write $A^{-k} = b(x,D) + R$, where R is smoothing. Then for the Schwartz-kernel K of A^{-k} we have

$$\mathsf{K}(x,y) = \mathsf{K}_0(x,y) + \mathsf{K}_1(x,y),$$

where $\mathsf{K}_1 \in \mathscr{S}(\mathbb{R}^n \times \mathbb{R}^n)$ is the Schwartz-kernel of R, and K_0 is the Schwartz-kernel of $b(x,D)$, given by

$$\mathsf{K}_0(x,y) = (2\pi)^{-n} \int e^{i\langle x-y, \xi\rangle} b(x,\xi) d\xi \in \mathscr{S}'(\mathbb{R}^n \times \mathbb{R}^n)$$

(in the sense of oscillatory integrals, see Helffer [17] or Shubin [67]). Let us choose in the first place the integer k so that $k\mu > n$. Then $\mathsf{K}_0 \in C(\mathbb{R}^n \times \mathbb{R}^n) \cap L^\infty(\mathbb{R}^n \times \mathbb{R}^n)$. If k is picked so that $k\mu > n+1$, we have in addition that

$$\begin{aligned} y_j \mathsf{K}_0(x,y) &= (2\pi)^{-n} \int y_j e^{i\langle x-y,\xi\rangle} b(x,\xi) d\xi \\ &= -(2\pi)^{-n} \int D_{\xi_j}(e^{i\langle x-y,\xi\rangle}) b(x,\xi) d\xi + x_j (2\pi)^{-n} \int e^{i\langle x-y,\xi\rangle} b(x,\xi) d\xi \\ &= (2\pi)^{-n} \int e^{i\langle x-y,\xi\rangle} \Big((D_{\xi_j} b)(x,\xi) + x_j b(x,\xi) \Big) d\xi, \end{aligned}$$

from which we deduce that $y_j \mathsf{K}_0 \in L^\infty(\mathbb{R}^n \times \mathbb{R}^n)$. By the same token, we may finally conclude that there is $k_0 = k_0(n)$ such that picking $k = k_0$ gives

$$x^\alpha y^\beta \mathsf{K}_0 \in C(\mathbb{R}^n \times \mathbb{R}^n) \cap L^\infty(\mathbb{R}^n \times \mathbb{R}^n),\ \forall \alpha, \beta \in \mathbb{Z}_+^n,\ |\alpha| + |\beta| \le 2(n+1),$$

whence $\mathsf{K}_0 \in L^2(\mathbb{R}^n \times \mathbb{R}^n)$, and therefore also $\mathsf{K} \in L^2(\mathbb{R}^n \times \mathbb{R}^n)$. Since $\{\varphi_j\}_{j\ge1}$ and $\{\bar{\varphi}_j\}_{j\ge1}$ are orthonormal systems in L_x^2 and L_y^2, respectively, we get that $\varphi_j(y)^* \otimes \varphi_{j'}(x) (= \varphi_{j'}(x)\overline{\varphi_j(y)}$ in the scalar case) is an orthonormal system in $L_{x,y}^2$. The Fourier coefficients of K with respect to this basis are given by

$$\begin{aligned} a_{jj'} &:= \iint \mathsf{K}(x,y)\varphi_j(y)\overline{\varphi_{j'}(x)}dxdy = (A^{-k_0}\varphi_j, \varphi_{j'}) \\ &= \lambda_j^{-k_0}(\varphi_j, \varphi_{j'}) = \lambda_j^{-k_0}\delta_{jj'}. \end{aligned}$$

Hence

$$\iint |\mathsf{K}(x,y)|^2 dxdy = \sum_{j,j'=1}^{\infty} |a_{jj'}|^2 = \sum_{j=1}^{\infty} \frac{1}{\lambda_j^{2k_0}} < +\infty,$$

and this concludes the proof. □

It is useful to define also the trace of $t \longmapsto e^{-itA}$, that is the trace of the Schrödinger group (which is clearly important in its own right, and basic when studying Poisson-like relations). We have the following proposition.

Proposition 5.4.5. *The map*

$$\mathscr{S}(\mathbb{R}) \ni \phi \longmapsto \mathrm{Tr}\,\hat{\phi}(A),\ \hat{\phi}(t) = \int e^{-itx}\phi(x)dx,$$

defines a distribution in $\mathscr{S}'(\mathbb{R})$, *denoted by* $\mathrm{Tr}\, e^{-itA}$.

Proof. The map $\phi \longmapsto \hat{\phi}$ is continuous in $\mathscr{S}$. It suffices therefore to prove that for some $C > 0$ we have

$$|\mathrm{Tr}\,\hat{\phi}(A)| \le C|\hat{\phi}|_{p,q}, \tag{5.11}$$

where $|\hat{\phi}|_{p,q}$ is some suitable $\mathscr{S}$-seminorm of $\hat{\phi}$. By Proposition 5.4.4 (alternatively, in case $\mu = 2\nu$ by Theorem 4.3.4, Remark 4.3.5, Theorem 4.4.1 and Remark 4.4.2, we may also use the behavior $\lambda_j \approx j^{\nu/n}$ as $j \to +\infty$), there exists $r \in \mathbb{N}$ sufficiently large such that $\sum_{j\ge1}(1+\lambda_j^2)^{-r} < +\infty$. Hence

$$\begin{aligned} \Big|\sum_{j\ge1} \hat{\phi}(\lambda_j)\Big| &= \Big|\sum_{j\ge1}(1+\lambda_j^2)^{-r}(1+\lambda_j^2)^r\hat{\phi}(\lambda_j)\Big| \\ &\le \Big(\sum_{j\ge1}(1+\lambda_j^2)^{-r}\Big)\sup_{t\in\mathbb{R}}|(1+t^2)^r\hat{\phi}(t)|, \end{aligned}$$

which proves (5.11) and the proposition. □

5.5 $f(A)$ as a Pseudodifferential Operator

The following theorem, which concludes the chapter, is also useful (see Helffer [17] and Robert [63–65]). It is obtained through the calculus with parameters.

Theorem 5.5.1. *Let* $0 < A = A^* \in \mathrm{OP}S_{\mathrm{cl}}(m^\mu, g)$, $\mu \in \mathbb{N}$, *be a scalar* **elliptic** GPDO *of order* μ. *Let* $f\colon (-c, +\infty) \longrightarrow \mathbb{C}$ *satisfy, for some* $r \in \mathbb{R}$, *the inequalities*

$$\forall k \in \mathbb{Z}_+,\ \exists C_k > 0,\ \text{such that}\ |f^{(k)}(\lambda)| \leq C_k(1 + |\lambda|)^{r-k}.$$

Then the operator-function $f(A) = F^{\mathrm{w}}(x, D) \in \mathrm{OP}S(m^{r\mu}, g)$, *with*

$$F \sim \sum_{j=0}^{\mu} F_{r\mu-j},\quad F_{r\mu-j} \in S(m^{r\mu-j}, g),$$

and

- $F_{r\mu} = f(a_\mu)$;
- $F_{r\mu-1} = a_{\mu-1} f'(a_\mu)$ *(and hence it is equal to 0, for in our case* $a_{\mu-1} = 0$*)*;
- $F_{r\mu-j} = \sum_{k=1}^{j} \frac{d_{jk}(a)}{k!} f^{(k)}(a_\mu)$, $j \geq 2$, *where the* $d_{jk} \in S(m^{k\mu-j}, g)$ *and depend only on the symbol* a.

Furthermore, if f *is classic, that is* $f(\lambda) \sim \sum_{j\geq 0} c_j \lambda^{r-j}$, $\lambda > 0$, *then* $f(A) \in \mathrm{OP}S_{\mathrm{cl}}(m^{r\mu}, g)$.

Remark 5.5.2. In the matrix-valued case we shall need the result of Theorem 5.5.1 only for the *complex power* A^z, $z \in \mathbb{C}$, of an **elliptic** system of GPDOs $0 < A = A^* \in \mathrm{OP}S_{\mathrm{cl}}(m^\mu, g; \mathsf{M}_N)$. By a result due to Robert [63], one has that A^z belongs to the class $\mathrm{OP}S_{\mathrm{cl}}(m^{\mu \operatorname{Re} z}, g; \mathsf{M}_N)$. △

Corollary 5.5.3. *In particular, for* $Q_{(\alpha,\beta)} \in S_{\mathrm{cl}}(m^2, g; \mathsf{M}_2)$, *which is globally positive elliptic for* $\alpha, \beta > 0$ *and* $\alpha\beta > 1$, *Remark 5.5.2 holds true.*

5.6 Notes

The reader is addressed to Helffer's book [17] for the functional calculus of global pseudodifferential operators (see also the paper by Helffer and Robert [19]).

Chapter 6
The Spectral Counting Function $N(\lambda)$ and the Behavior of the Eigenvalues: Part 2

In this chapter we shall describe the parametrix approximation of e^{-tA}, A a positive elliptic global polynomial differential system, which will then be used, through Karamata's Tauberian theorem (proved in Section 6.2), to compute the leading coefficient of the asymptotic behavior for large eigenvalues of the spectral counting function, in terms of the symbol of the system.

6.1 A Parametrix Approximation of e^{-tA}

In this section, following Parenti-Parmeggiani [50], we consider an $N \times N$ **elliptic** system of GPDOs $0 < A = A^* \in \mathrm{OP}S_{\mathrm{cl}}(m^{\mu}, g; \mathsf{M}_N)$, $\mu \in \mathbb{N}$, and introduce a class of symbols which is suitable for constructing a pseudodifferential approximation of e^{-tA}. Recall that $\overline{\mathbb{R}}_+ = [0, +\infty)$.

Definition 6.1.1. Let $r \in \mathbb{R}$. By $S(\mu, r)$ we denote the set of all smooth maps $b\colon \overline{\mathbb{R}}_+ \times \mathbb{R}^n \times \mathbb{R}^n \longrightarrow \mathsf{M}_N$ satisfying the following estimates: For any given $\alpha \in \mathbb{Z}_+^{2n}$ and any given $p, j \in \mathbb{Z}_+$ there exists $C > 0$ such that

$$\sup_{t \in \overline{\mathbb{R}}_+} \left| t^p \left(\frac{d}{dt}\right)^j \partial_X^{\alpha} b(t, X) \right| \leq C m(X)^{r - |\alpha| + (j-p)\mu}. \tag{6.1}$$

For $b \in S(\mu, r)$ we then consider the pseudodifferential operator

$$b^{\mathrm{w}}(t, x, D)u(x) = (2\pi)^{-n} \iint e^{i\langle x-y, \xi\rangle} b(t, \frac{x+y}{2}, \xi) u(y) dy d\xi, \, u \in \mathscr{S}(\mathbb{R}^n; \mathbb{C}^N),$$

and we shall say that $B \in \mathrm{OP}S(\mu, r)$ if $B = b^{\mathrm{w}}(t, x, D) + R$, where R is smoothing. In this setting, a smoothing operator R is any **continuous** map

$$R\colon \mathscr{S}'(\mathbb{R}^n; \mathbb{C}^N) \longrightarrow \mathscr{S}(\overline{\mathbb{R}}_+; \mathscr{S}(\mathbb{R}^n; \mathbb{C}^N)). \tag{6.2}$$

A. Parmeggiani, *Spectral Theory of Non-Commutative Harmonic Oscillators: An Introduction*, Lecture Notes in Mathematics 1992,
DOI 10.1007/978-3-642-11922-4_6, © Springer-Verlag Berlin Heidelberg 2010

Notice in particular that for the time-dependent seminorms of $b \in S(\mu,r)$ we have

$$m(X)^{|\alpha|}|\partial_t^j \partial_X^\alpha b(t,X)| \leq C_{\alpha,j,p} m(X)^{r+(j-p)\mu} t^{-p}, \ \forall X \in \mathbb{R}^{2n}, \ \forall t \geq 1,$$

that is, for any given $k,j,p \in \mathbb{Z}_+$ there exists $C = C_{k,j,p} > 0$ such that (see (3.1))

$$\|\partial_t^j b(t,\cdot)\|_{k,S(m^{r+(j-p)\mu},g)} \leq Ct^{-p}, \ \forall t \geq 1. \tag{6.3}$$

We now introduce the "classical operators". The key point here is to consider the correct homogeneity properties.

The basic example to bear in mind is the matrix $e^{-ta_\mu(x,\xi)}$. For instance, the bound for $t^p e^{-ta_\mu(X)}$ appearing in (6.1) follows from writing

$$t^p e^{-ta_\mu(X)} = a_\mu(X)^{-p}(ta_\mu(X))^p e^{-ta_\mu(X)},$$

and from the inequality $\tau^p e^{-\tau} \leq (p/e)^p$, all $\tau > 0$ (see (5.6)). The other bounds follow by the same considerations.

Definition 6.1.2. We say that the operator $B \in \mathrm{OP}S(\mu,r)$, $B = b^{\mathrm{w}} + R$ is **classical**, and write $B \in \mathrm{OP}S_{\mathrm{cl}}(\mu,r)$, if there exists a sequence of functions $b_{r-2j} = b_{r-2j}(t,X)$, $j \geq 0$, $t \geq 0$ and $X \neq 0$, such that:

(i) One has the homogeneity

$$b_{r-2j}(t,\tau X) = \tau^{r-2j} b_{r-2j}(\tau^\mu t, X), \ \forall \tau > 0, \ \forall j \geq 0; \tag{6.4}$$

(ii) The function

$$\mathbb{R}^{2n} \setminus \{0\} \ni X \longmapsto b_{r-2j}(\cdot,X) \in \mathscr{S}(\overline{\mathbb{R}}_+; \mathsf{M}_N)$$

is smooth for all $j \geq 0$;

(iii) For all $\nu \geq 1$

$$b(t,X) - \sum_{j=0}^{\nu-1} \chi(X) b_{r-2j}(t,X) \in S(\mu, r-2\nu),$$

where χ is an excision function.

We call $b_r = \sigma_r(B)$ the **principal symbol** of B.

Of course, **semi-regular classical** symbols are defined accordingly (see Remark 3.2.4).

Note that

$$\sigma_r(B) \equiv 0 \Longrightarrow B \in \mathrm{OP}S_{\mathrm{cl}}(\mu, r-2),$$

and that given any smooth map $b\colon \mathbb{R}^{2n}\setminus\{0\} \longrightarrow \mathscr{S}(\overline{\mathbb{R}}_+;\mathsf{M}_N)$, with the homogeneity property

$$b(t,\tau X) = \tau^r b(\tau^\mu t, X),\ \ \forall\tau>0,$$

then there exists $B\in \mathrm{OP}S_{\mathrm{cl}}(\mu,r)$ such that $\sigma_r(B)=b$.

One has the following results (whose proofs are left as an exercise to the reader) that are useful to the parametrix construction.

Lemma 6.1.3.

- *One has*

$$B\in \mathrm{OP}S_{\mathrm{cl}}(\mu,r),\ \ C\in \mathrm{OP}S_{\mathrm{cl}}(\mu,r') \Longrightarrow BC\in \mathrm{OP}S_{\mathrm{cl}}(\mu,r+r'),$$

with

$$\sigma_{r+r'}(BC)(t,X) = \sigma_r(B)(t,X)\sigma_{r'}(C)(t,X).$$

- *One has*

$$B\in \mathrm{OP}S_{\mathrm{cl}}(\mu,r), P\in \mathrm{OP}S_{\mathrm{cl}}(m^\ell,g;\mathsf{M}_N) \Longrightarrow PB,\ BP\in \mathrm{OP}S_{\mathrm{cl}}(\mu,r+\ell),$$

with

$$\sigma_{r+\ell}(PB)(t,X) = p_\ell(X)\sigma_r(B)(t,X),$$

and likewise for $\sigma_{r+\ell}(BP)$.

- *For $B\in \mathrm{OP}S_{\mathrm{cl}}(\mu,r)$ consider*

$$\frac{d}{dt}B\colon \mathscr{S}(\mathbb{R}^n;\mathbb{C}^N) \longrightarrow \mathscr{S}(\overline{\mathbb{R}}_+;\mathscr{S}(\mathbb{R}^n;\mathbb{C}^N)),\ \ u\longmapsto \frac{d}{dt}(Bu).$$

Then

$$\frac{d}{dt}B\in \mathrm{OP}S_{\mathrm{cl}}(\mu,\mu+r),\ \text{with}\ \sigma_{\mu+r}(\frac{d}{dt}B)(t,X) = \frac{d}{dt}\big(\sigma_r(B)\big)(t,X).$$

- *For $B\in \mathrm{OP}S_{\mathrm{cl}}(\mu,r)$ consider*

$$\int_0^\infty Bdt\colon \mathscr{S}(\mathbb{R}^n;\mathbb{C}^N) \longrightarrow \mathscr{S}(\mathbb{R}^n;\mathbb{C}^N),\ \ u\longmapsto \int_0^\infty (Bu)dt.$$

Then

$$\int_0^\infty Bdt\in \mathrm{OP}S_{\mathrm{cl}}(m^{r-\mu},g;\mathsf{M}_N),$$

with principal symbol

$$X\longmapsto \int_0^\infty \sigma_r(B)(t,X)dt.$$

We are now in a position to prove the following theorem about the existence of a parametrix approximation of e^{-tA}. The parametrix we are interested in is that of the operator $d/dt+A$.

Theorem 6.1.4. *There exists* $U_A \in \mathrm{OPS}_{\mathrm{cl}}(\mu,0)$ *such that*

$$(i) \qquad \frac{d}{dt}U_A + AU_A \colon \mathscr{S}'(\mathbb{R}^n;\mathbb{C}^N) \longrightarrow \mathscr{S}(\overline{\mathbb{R}}_+;\mathscr{S}(\mathbb{R}^n;\mathbb{C}^N))$$

is smoothing, and

$$(ii) \qquad U_A\big|_{t=0} - I \colon \mathscr{S}'(\mathbb{R}^n;\mathbb{C}^N) \longrightarrow \mathscr{S}(\mathbb{R}^n;\mathbb{C}^N)$$

is smoothing. One has

$$\sigma_0(U_A)(t,X) = e^{-ta_\mu(X)}.$$

Proof. Let $b_0(t,X) = e^{-ta_\mu(X)}$, and let $B_0 \in \mathrm{OPS}_{\mathrm{cl}}(\mu,0)$ with $\sigma_0(B_0) = b_0$. By Lemma 6.1.3 we have

$$\frac{d}{dt}B_0 + AB_0 \in \mathrm{OPS}_{\mathrm{cl}}(\mu,\mu-2),$$

with principal symbol $r_{\mu-2}(t,X)$, and

$$B_0\big|_{t=0} - I \in \mathrm{OPS}_{\mathrm{cl}}(m^{-2},g;\mathsf{M}_N),$$

with principal symbol $p_{-2}(X)$. We next look for a symbol $b_{-2}(t,X)$, positively homogeneous of degree -2 (in the sense of (6.4)), such that

$$\begin{cases} \dfrac{d}{dt}b_{-2} + a_\mu b_{-2} = -r_{\mu-2}, \\[2ex] b_{-2}\big|_{t=0} = -p_{-2}. \end{cases} \tag{6.5}$$

The solution of (6.5),

$$b_{-2}(t,X) = -e^{-ta_\mu(X)}p_{-2}(X) - \int_0^t e^{-(t-t')a_\mu(X)}r_{\mu-2}(t',X)dt', \tag{6.6}$$

is easily seen to be smooth and have the required homogeneity properties. Taking $B_{-2} \in \mathrm{OPS}_{\mathrm{cl}}(\mu,-2)$ with $\sigma_{-2}(B_{-2}) = b_{-2}$, gives

$$\frac{d}{dt}(B_0 + B_{-2}) + A(B_0 + B_{-2}) \in \mathrm{OPS}_{\mathrm{cl}}(\mu,\mu-4),$$

and

$$(B_0 + B_{-2})\big|_{t=0} - I \in \mathrm{OPS}_{\mathrm{cl}}(m^{-4},g;\mathsf{M}_N).$$

Iterating the above procedure gives a formal series

$$\sum_{k\geq 0} B_{-2k}, \ B_{-2k} \in \mathrm{OPS}_{\mathrm{cl}}(\mu,-2k).$$

We may hence take, by an adaptation of Proposition 3.2.15, an operator $U_A \in \mathrm{OPS}_{\mathrm{cl}}(\mu,0)$ for which

$$U_A - \sum_{k=0}^{\nu-1} B_{-2k} \in \mathrm{OPS}(\mu,-2\nu),\ \forall \nu \geq 1,$$

and therefore obtain the required parametrix. □

Hence, from Lemma 5.1.1 and Theorem 6.1.4 we have that

$$R(t) := e^{-tA} - U_A(t) \colon \mathscr{S}'(\mathbb{R}^n;\mathbb{C}^N) \longrightarrow \mathscr{S}(\overline{\mathbb{R}}_+;\mathscr{S}(\mathbb{R}^n;\mathbb{C}^N)) \textbf{ is continuous} \tag{6.7}$$

(i.e., it is a smoothing operator), and

$$\left(e^{-tA} - U_A(t)\right)\Big|_{t=0} \colon \mathscr{S}'(\mathbb{R}^n;\mathbb{C}^N) \longrightarrow \mathscr{S}(\mathbb{R}^n;\mathbb{C}^N) \textbf{ is smoothing.} \tag{6.8}$$

Remark 6.1.5. In the applications of Theorem 6.1.4 we shall always consider a parametrix approximation of e^{-tA} where $b_{-2j}\big|_{t=0} = 0$ for $j \geq 1$,

$$B_{-2j}(t) = (\chi b_{-2j})^{\mathrm{w}}(t,x,D),$$

where χ is our usual excision function, and hence consider the symbol $c_A(t,X)$ of $U_A(t)$, i.e. $U_A(t) = c_A^{\mathrm{w}}(t,x,D)$, given by

$$c_A(t,X) = \sum_{j\geq 0} \chi_j(X) b_{-2j}(t,X),$$

where $\chi_0(X) = \chi(X)$ and $\chi_j(X) = \chi(X/R_j)$, $j \geq 1$, with $R_j \nearrow +\infty$, as $j \to +\infty$, sufficiently fast (in analogy with the R_j constructed in Proposition 3.2.15). We shall write

$$U_A(t) \sim \sum_{j\geq 0} B_{-2j}(t). \tag{6.9}$$

△

In order to exploit the parametrix approximation $U_A(t)$ of e^{-tA} to study $N_A(\lambda)$ as $\lambda \to +\infty$, we have to recall the Karamata theorem.

6.2 The Karamata Theorem

Recall that the Euler Gamma function $\Gamma(x) = \int_0^\infty e^{-t} t^{x-1} dt$, $x > 0$, fulfills the relation $\Gamma(x+1) = x\Gamma(x)$.

Also, following Halmos' book on Measure Theory [16], given a Borel measure μ over a measure space X, a measurable space Y and a measurable transformation

$T: X \longrightarrow Y$, the T**-image measure** μ_T of Y is defined for any Borel subset $F \subset Y$ by

$$\mu_T(F) = \mu(T^{-1}F).$$

Hence, for any given μ_T-measurable function $f: Y \longrightarrow [0, +\infty]$

$$\int_F f(y) d\mu_T(y) = \int_{T^{-1}F} f(Tx) d\mu(x). \tag{6.10}$$

Theorem 6.2.1 (Karamata). *Let μ be a positive (locally finite) Borel measure on $[0, +\infty)$, and let $\alpha > 0$. Then*

$$\int_0^\infty e^{-t\lambda} d\mu(\lambda) \sim at^{-\alpha}, \text{ as } t \to 0+, \tag{6.11}$$

implies

$$\int_0^x d\mu(\lambda) \sim \frac{a}{\Gamma(\alpha+1)} x^\alpha, \text{ as } x \to +\infty. \tag{6.12}$$

Proof. For $t > 0$ define the measure $d\mu_t(\lambda)$ by $\mu_t(A) = t^\alpha \mu(t^{-1}A)$. Let $d\nu(\lambda) = \alpha\lambda^{\alpha-1} d\lambda$, $\lambda > 0$. Then $\nu_t = \nu$. In fact, in general we have by definition that, given any Borel set $A \subset [0, +\infty)$,

$$\mu_t(A) = \int_A d\mu_t(\lambda) = t^\alpha \mu(t^{-1}A) = t^\alpha \int_{t^{-1}A} d\mu(\lambda),$$

and that, by (6.10), for any given measurable function $f: [0, +\infty) \longrightarrow [0, +\infty]$

$$\int_A f(\lambda) d\mu_t(\lambda) = t^\alpha \int_{t^{-1}A} f(t\sigma) d\mu(\sigma). \tag{6.13}$$

Hence, specializing to the case of ν_t gives

$$\nu_t(A) = \int_A d\nu_t(\lambda) = t^\alpha \nu(t^{-1}A) = \alpha t^\alpha \int_{t^{-1}A} \lambda^{\alpha-1} d\lambda = \nu(A).$$

With $b = a/\Gamma(\alpha+1)$, we have that, by virtue of (6.13), hypothesis (6.11) can be rewritten as

$$\lim_{t \to 0+} \int_0^\infty e^{-\lambda} d\mu_t(\lambda) = b \int_0^\infty e^{-\lambda} d\nu(\lambda), \tag{6.14}$$

and also, since

$$t^\alpha \int_0^{1/t} d\mu(\lambda) = \int_0^1 d\mu_t(\lambda),$$

that the desired conclusion can be rewritten (just by putting $x = 1/t$) as

$$\lim_{t \to 0+} \int_0^1 d\mu_t(\lambda) = b \int_0^1 d\nu(\lambda). \tag{6.15}$$

Suppose we know that

$$\lim_{t\to 0+}\int_0^\infty f(\lambda)d\mu_t(\lambda) = b\int_0^\infty f(\lambda)d\nu(\lambda),\ \forall f\in C_0([0,+\infty)), \tag{6.16}$$

where $C_0([0,+\infty))$ denotes the space of continuous functions compactly supported in $[0,+\infty)$. Since ν and the μ_t, $t\in(0,1]$, are regular measures, we have that for any given open subset V of $[0,+\infty)$,

$$\nu(V) = \sup\Big\{\int_{[0,+\infty)} f(\lambda)d\nu(\lambda);\ f\in C_0([0,+\infty)),\ 0\le f\le \mathbf{1}_V\Big\}, \tag{6.17}$$

where $\mathbf{1}_V$ is the characteristic function of V (hence in particular $\operatorname{supp} f\subset V$), and for any given compact $K\subset[0,+\infty)$,

$$\nu(K) = \inf\Big\{\int_{[0,+\infty)} f(\lambda)d\nu(\lambda);\ f\in C_0([0,+\infty)),\ \mathbf{1}_K\le f\le 1\Big\}, \tag{6.18}$$

hence $f\big|_K\equiv 1$, and likewise for the measures μ_t. Since (6.16) says that $\mu_t\to\nu$ weakly as measures on $[0,+\infty)$ as $t\to 0+$, we have from (6.17) that

$$\begin{aligned} b\nu((0,1)) &= \sup\Big\{b\int_{[0,+\infty)} fd\nu;\ f\in C_0([0,+\infty)),\ 0\le f\le \mathbf{1}_{(0,1)}\Big\}\\ &\le \liminf_{t\to 0+}\sup\Big\{\int_{[0,+\infty)} fd\mu_t;\ f\in C_0([0,+\infty)),\ 0\le f\le \mathbf{1}_{(0,1)}\Big\}\\ &= \liminf_{t\to 0+}\mu_t((0,1)) \le \liminf_{t\to 0+}\mu_t([0,1]), \end{aligned}$$

and likewise, from (6.18), that

$$b\nu([0,1]) \ge \limsup_{t\to 0+}\mu_t([0,1]).$$

Since $\nu((0,1)) = \nu([0,1])$ (for $\nu(\{0\}) = \nu(\{1\}) = 0$) we have (a particular case of Prohorov's theorem)

$$\lim_{t\to 0+}\mu_t([0,1]) = b\nu([0,1]),$$

and this proves that (6.16) implies (6.15).

Next, by (6.14) the measures $e^{-\lambda}d\mu_t(\lambda)$ are **uniformly bounded**, for $t\in(0,1]$, so that (6.16) follows from

$$\lim_{t\to 0+}\int_0^\infty g(\lambda)e^{-\lambda}d\mu_t(\lambda) = b\int_0^\infty g(\lambda)e^{-\lambda}d\nu(\lambda), \tag{6.19}$$

for a dense (in the sup-norm) subset of g in $C_\infty([0,+\infty))$, the space of continuous functions that vanish at infinity. But (6.19) holds for $g(\lambda) = e^{-n\lambda}$ by hypothesis (6.11). Since polynomials in $e^{-\lambda}$ are dense in $C_\infty([0,+\infty))$ by the Stone-Weierstrass Theorem, (6.19) holds and the theorem is proven. □

6.3 Use of the Parametrix Approximation of e^{-tA} for Obtaining the Weyl Asymptotics of $N(\lambda)$

We may now exploit the existence of the parametrix approximation of e^{-tA} to study, through Karamata's theorem, the Weyl-asymptotics of $N(\lambda)$ as $\lambda \to +\infty$. Hence, let

$$\mathrm{Spec}(A) = \{\lambda_j\}_{j\geq 1},\ 0 < \lambda_1 \leq \lambda_2 \leq \ldots \to +\infty,$$

repeated according to multiplicity, and let $N(\lambda) = \sharp\{j \in \mathbb{N};\ \lambda_j < \lambda\}$. Then

$$\sum_{j\geq 1} e^{-t\lambda_j} = \int_0^{\infty} e^{-t\lambda} dN(\lambda) = \mathrm{Tr}\, e^{-tA}.$$

Actually, from Theorem 6.1.4, (6.7) and Remark 6.1.5 we have

$$\mathrm{Tr}\, e^{-tA} = \mathrm{Tr}\, U_A(t) + \mathrm{Tr}\, R(t),\ \ t > 0, \tag{6.20}$$

where

$$\mathrm{Tr}\, U_A(t) = \sum_{j\geq 0} \mathrm{Tr}\, B_{-2j}(t) \tag{6.21}$$

(we shall specify in a moment in what sense the $=$ sign holds). Note that since the operator $R(t)$ is smoothing (see (6.2)), we have

$$0 \leq t \longmapsto \mathrm{Tr}\, R(t) \in \mathscr{S}(\overline{\mathbb{R}}_+;\mathbb{C}). \tag{6.22}$$

Now,

$$\mathrm{Tr}\, B_{-2j}(t) = (2\pi)^{-n} \int_{\mathbb{R}^{2n}} \chi(X)\mathrm{Tr}\, b_{-2j}(t,X) dX.$$

Hence, upon denoting by $\mathbb{S}^{2n-1}$ the unit sphere in $\mathbb{R}^{2n}$, using polar coordinates $0 \neq X = |X|\dfrac{X}{|X|} = \rho\omega$, where $\rho \in \mathbb{R}_+$ and $\omega \in \mathbb{S}^{2n-1}$, gives

$$\mathrm{Tr}\, B_{-2j}(t) = (2\pi)^{-n} \int_0^{+\infty}\int_{\mathbb{S}^{2n-1}} \chi(\rho\omega)\mathrm{Tr}\, b_{-2j}(t,\rho\omega)\rho^{2n-1} d\rho d\omega, \tag{6.23}$$

where $d\omega$ is the induced Riemannian measure on $\mathbb{S}^{2n-1}$.

In general, if $b \in S(\mu,-2\nu)$, with $\nu > n$, then by the estimates (6.1) we have that **uniformly** in $t \geq 0$,

$$\left|(2\pi)^{-n} \int_0^{+\infty}\int_{\mathbb{S}^{2n-1}} b(t,\rho\omega)\rho^{2n-1} d\rho d\omega\right| \lesssim \int_0^{+\infty} \frac{\rho^{2n-1}}{(1+\rho)^{2\nu}} d\rho < +\infty, \tag{6.24}$$

from which it follows that

$$0 \leq t \longmapsto f_b(t) := (2\pi)^{-n} \int_0^{+\infty}\int_{\mathbb{S}^{2n-1}} b(t,\rho\omega)\rho^{2n-1} d\rho d\omega$$

is **continuous and bounded** on $\overline{\mathbb{R}}_+$. Hence

$$0 \le t \longmapsto \operatorname{Tr} U_A(t) - \sum_{j=0}^{\nu} \operatorname{Tr} B_{-2j}(t) \in C(\overline{\mathbb{R}}_+) \cap L^\infty(\overline{\mathbb{R}}_+), \ \forall \nu > n. \tag{6.25}$$

More generally, given any $p, k \in \mathbb{Z}_+$, we have

$$|t^p \partial_t^k f_b(t)| < +\infty, \ \forall t \ge 0,$$

provided $b \in S(\mu, -2\nu)$ with $2\nu > 2n - (p-k)\mu$. Hence, on defining

$$\mathscr{S}_{p,k}(\overline{\mathbb{R}}_+) := \{ f \in C^k(\overline{\mathbb{R}}_+);\ t^p f^{(r)} \in L^\infty(\overline{\mathbb{R}}_+), r = 0, 1, \ldots, k \}, \tag{6.26}$$

for any given $p, k \in \mathbb{Z}_+$ there exists ν with $2\nu > 2n - (p-k)\mu$ such that

$$0 \le t \longmapsto \operatorname{Tr} U_A(t) - \sum_{j=0}^{\nu} \operatorname{Tr} B_{-2j}(t) \in \mathscr{S}_{p,k}(\overline{\mathbb{R}}_+), \tag{6.27}$$

and this is the meaning of the $=$ sign in (6.21).

Next, by the homogeneity of the b_{-2j} we may write ($\operatorname{supp} \chi \subset \{|X| \ge 1/2\}$, say)

$$\operatorname{Tr} B_{-2j}(t) = (2\pi)^{-n} \int_{1/2}^{+\infty} \int_{\mathbb{S}^{2n-1}} \chi(\rho\omega) \mathsf{Tr}\, b_{-2j}(\rho^\mu t, \omega) \rho^{2(n-j)-1} d\rho d\omega. \tag{6.28}$$

Since $\mathsf{Tr}\, b_{-2j}(\cdot, \omega) \in \mathscr{S}(\overline{\mathbb{R}}_+, \mathbb{C})$ (which is also the case for the remainder in the asymptotic expansion of $\sigma_0(U_A)$, by virtue of (6.1)), the integral in ρ converges for all $t > 0$, and hence the whole integral in $d\rho d\omega$ converges absolutely.

Thus

$$|\operatorname{Tr} B_{-2j}(t)| < +\infty, \quad \forall t > 0, j \ge 0,$$

and

$$\left| (2\pi)^{-n} \int_0^{+\infty} \int_{\mathbb{S}^{2n-1}} \mathsf{Tr}\, b_{-2j}(\rho^\mu t, \omega) \rho^{2(n-j)-1} d\rho d\omega \right| < +\infty, \ \forall t > 0, 0 \le j < n.$$

Furthermore, it follows from (6.23) and (6.24) that *if* $j \ge n+1$ *in* $\operatorname{Tr} B_{-2j}(t)$ *the whole integral in* $d\rho d\omega$ *converges also for* $t = 0$.

We therefore have from (6.22), (6.25) and the fact (that we have just seen) that for $j \ge n+1$ the terms $\operatorname{Tr} B_{-2j}(t)$ are non-singular at $t = 0$, that the singularity of $\operatorname{Tr} e^{-tA}$ at $t = 0$, which of course must be present, is necessarily contained in

$$\sum_{j=0}^{n} \operatorname{Tr} B_{-2j}(t). \tag{6.29}$$

Hence, to study the singularity at $t = 0$ of $\mathrm{Tr}\, e^{-tA}$, let us at first consider a term $\mathrm{Tr}\, B_{-2j}(t)$ of (6.29) with $0 \leq j < n$.

Write for $t > 0$

$$\mathrm{Tr}\, B_{-2j}(t) = \beta'_{-2j}(t) - \beta''_{-2j}(t),$$

where

$$\beta'_{-2j}(t) := (2\pi)^{-n} \int_0^{+\infty} \int_{\mathbb{S}^{2n-1}} \mathsf{Tr}\, b_{-2j}(\rho^{\mu} t, \omega) \rho^{2(n-j)-1} d\rho d\omega$$

and

$$\beta''_{-2j}(t) := (2\pi)^{-n} \int_0^{+\infty} \int_{\mathbb{S}^{2n-1}} (1 - \chi(\rho\omega)) \mathsf{Tr}\, b_{-2j}(\rho^{\mu} t, \omega) \rho^{2(n-j)-1} d\rho d\omega.$$

Since $\rho\omega \in \mathrm{supp}(1-\chi) \Longrightarrow \rho \in [0,1]$, it follows that we may rewrite

$$\beta''_{-2j}(t) = (2\pi)^{-n} \int_0^{1} \int_{\mathbb{S}^{2n-1}} (1 - \chi(\rho\omega)) \mathsf{Tr}\, b_{-2j}(\rho^{\mu} t, \omega) \rho^{2(n-j)-1} d\rho d\omega,$$

whence it follows that

$$\beta''_{-2j} \in C^{\infty}(\overline{\mathbb{R}}_{+}) \cap L^{\infty}(\overline{\mathbb{R}}_{+}),$$

with the derivatives of all orders bounded on $\overline{\mathbb{R}}_{+}$.

As regards β'_{-2j}, switching to the variable $s = t^{1/\mu}\rho$ gives

$$\begin{aligned}\beta'_{-2j}(t) &= t^{-2(n-j)/\mu} (2\pi)^{-n} \int_0^{+\infty} \int_{\mathbb{S}^{2n-1}} \mathsf{Tr}\, b_{-2j}(s^{\mu}, \omega) s^{2(n-j)-1} ds d\omega \\ &= c_{-2j,n}(\mu) t^{-2(n-j)/\mu},\end{aligned}$$

where

$$c_{-2j,n}(\mu) := (2\pi)^{-n} \int_0^{+\infty} \int_{\mathbb{S}^{2n-1}} \mathsf{Tr}\, b_{-2j}(s^{\mu}, \omega) s^{2(n-j)-1} ds d\omega \tag{6.30}$$

is such that $|c_{-2j,n}(\mu)| < +\infty$, because of the fact that

$$\mathsf{Tr}\, b_{-2j}(\cdot, \omega) \in \mathscr{S}(\overline{\mathbb{R}}_{+}) \quad \text{and} \quad 0 \leq j \leq n-1.$$

Notice that, by the homogeneity, we may also write

$$c_{-2j,n}(\mu) = (2\pi)^{-n} \int_0^{+\infty} \int_{\mathbb{S}^{2n-1}} \mathsf{Tr}\, b_{-2j}(1, s\omega) s^{2n-1} ds d\omega, \ 0 \leq j \leq n-1. \tag{6.31}$$

It therefore follows that

$$\mathrm{Tr}\, B_{-2j}(t) = t^{-2(n-j)/\mu} \big(c_{-2j,n}(\mu) + o(1)\big), \text{ as } t \to 0+,\ 0 \leq j \leq n-1.$$

As regards the term $\mathrm{Tr}B_{-2n}(t)$, again by virtue of the fact that $\mathrm{Tr}b_{-2j}(\cdot,\omega)\in\mathscr{S}(\overline{\mathbb{R}}_+)$, we have that for any given $\varepsilon>0$ we may find $C_\varepsilon>0$ such that for all $t>0$

$$\begin{aligned}|\mathrm{Tr}B_{-2n}(t)| &= \left|(2\pi)^{-n}\int_{1/2}^{+\infty}\int_{\mathbb{S}^{2n-1}}\chi(\rho\omega)\mathrm{Tr}b_{-2n}(\rho^\mu t,\omega)\rho^{-1}d\rho d\omega\right|\\ &= t^{-\varepsilon}\left|(2\pi)^{-n}\int_{1/2}^{+\infty}\int_{\mathbb{S}^{2n-1}}\chi(\rho\omega)\rho^{\varepsilon\mu}t^{\varepsilon}\,\mathrm{Tr}b_{-2n}(\rho^\mu t,\omega)\rho^{-1-\varepsilon\mu}d\rho d\omega\right|\\ &\le C_\varepsilon/t^\varepsilon. \end{aligned}\tag{6.32}$$

It follows that as $t\to 0+$

$$\sum_{j=0}^{n}\mathrm{Tr}B_{-2j}(t)=t^{-2n/\mu}\Big[\sum_{j=0}^{n-1}\Big(c_{-2j,n}(\mu)t^{2j/\mu}+o(t^{2j/\mu})\Big)+t^{2n/\mu}\mathrm{Tr}B_{-2n}(t)\Big],$$

so that, by (6.32),

$$\sum_{j=0}^{n}\mathrm{Tr}B_{-2j}(t)=t^{-2n/\mu}\big(c_{0,n}(\mu)+o(1)\big),\ \text{ as } t\to 0+.$$

This shows (using (6.24) to control the remainder term) that the leading singularity at $t=0$ of $\mathrm{Tr}e^{-tA}$ is exactly given by the term $t^{-2n/\mu}$ with coefficient

$$c_{0,n}(\mu)=(2\pi)^{-n}\int_0^{+\infty}\int_{\mathbb{S}^{2n-1}}\mathrm{Tr}\,b_0(s^\mu,\omega)s^{2n-1}ds d\omega. \tag{6.33}$$

Since

$$\mathrm{Tr}\,b_0(s^\mu,\omega)=\mathrm{Tr}\,e^{-s^\mu a_\mu(\omega)}=\sum_{j=1}^{N}e^{-s^\mu\ell_j(\omega)},$$

where $0<\ell_1(\omega),\dots,\ell_N(\omega)$, $\omega\in\mathbb{S}^{2n-1}$, are the (possibly repeated) eigenvalues of the $N\times N$ Hermitian matrix $a_\mu(\omega)$, labelled in such a way to be represented by continuous functions on $\mathbb{S}^{2n-1}$. Therefore

$$\begin{aligned}c_{0,n}(\mu)&=(2\pi)^{-n}\sum_{j=1}^{N}\int_0^{+\infty}\int_{\mathbb{S}^{2n-1}}e^{-s^\mu\ell_j(\omega)}s^{2n-1}dsd\omega\\ &=(2\pi)^{-n}\frac{\Gamma(\frac{2n}{\mu})}{\mu}\sum_{j=1}^{N}\int_{\mathbb{S}^{2n-1}}\frac{1}{\ell_j(\omega)^{2n/\mu}}d\omega,\end{aligned}$$

where we have used

$$\int_0^{+\infty}e^{-s^\mu}s^{2n-1}ds=\Gamma(\frac{2n}{\mu})/\mu.$$

We have thus proved the following theorem.

Theorem 6.3.1. *Let $0 < A = A^* \in \mathrm{OPS}_{\mathrm{cl}}(m^\mu, g; \mathsf{M}_N)$ be an elliptic $N \times N$ system of* GPDO*s of order $\mu \in \mathbb{N}$ in $\mathbb{R}^n$. Then*

$$\mathrm{Tr}\, e^{-tA} \sim \frac{\Gamma(\frac{2n}{\mu})/\mu}{(2\pi)^n} t^{-2n/\mu} \sum_{j=1}^{N} \int_{\mathbb{S}^{2n-1}} \frac{1}{\ell_j(\omega)^{2n/\mu}} d\omega, \text{ as } t \to 0+,$$

where $0 < \ell_1(\omega), \ldots, \ell_N(\omega)$, $\omega \in \mathbb{S}^{2n-1}$, are the (possibly repeated) eigenvalues of the $N \times N$ Hermitian matrix $a_\mu(\omega)$, labelled in such a way to be represented by continuous functions on $\mathbb{S}^{2n-1}$. Hence, Karamata's theorem gives for the spectral counting function

$$N(\lambda) \sim \frac{\Gamma(\frac{2n}{\mu})/\mu}{(2\pi)^n \Gamma(\frac{2n}{\mu}+1)} \left(\sum_{j=1}^{N} \int_{\mathbb{S}^{2n-1}} \frac{1}{\ell_j(\omega)^{2n/\mu}} d\omega \right) \lambda^{2n/\mu}, \text{ as } \lambda \to +\infty. \quad (6.34)$$

In particular, when A is an elliptic NCHO*, we have*

$$N(\lambda) \sim \frac{1}{2n(2\pi)^n} \left(\int_{\mathbb{S}^{2n-1}} \mathsf{Tr}(a_2(\omega)^{-n}) d\omega \right) \lambda^n, \text{ as } \lambda \to +\infty. \quad (6.35)$$

Corollary 6.3.2. *For the* NCHO *$Q^{\mathrm{w}}_{(\alpha,\beta)}(x,D)$, $\alpha, \beta > 0$ and $\alpha\beta > 1$, we have*

$$N(\lambda) \sim \frac{1}{4\pi} \left(\int_{\mathbb{S}^1} \mathsf{Tr}(Q_{(\alpha,\beta)}(\omega)^{-1}) d\omega \right) \lambda = \frac{\alpha+\beta}{\sqrt{\alpha\beta(\alpha\beta-1)}} \lambda, \text{ as } \lambda \to +\infty. \quad (6.36)$$

Remark 6.3.3. Given a positive function a_μ, positively homogeneous of degree $\mu > 0$ in $\mathbb{R}^{2n} \setminus \{0\}$, we have

$$\iint_{a_\mu(x,\xi) \le 1} dx d\xi = \frac{(2\pi)^{-n}}{2n} \int_{\mathbb{S}^{2n-1}} \frac{1}{a_\mu(\omega)^{2n/\mu}} d\omega.$$

Hence, using the homogeneity of degree μ of the eigenvalues ℓ_j, the main term in (6.34) (i.e. the coefficient of $\lambda^{2n/\mu}$) may also be written as

$$\frac{2n}{\mu} \frac{\Gamma(\frac{2n}{\mu})}{\Gamma(\frac{2n}{\mu}+1)} \sum_{j=1}^{N} \iint_{\ell_j(x,\xi) \le 1} dx d\xi, \quad (6.37)$$

and the one in (6.36) as

$$\mathrm{Vol}\left(\left\{ (x,\xi) \in \mathbb{R} \times \mathbb{R};\ \frac{\det Q_{(\alpha,\beta)}(x,\xi)}{\mathsf{Tr}\, Q_{(\alpha,\beta)}(x,\xi)} \le 1 \right\} \right). \quad (6.38)$$

△

6.4 Remarks on the Heuristics on $N(\lambda)$ and $\zeta_A(s)$

It is worth noting that information on $N(\lambda)$ can be obtained from information on $\zeta_A(s)$ through Ikehara's theorem. Here we follow Shubin [67, p. 120].

Theorem 6.4.1. *Let $N(\lambda)$ be non-decreasing and equal to 0 for $\lambda < \lambda_1$. Suppose that*

$$\zeta_A(s) = \int_{\lambda_1}^{+\infty} \lambda^{-s} dN(\lambda)$$

converges for $\operatorname{Re} s > k_0$, *for some $k_0 > 0$, and that*

$$\zeta_A(s) - \frac{A}{s-k_0}$$

can be extended as a continuous function to the closed half-plane $\operatorname{Re} s \geq k_0$. *Then*

$$N(\lambda) \sim \frac{A}{k_0}\lambda^{k_0}, \quad as\ \lambda \to +\infty.$$

What we have behind scenes is the following. Since the spectrum of A is made of eigenvalues $0 < \lambda_1 \leq \lambda_2 \leq \ldots \to +\infty$, considering the Stiltjes integral

$$\zeta_A(s) = \int_0^{+\infty} \lambda^{-s} dN(\lambda)$$

gives the spectral zeta function $\zeta_A(s) = \sum_{j\geq 1} \lambda_j^{-s}$, which we already know to be holomorphic for $\operatorname{Re} s > 2n/\mu$. Suppose now that

$$N(\lambda) = c_1\lambda^{\alpha_1} + c_2\lambda^{\alpha_2} + \ldots + c_k\lambda^{\alpha_k} + O(\lambda^{\alpha_{k+1}}), \text{ as } \lambda \to +\infty,$$

where $\operatorname{Re}\alpha_1 > \operatorname{Re}\alpha_2 > \ldots > \operatorname{Re}\alpha_k > \operatorname{Re}\alpha_{k+1}$. Then

$$\zeta_A(s) = \sum_{j=1}^{k} c_j \int_{\lambda_1}^{+\infty} \lambda^{-s} dN(\lambda) + f_k(s),$$

where f_k is holomorphic for $\operatorname{Re} s > \operatorname{Re}\alpha_{k+1}$. Since for $\operatorname{Re} s > \operatorname{Re}\alpha_j$

$$\int_{\lambda_1}^{+\infty} \lambda^{-s} d(\lambda^{\alpha_j}) = \alpha_j \int_{\lambda_1}^{+\infty} \lambda^{\alpha_j - s - 1} d\lambda = \frac{\alpha_j}{\alpha_j - s}[\lambda^{\alpha_j - s}]_{\lambda_1}^{+\infty} = \frac{\alpha_j}{s-\alpha_j}\lambda_1^{\alpha_j - s},$$

we obtain, for $\operatorname{Re} s > \operatorname{Re}\alpha_{k+1}$,

$$\zeta_A(s) = \sum_{j=1}^{k} c_j \frac{\lambda_1^{\alpha_j - s}\alpha_j}{s-\alpha_j} + f_k(s),$$

so that for $s = \alpha_j$ we have simple poles with residue $c_j\alpha_j$, $1 \le j \le k$. Conversely, knowing the poles of ζ_A gives information on the aymptotic behavior of $N(\lambda)$, as $\lambda \to +\infty$. However, things are very delicate and difficult even in the scalar case (be it on a compact boundaryless Riemannian manifold or the whole $\mathbb{R}^n \times \mathbb{R}^n$; see Ivrii [34] and references therein).

6.5 Notes

The proof of Karamata's Theorem 6.2.1 follows an unpublished idea of M. Aizenman.

Corollary 6.3.2 was obtained by Ichinose and Wakayama in [31] as a consequence, through Ikehara's theorem, of their result about the meromorphic continuation of the spectral ζ-function associated with $Q^{\mathrm{w}}_{(\alpha,\beta)}(x,D)$ (where $\alpha,\beta > 0$, $\alpha\beta > 1$), and later also by Parmeggiani in [53] (by Fourier integral operator methods) as a consequence of a theorem about the singularity at $t = 0$ of the Fourier transform of the spectral density $\sum_{j\ge 1}\delta(\lambda - \lambda_j)$. Formula (6.38) for the main coefficient in the asymptotics for $N(\lambda)$ was observed by Parmeggiani in [53], where the relation between that coefficient and the periods of the bicharacteristics curves associated with the eigenvalues of $Q_{(\alpha,\beta)}(x,\xi)$ (they are all periodic, as it will be seen below) was also shown.

It is worth mentioning also that Taniguchi in [68] approaches the construction of the heat-semigroup associated with $Q^{\mathrm{w}}_{(\alpha,\beta)}(x,D)$, for $\alpha,\beta > 0$, $\alpha\beta > 1$, by probabilistic methods. This method is deep and powerful, and should be extended to general NCHOs, to provide further geometrical understanding of the behavior of the eigenvalues of a NCHO in terms of "classical" quantities (i.e. related to the classical dynamics associated with the bicharacteristics).

Chapter 7
The Spectral Zeta Function

In the first section of this chapter we will briefly recall, addressing the reader to Robert's paper [63] (see also Aramaki [1]), the construction of the spectral zeta function $\zeta_A(s) = \operatorname{Tr} A^{-s}$, where $0 < A = A^* \in \mathrm{OP}S_{\mathrm{cl}}(m^\mu, g; \mathsf{M}_N)$ is an **elliptic** $N \times N$ system of GPDOs in $\mathbb{R}^n$ of order $\mu \in \mathbb{N}$. We will then prove a theorem about the meromorphic continuation of the spectral zeta function of an elliptic NCHO $A = A^* > 0$ in $\mathbb{R}^n$ (Theorem 7.2.1) by using the parametrix approximation of the heat-semigroup e^{-tA} constructed in Chapter 6, Section 6.1, from which we immediately deduce a corollary (Corollary 7.2.8) for our NCHO $Q^{\mathrm{w}}_{(\alpha,\beta)}(x,D)$ ($\alpha, \beta > 0$ and $\alpha\beta > 1$) which gives part of the result of Ichinose and Wakayama (Theorem 7.3.1 below; see [31]), whose proof we will sketch in the final section.

7.1 Robert's Construction of ζ_A by Complex Powers

Let $0 < A = A^* \in \mathrm{OP}S_{\mathrm{cl}}(m^\mu, g; \mathsf{M}_N)$ be an **elliptic** $N \times N$ system of GPDOs in $\mathbb{R}^n$ of order $\mu \in \mathbb{N}$. Recall from Remark 3.3.20 that the construction of the complex powers of A may be cast in the Weyl-Hörmander calculus with parameters. So, suppose the principal symbol a_μ of A be positive as an Hermitian matrix, that is $a_\mu = a_\mu^* > 0$, and let

$$\Lambda = \{z \in \mathbb{C};\ \arg z \in [\theta, 2\pi - \theta]\},$$

for some $\theta \in (0, \pi/4)$. Recalling the weight $m_\lambda(X) = (1 + |X|^2 + |\lambda|^{2/\mu})^{1/2}$ and the metric $g_{\lambda,X} = m_\lambda(X)^{-2}|dX|^2$, we have that

$$A - \lambda \in \mathrm{OP}S_{\mathrm{cl}}(m_\lambda^\mu, g_\lambda; \lambda \in \Lambda; \mathsf{M}_N),$$

and $A - \lambda$ is **elliptic** for $\lambda \in \Lambda$:

$$|a_\mu(X) - \lambda| \geq C m_\lambda(X)^\mu,\ \ |X|^2 + |\lambda|^{2/\mu} \gtrsim 1,\ \lambda \in \Lambda,\ X \in \mathbb{R}^{2n}.$$

Note that

$$a_\mu(tX) - (t^\mu \lambda) = t^\mu (a_\mu(X) - \lambda).$$

A. Parmeggiani, *Spectral Theory of Non-Commutative Harmonic Oscillators: An Introduction*, Lecture Notes in Mathematics 1992,
DOI 10.1007/978-3-642-11922-4_7, © Springer-Verlag Berlin Heidelberg 2010

Hence it suffices to check ellipticity only when $|X|^2 + |\lambda|^{2/\mu} = 1$ (recall that we are dealing with systems of GPDOs in $\mathbb{R}^n$).

One next constructs a parametrix $B_\lambda \in \mathrm{OP}S_{\mathrm{cl}}(m_\lambda^{-\mu}, g_\lambda; \lambda \in \Lambda; \mathsf{M}_N)$ of $A_\lambda = A - \lambda$, such that

$$B_\lambda A_\lambda = I + R_\lambda,\ A_\lambda B_\lambda = I + R'_\lambda,\ R_\lambda, R'_\lambda \in \mathrm{OP}S(m_\lambda^{-\infty}, g_\lambda; \lambda \in \Lambda; \mathsf{M}_N).$$

Starting from the symbol

$$b(X;\lambda) \sim \sum_{j\geq 0} b_{-\mu-2j}(X;\lambda)$$

of B_λ one then takes

$$a_z(X) \sim \frac{1}{2\pi i}\int_\gamma \lambda^z b(X;\lambda)d\lambda,$$

where $\gamma \subset \mathbb{C}$ is the curve

$$\{z \in \mathbb{C};\ |z| = c, |\arg z| \in [\theta', 2\pi - \theta']\} \cup \{z \in \mathbb{C};\ \arg z = \theta', |z| \geq c\}$$
$$\cup \{z \in \mathbb{C};\ \arg z = 2\pi - \theta', |z| \geq c\},$$

for some fixed $\theta' \in (\theta, \pi/4)$ and $c > 0$, oriented in such a way that the circle-part is clockwise oriented. If $a_{z,N}$ is any truncation of the asymptotics of a_z, with remainder $r_{z,N}$, one defines

$$A^z = a_{z,N}^{\mathrm{w}}(x,D) + r_{z,N}^{\mathrm{w}}(x,D), \ \text{ when } \operatorname{Re} z < 0,$$

and for $\operatorname{Re} z < \ell$, $\ell \in \mathbb{Z}_+$,

$$A^z = A^{z-\ell}A^\ell.$$

It follows that

$$A^z \in \mathrm{OP}S_{\mathrm{cl}}(m^{\mu \operatorname{Re} z}, g; \mathsf{M}_N).$$

Robert then proved the following result (later generalized by Aramaki to some infinite-dimensional situations), that we state for *classical* symbols (Robert's statement holds for more general polyhomogenous classes).

Theorem 7.1.1. *Let $\mathsf{K}^{(z)}$ be the Schwartz kernel of A^z. Then*

$$\zeta_A(s) = \operatorname{Tr} A^{-s} = \int_{\mathbb{R}^n} \operatorname{Tr} \mathsf{K}^{(-s)}(x,x)dx,$$

is holomorphic in $\{s \in \mathbb{C};\ \operatorname{Re} s > 2n/\mu\}$, and can be extended as a meromorphic function in $\mathbb{C}$, with at most **simple** *poles belonging to the sequence $s_j = \dfrac{2n}{\mu} - \dfrac{2j}{\mu}$, $j \in \mathbb{Z}_+$, with residue*

$$\mathrm{Res}(\zeta_A, s_j) = \frac{\mu}{i(2\pi)^{n+1}} \int_{\mathbb{S}^{2n-1}} \int_\gamma \lambda^{-s_j} \mathrm{Tr}\, b_{-\mu-2j}(\omega,\lambda) d\lambda d\omega.$$

The function ζ_A is holomorphic in 0 with value

$$\zeta_A(0) = \frac{1}{\mu(2\pi)^n} \int_{\mathbb{S}^{2n-1}} \int_0^{+\infty} \mathrm{Tr}\, b_{-\mu-2n}(\omega,-\lambda) d\lambda d\omega,$$

and since in our case A is a system of GPDOs*, we have that ζ_A is holomorphic in $-j$, $j \in \mathbb{N}$, with value*

$$\zeta_A(-j) = \frac{(-1)^j}{\mu(2\pi)^n} \int_{\mathbb{S}^{2n-1}} \int_0^{+\infty} \lambda^j \mathrm{Tr}\, b_{-\mu-2n-j\mu}(\omega,-\lambda) d\lambda d\omega,$$

hence it is surely 0 when $j\mu$ is not even, for in this case $b_{-\mu-2n-j\mu} = 0$. Note that $\zeta_A(-j) = \mathrm{Tr} A^j$, so that, since in our case A is a system of GPDOs*, the value $\zeta_A(-j)$ is the trace of a* **local** *operator.*

In the case $A = Q^{\mathrm{w}}_{(\alpha,\beta)}(x,D)$, we have $\mu = 2$ and $n = 1$, so that ζ_Q is meromorphic in $\mathbb{C}$, with at most simple poles that belong to the sequence $s_j = 1 - j$, $j \in \mathbb{Z}_+$. But, as we shall see below, by the theorem of Ichinose and Wakayama, or Corollary 7.2.8 to Theorem 7.2.1, we have that ζ_Q has **only one** simple pole at $s = 1$. Moreover, Ichinose and Wakayama prove in addition the remarkable result that $\zeta_Q(-2j) = 0$, $j \in \mathbb{Z}_+$, i.e. that ζ_Q has, beyond the zero at $s = 0$, the same "trivial zeros" of the Riemann zeta function (the negative even integers $-2k$, $k \in \mathbb{N}$).

The method of Robert to construct the complex powers of a NCHO (i.e. through a parametrix of the resolvent operator) seems to be not well-suited if one is willing to keep the expressions of the terms of the parametrix B_λ as much explicit as possible, as one can see already in the basic case of $Q^{\mathrm{w}}_{(\alpha,\beta)}(x,D)$. Since in a NCHO many symmetries are present (some of them are hidden because of the non-commutativity of the variables x and ξ, and because of the matrix-nature of the system), one should find the way to exploit them to simplify computations (we made some computations, and we could only prove that $\zeta_Q(0) = 0$).

Exercise 7.1.2. Upon computing a few terms in the symbol of the parametrix for $H - \lambda$, $H = p_0^{\mathrm{w}}(x,D)$, $n = 1$, calculate the residues of ζ_H and its value at $-j$, $j \in \mathbb{Z}_+$, using Robert's theorem. (See Remark 4.4.5.) △

In the next section we shall give a proof of the meromorphic continuation of the spectral zeta function ζ_A of an **elliptic** $N \times N$ NCHO $0 < A = A^* \in \mathrm{OP}S_{\mathrm{cl}}(m^2, g; \mathsf{M}_N)$, with positive principal symbol $a_2 = a_2^* > 0$, by using the parametrix approximation of e^{-tA} constructed in Section 6.1, which will yield Robert's result in the special, yet meaningful, case of a second order system of GPDOs.

7.2 The Meromorphic Continuation of ζ_A via the Parametrix Approximation of e^{-tA}

Let $0 < A = A^* \in \mathrm{OP}S_{\mathrm{cl}}(m^2, g; \mathsf{M}_N)$ be an elliptic $N \times N$ NCHO (i.e. an elliptic $N \times N$ second order GPD system) with positive principal symbol $a_2 = a_2^* > 0$. Let $\zeta_A(s)$ be the corresponding spectral zeta function. We have the following theorem.

Theorem 7.2.1. *There exist constants $c_{-2j,n}$, $0 \le j \le n-1$, and constants C_j, $j \ge n$, such that, for any given integer $\nu \in \mathbb{Z}_+$ with $\nu \ge n$,*

$$\zeta_A(s) = \frac{1}{\Gamma(s)}\Big[\sum_{j=0}^{n-1} \frac{c_{-2j,n}}{s-n+j} + \sum_{j=n}^{\nu} \frac{C_j}{s-n+j} + H_\nu(s)\Big], \tag{7.1}$$

where $\Gamma(s)$ is the Euler gamma function, and H_ν is holomorphic in the region $\mathrm{Re}\, s > -(\nu - n) - 1$. *Consequently, the spectral zeta function $\zeta_A(s)$ is meromorphic in the whole complex plane $\mathbb{C}$ with at most simple poles at $s = n, n-1, n-2, \ldots, 1$. In particular, when $n = 1$, ζ_A has only one simple pole at $s = 1$. One has (recall (6.31))*

$$c_{-2j,n} = (2\pi)^{-n} \int_0^{+\infty} \int_{\mathbb{S}^{2n-1}} \mathsf{Tr}\, b_{-2j}(1, \rho\omega)\rho^{2n-1} d\rho\, d\omega, \; 0 \le j \le n-1, \tag{7.2}$$

where the b_{-2j} are the terms in the symbol of the parametrix $U_A \in \mathrm{OP}S_{\mathrm{cl}}(2,0)$ of Theorem 6.1.4, Remark 6.1.5 and (6.9),

$$U_A \sim \sum_{j \ge 0} B_{-2j}.$$

Moreover,

$$c_{0,n} = (2\pi)^{-n} \int_0^{+\infty} \int_{\mathbb{S}^{2n-1}} \mathsf{Tr}(e^{-a_2(\rho\omega)})\rho^{2n-1} d\rho\, d\omega$$

is the leading coefficient in the Weyl asymptotics for $N_A(\lambda)$ (see (6.33) and Theorem 6.3.1).

Proof. By the properties of the semigroup $0 \le t \mapsto e^{-tA}$ we may use the Mellin transform and write (recall Corollary 4.4.4 with $\mu = 2$)

$$A^{-s} = \frac{1}{\Gamma(s)} \int_0^{+\infty} t^{s-1} e^{-tA} dt, \; \mathrm{Re}\, s > 2n/2 = n, \tag{7.3}$$

so that

$$\zeta_A(s) = \mathrm{Tr}\, A^{-s} = \frac{1}{\Gamma(s)} \int_0^{+\infty} t^{s-1} \mathrm{Tr}\, e^{-tA}\, dt. \tag{7.4}$$

Let hence $U_A \sim \sum_{j \ge 0} B_{-2j} \in \mathrm{OP}S_{\mathrm{cl}}(2,0)$ be the parametrix approximation of e^{-tA} constructed in Theorem 6.1.4 and Remark 6.1.5.

Next write

$$\zeta_A(s) = \frac{1}{\Gamma(s)}\left(\int_0^1 + \int_1^{+\infty}\right) t^{s-1} \mathrm{Tr}\, e^{-tA}\, dt =: Z_0(s) + Z_\infty(s). \tag{7.5}$$

In the first place we claim that $Z_\infty(s)$ is holomorphic in $\mathbb{C}$. In fact, on the one hand, since $t \mapsto \mathrm{Tr}\, R(t)$ is rapidly decreasing for $t \to +\infty$ (where $R(t) = e^{-tA} - U_A(t)$, see (6.7)), we have that for all $p \in \mathbb{N}$ and for all $t \geq 1$

$$|\mathrm{Tr}\, R(t)| \lesssim t^{-p}.$$

On the other, given any $\nu \geq 0$ and any symbol $b \in S(2, -2\nu)$, we have by (6.1) that for all $t \geq 1$ and all $p \in \mathbb{N}$

$$\begin{aligned}
&\left|(2\pi)^{-n} \int_0^{+\infty} \int_{\mathbb{S}^{2n-1}} \mathrm{Tr}\, b(t, \rho\omega) \rho^{2n-1} d\rho\, d\omega\right| \\
&= t^{-p} \left|(2\pi)^{-n} \int_0^{+\infty} \int_{\mathbb{S}^{2n-1}} t^p\, \mathrm{Tr}\, b(t, \rho\omega) \rho^{2n-1} d\rho\, d\omega\right| \\
&\lesssim t^{-p} \int_0^{+\infty} \frac{\rho^{2n-1}}{(1+\rho)^{2\nu+2p}} d\rho \lesssim t^{-p}.
\end{aligned}$$

It thus follows that for all $p \in \mathbb{N}$ and for all $t \geq 1$

$$|\mathrm{Tr}\, U_A(t)| \lesssim t^{-p}.$$

In conclusion, since

$$\mathrm{Tr}\, e^{-tA} = \mathrm{Tr}\, U_A(t) + \mathrm{Tr}\, R(t),$$

for every $p \geq 1$ there exists $C_p > 0$ such that

$$|\mathrm{Tr}\, e^{-tA}| \leq C_p t^{-p}, \quad \forall t \geq 1,$$

which proves the claim, since the term $1/\Gamma(s)$ is already holomorphic in $\mathbb{C}$.

Therefore the crux of the matter lies in the study of the function $Z_0(s)$. It is important to be a little more precise on the form of the b_{-2j}. First of all, let us rewrite the composition formula (3.3) as

$$(a \sharp b)(X) \sim a(X) b(X) + \sum_{j \geq 1} \frac{1}{j!} \left(\frac{-i}{2}\right)^j \{a, b\}_{(j)}(X),$$

where $\{\cdot,\cdot\}_{(1)} = \{\cdot,\cdot\}$ is the Poisson bracket. Then, from the construction of the parametrix approximation of e^{-tA}, and using the fact that the term a_0 in the symbol of A is a **constant** $N \times N$ Hermitian matrix, we have

$$r_{-2j} = a_0 b_{-2j} + \frac{1}{2}\left(\frac{-i}{2}\right)^2 \{a_2, b_{-2(j-1)}\}_{(2)} - \frac{i}{2}\{a_2, b_{-2j}\}, \ j \geq 0 \tag{7.6}$$

(where we set $b_2 \equiv 0$).

Recall hence, by (6.6), that

$$\begin{cases} b_0(t,X) = e^{-ta_2(X)}, \\ b_{-2(j+1)}(t,X) = -\int_0^t e^{-(t-t')a_2(X)} r_{-2j}(t',X)dt', \ j \geq 0. \end{cases} \tag{7.7}$$

To have a better understanding of the terms (6.28), $j \geq 0$, we need to control the behavior of the b_{-2j} as $t \to 0+$. This is provided by the following technical proposition, whose proof is postponed to the end of the section.

Proposition 7.2.2. *For any given $j \geq 0$ we have*

$$b_{-2j}(t,\omega) = O(t^j), \ t \to 0+,$$

and for all $\alpha, \beta \in \mathbb{Z}_+^n$, with $|\alpha| = 2k+1$, $k \geq 0$, and $|\beta| \leq 1$ we have

$$\partial_X^{\alpha+\beta} b_{-2j}(t,\omega) = O(t^{j+k+1}), \ t \to 0+,$$

where the constants in $O(\cdot)$ do not depend on $\omega \in \mathbb{S}^{2n-1}$.

Now, taking the proposition for granted, recall that by the homogeneity of the b_{-2j} (see (6.28)), for $t > 0$

$$\begin{aligned} \mathrm{Tr} B_{-2j}(t) &= (2\pi)^{-n} \int_0^{+\infty} \int_{\mathbb{S}^{2n-1}} \chi(\rho\omega) \mathrm{Tr}\, b_{-2j}(t,\rho\omega) \rho^{2n-1} d\rho d\omega \\ &= (2\pi)^{-n} \int_{1/2}^{+\infty} \int_{\mathbb{S}^{2n-1}} \chi(\rho\omega) \mathrm{Tr}\, b_{-2j}(\rho^2 t,\omega) \rho^{2(n-j)-1} d\rho d\omega. \end{aligned}$$

Consider

$$c_{-2j,n} = (2\pi)^{-n} \int_0^{+\infty} \int_{\mathbb{S}^{2n-1}} \mathrm{Tr}\, b_{-2j}(\rho^2,\omega) \rho^{2(n-j)-1} d\rho d\omega.$$

We claim that

$$|c_{-2j,n}| < +\infty, \ \forall j \in \mathbb{Z}_+.$$

In fact, the integral is convergent at $\rho = +\infty$ for all j since $\mathrm{Tr}\, b_{-2j}(\cdot,\omega)$ is a Schwartz function, it is clearly convergent at $\rho = 0$ for $0 \leq j \leq n-1$, and finally it is convergent at $\rho = 0$ also when $j \geq n$, for the singularity at 0 of the factor $\rho^{2(n-j)-1}$ is compensated by $\mathrm{Tr}\, b_{-2j}(t,\omega) = O(t^j)$ as $t \to 0+$.

Define now the functions

$$f_j(t) := -(2\pi)^{-n} \int_0^1 \int_{\mathbb{S}^{2n-1}} (1-\chi(\rho\omega)) \mathrm{Tr}\, b_{-2j}(t,\rho\omega) \rho^{2n-1} d\rho d\omega, \ j \in \mathbb{Z}_+. \tag{7.8}$$

Then $f_j \in C^\infty([0,+\infty);\mathbb{C})$, for all $j \in \mathbb{Z}_+$, and by Proposition 7.2.2

$$f_j(t) = O(t^j),\ t \to 0+. \tag{7.9}$$

It follows that

$$\operatorname{Tr} B_{-2j}(t) = c_{-2j,n} t^{-(n-j)} + f_j(t) = c_{-2j,n} t^{-(n-j)} + O(t^j),\ \ t \to 0+, \tag{7.10}$$

for all $j \geq 0$, and that (by the proof of Proposition 3.2.15 adapted to the present setting),

$$\operatorname{Tr} U_A(t) - \sum_{j=0}^{\nu} \operatorname{Tr} B_{-2j}(t) =: \operatorname{Tr} R_{\nu+1}(t) = O(t^{\nu+1}),\ \ t \to 0+,\ \forall \nu \in \mathbb{Z}_+. \tag{7.11}$$

However, the information contained in (7.10) alone is not yet sufficient to obtain the meromorphic continuation of ζ_A, and we need a better control of f_j. Notice that for all $j,k \in \mathbb{Z}_+$, writing $\partial_t^k f_j(t) = f_j^{(k)}(t)$,

$$f_j^{(k)}(t) = -(2\pi)^{-n} \int_0^1 \int_{\mathbb{S}^{2n-1}} (1 - \chi(\rho\omega)) \operatorname{Tr} \partial_t^k b_{-2j}(t, \rho\omega) \rho^{2n-1} d\rho d\omega,$$

so that $f_j^{(k)}(0)$ is finite and can be computed through (7.6), and through the differential equations (7.20) and (7.24) below used to construct the b_{-2j}. Note, in particular, that

$$f_0^{(k)}(0) = (-1)^{k+1}(2\pi)^{-n} \int_0^1 \int_{\mathbb{S}^{2n-1}} (1 - \chi(\rho\omega)) \operatorname{Tr} a_2(\rho\omega)^k \rho^{2n-1} d\rho d\omega.$$

It is useful now to have the following elementary result.

Lemma 7.2.3. *Let $f \in C^\infty([0,1];\mathbb{C})$. Define, for $s \in \mathbb{C}$, $\operatorname{Re} s > 0$,*

$$F(s) = \int_0^1 t^{s-1} f(t) dt.$$

Then, for any given $\nu \in \mathbb{Z}_+$,

$$F(s) = \sum_{k=0}^{\nu} \frac{f^{(k)}(0)}{k!} \frac{1}{s+k} + F_\nu(s), \tag{7.12}$$

where

$$F_\nu(s) = \frac{1}{\nu!} \int_0^1 t^{s+\nu} \Big(\int_0^1 (1-\tau)^\nu f^{(\nu+1)}(t\tau) d\tau \Big) dt \tag{7.13}$$

is holomorphic for $\operatorname{Re} s > -\nu - 1$. *Hence F can be meromorphically continued to the whole complex plane, with at most simple poles at the non-positive integers* $s = -k$, $k \in \mathbb{Z}_+$, *and*

$$\operatorname{Res}(F, -k) = \frac{f^{(k)}(0)}{k!}. \tag{7.14}$$

In particular, if $f(t) = O(t^j)$ *as* $t \to 0+$ *for some* $j \in \mathbb{Z}_+$, *then* $F(s)$ *is meromorphic on* $\mathbb{C}$ *with at most simple poles at* $s = -j - k$, $k \in \mathbb{Z}_+$.

Proof (of Lemma 7.2.3). By Taylor's formula, for any given $\nu \in \mathbb{Z}_+$,

$$f(t) = \sum_{k=0}^{\nu} \frac{f^{(k)}(0)}{k!} t^k + \frac{t^{\nu+1}}{\nu!} \int_0^1 (1-\tau)^\nu f^{(\nu+1)}(t\tau) d\tau.$$

Hence, if $\nu = 0$ and $\operatorname{Re} s > 0$ we write $f(t) = f(0) + t \int_0^1 f'(t\tau) d\tau$ and have

$$F(s) = \frac{f(0)}{s} + \int_0^1 t^s \Big(\int_0^1 f'(t\tau) d\tau \Big) dt,$$

where

$$\int_0^1 t^s \Big(\int_0^1 f'(t\tau) d\tau \Big) dt$$

is holomorphic for $\operatorname{Re} s > -1$. This shows that F can be meromorphically extended for $\operatorname{Re} s > -1$, with possibly a simple pole at $s = 0$ with residue $f(0)$. Iterating the process using Taylor's formula gives (7.12) to (7.14).

If now $f(t) = O(t^j)$ as $t \to 0+$ for some $j \in \mathbb{Z}_+$, we have that $F(s)$ is holomorphic for $\operatorname{Re} s > -j$, and meromorphic on $\mathbb{C}$ with at most simple poles at $s = -j - k$, $k \in \mathbb{Z}_+$. This completes the proof of the lemma. □

We next apply Lemma 7.2.3 to the functions f_j, so that for any given $\nu \in \mathbb{Z}_+$ we may write, by (7.9),

$$F_j(s) := \int_0^1 t^{s-1} f_j(t) dt = \sum_{k=0}^{\nu} \frac{f_j^{(j+k)}(0)}{(j+k)!} \frac{1}{s+j+k} + F_{j,\nu}(s),$$

where $F_{j,\nu}$ is holomorphic for $\operatorname{Re} s > -j - \nu - 1$.

Using this in (7.10) gives the following lemma.

Lemma 7.2.4. *For each* $j \geq 0$, *for any given* $\nu \in \mathbb{Z}_+$,

$$s \longmapsto \int_0^1 t^{s-1} \operatorname{Tr} B_{-2j}(t) dt = \frac{c_{-2j,n}}{s-(n-j)} + \sum_{k=0}^{\nu} \frac{f_j^{(j+k)}(0)}{(j+k)!} \frac{1}{s+j+k} + F_{j,\nu}(s),$$

where $F_{j,k}$ *is holomorphic for* $\operatorname{Re} s > -j - \nu - 1$.

Analogously, from (6.22) and Lemma 7.2.3 we also have, with $f_R(t) := \mathrm{Tr}R(t)$, that for any given $\nu \in \mathbb{Z}_+$

$$\int_0^1 t^{s-1} f_R(t)dt = \sum_{k=0}^{\nu} \frac{f_R^{(k)}(0)}{k!} \frac{1}{s+k} + F_{R,\nu}(s),$$

where $F_{R,\nu}$ is holomorphic for $\mathrm{Re}\, s > -\nu - 1$.

We therefore obtain that for any given $\nu \in \mathbb{Z}_+$

$$Z_0(s) = \frac{1}{\Gamma(s)} \Big[\sum_{j=0}^{\nu} \int_0^1 t^{s-1} \mathrm{Tr}B_{-2j}(t)dt + \int_0^1 t^{s-1} \mathrm{Tr}R_{\nu+1}(t)dt + \int_0^1 t^{s-1} \mathrm{Tr}R(t)dt \Big].$$

Since the function $s \mapsto \int_0^1 t^{s-1} \mathrm{Tr}R_{\nu+1}(t)dt =: F_{\nu+1}(s)$ is holomorphic for $\mathrm{Re}\, s > -\nu - 1$, we thus obtain that, for any given $\nu \in \mathbb{Z}_+$ with $\nu \geq n$,

$$\begin{aligned} Z_0(s) &= \frac{1}{\Gamma(s)} \Big[\sum_{j=0}^{\nu} \frac{c_{-2j,n}}{s-n+j} + \sum_{j,k=0}^{\nu} \frac{f_j^{(j+k)}(0)}{(j+k)!} \frac{1}{s+j+k} \\ &\quad + \sum_{k=0}^{\nu} \frac{f_R^{(k)}(0)}{k!} \frac{1}{s+k} + \sum_{j=0}^{\nu} F_{j,\nu}(s) + F_{R,\nu}(s) + F_{\nu+1}(s) \Big] \\ &= \frac{1}{\Gamma(s)} \Big[\sum_{j=0}^{n-1} \frac{c_{-2j,n}}{s-n+j} + \sum_{j=n}^{\nu} \frac{C_j}{s-n+j} + \tilde{H}_\nu(s) \Big], \end{aligned} \tag{7.15}$$

with $\tilde{H}_\nu(s)$ holomorphic for $\mathrm{Re}\, s > -(\nu - n) - 1$. Since the function $1/\Gamma(s)$ is holomorphic in $\mathbb{C}$ and has zeros at the non-positive integers $-k$, $k \in \mathbb{Z}_+$, this proves the theorem. □

As a corollary of the proof of Theorem 7.2.1 one has the following expression of the coefficients C_j appearing in (7.15).

Corollary 7.2.5. *With the notation of the proof of Theorem 7.2.1 one has, for all* $\ell \geq 0$,

$$C_{\ell+n} = c_{-2(\ell+n),n} + \frac{1}{\ell!} \Big(\sum_{j=0}^{\ell} f_j^{(\ell)}(0) + f_R^{(\ell)}(0) \Big). \tag{7.16}$$

As another corollary of the proof of Theorem 7.2.1, we have the following result on $\zeta_A(0)$.

Corollary 7.2.6. *We have*

$$\zeta_A(0) = c_{-2n,n} = (2\pi)^{-n} \int_0^{+\infty} \int_{\mathbb{S}^{2n-1}} \mathrm{Tr}\, b_{-2n}(\rho^2, \omega) \rho^{-1} d\rho d\omega.$$

Proof. In the first place, due to the (simple) pole at $s = 0$ of $\Gamma(s)$, we have from (7.5), (7.15) and (7.16) that

$$\zeta_A(0) = \mathrm{Res}(\Gamma\zeta_A, 0) = c_{-2n,n} + f_0(0) + f_R(0).$$

Now, recalling that (see Remark 6.1.5)

$$c_A\Big|_{t=0} = \chi I_{\mathbb{C}^N},$$

we have

$$R(0) = (e^{-tA} - U_A(t))\Big|_{t=0} = (1-\chi)^{\mathrm{w}}(x,D) I_{\mathbb{C}^N}.$$

Hence

$$f_R(0) = \mathrm{Tr}\, R(0) = (2\pi)^{-n} \int_0^{+\infty} \int_{\mathbb{S}^{2n-1}} (1 - \chi(\rho\omega)) \mathsf{Tr}(I_{\mathbb{C}^N}) \rho^{2n-1} d\rho\, d\omega.$$

But we also have

$$f_0(0) = -(2\pi)^{-n} \int_0^{+\infty} \int_{\mathbb{S}^{2n-1}} (1 - \chi(\rho\omega)) \mathsf{Tr}(I_{\mathbb{C}^N}) \rho^{2n-1} d\rho\, d\omega,$$

whence

$$c_{-2n,n} + f_0(0) + f_R(0) = c_{-2n,n},$$

which proves the corollary. □

Analogously, by looking into the proof of Theorem 7.2.1, one may also compute the values $\zeta_A(-k)$, $k \in \mathbb{Z}_+$, by using

$$\zeta_A(-k) = \mathrm{Res}(\Gamma\zeta_A, -k),$$

and formulas (7.15) and (7.16). One needs to control $f_R^{(k)}(0)$. This may be achieved as follows. For any given $\nu \in \mathbb{Z}_+$, write again

$$U_A(t) = \sum_{j=0}^{\nu} B_{-2j}(t) + R_{\nu+1}(t).$$

Then

$$\left(\frac{d}{dt} + A\right) R(t) = -\left(\frac{d}{dt} + A\right) R_{\nu+1}(t), \tag{7.17}$$

whence

$$R(t) = e^{-tA} R(0) - \int_0^t e^{-(t-t')A} \left(\frac{d}{dt} + A\right) R_{\nu+1}(t') dt',$$

and for any given $k \geq 1$ one may compute from (7.17) by successive differentiations a formula for

$$f_R^{(k)}(0) = \frac{d^k}{dt^k}\mathrm{Tr}\,R\Big|_{t=0}.$$

Remark 7.2.7. To obtain explicitly the terms $\{a_2, b_{-2j}\}_{(\ell)}$ in order to compute the coefficients $c_{-2j,n}$, it is covenient to make use of the following formula (see, e.g., Rossmann's book [66, p. 15]): for $\alpha \in \mathbb{Z}_+^n$ with $|\alpha| = 1$,

$$\partial_X^\alpha e^{a_2(X)} = e^{a_2(X)} \frac{1-\exp(-\mathrm{ad}\,a_2(X))}{\mathrm{ad}\,a_2(X)} \partial_X^\alpha a_2(X), \tag{7.18}$$

where

$$\frac{1-\exp(-\mathrm{ad}\,a_2(X))}{\mathrm{ad}\,a_2(X)} := \sum_{k=0}^{\infty} \frac{(-1)^k}{(k+1)!}(\mathrm{ad}\,a_2(X))^k, \quad \mathrm{ad}\,a_2(X)b = [a_2(X), b].$$

Recently, M. Hitrik and I. Polterovich have given in [24] (see also [23]), nice formulas for resolvent expansions and trace regularizations for Schrödinger operators. A nice problem is that of using these formulas to compute the coefficients $c_{-2j,n}$ and therefore re-obtain the full result by Wakayama and Ichinose, Theorem 7.3.1 below. △

When the order of the system is μ, which then must be an even integer by Lemma 3.2.13, one can prove the analog of Theorem 7.2.1, that is, that ζ_A has a meromorphic continuation to the whole complex plane with at most simple poles at $s = 2(n-j)/\mu$, $j \in \mathbb{Z}_+$. However, in this case the analog of Proposition 7.2.2 is much more involved, and we leave the (non trivial) details to the interested reader. We just give the following formula for the terms $r_{\mu-2\nu}$, $\nu \geq 1$:

$$\begin{aligned} r_{\mu-2\nu} = & \sum_{j=\max(\nu-\frac{\mu}{2},0)}^{\nu-1} a_{\mu-2(\nu-j)} b_{-2j} \\ & + \sum_{k=0}^{\frac{\mu}{2}-1} \sum_{j=\max(\nu-\mu+k,0)}^{\nu-1-k} \frac{1}{(\nu-k-j)!}\left(\frac{-i}{2}\right)^{\nu-k-j} \{a_{\mu-2k}, b_{-2j}\}_{(\nu-k-j)}. \end{aligned} \tag{7.19}$$

Notice that ζ_A is still continuous at $s = 0$, and that the zero of the factor $1/\Gamma(s)$ at $s = -2k$ now cancels the pole at $s = 2(n-j)/\mu$ of $\Gamma\zeta_A$ only when j is such that $2n - 2j = -2\mu k$, i.e. $j = n + \mu k$. We hence expect, for $\mu > 2$, a sequence of simple poles also for $\mathrm{Re}\,s < 0$.

Specializing to the case $A = Q^{\mathrm{w}}_{(\alpha,\beta)}(x,D)$, we have the following corollary of Theorem 7.2.1.

Corollary 7.2.8. *Consider an elliptic* NCHO $Q^{\mathrm{w}}_{(\alpha,\beta)}(x,D)$ *($\alpha,\beta>0$ and $\alpha\beta>1$). Then the associated spectral zeta function ζ_Q can be meromorphically continued to the whole complex place $\mathbb{C}$ with only one simple pole at $s=1$.*

We close this section by providing the proof of Proposition 7.2.2.

Proof (of Proposition 7.2.2). We won't be writing the dependence on ω, and we will write $b^{(\ell)}_{-2j}$ for a generic $\partial^{\alpha}_X b_{-2j}$ with $|\alpha|=\ell$.

Denote by $E(a_2^{(2)},b_{-2j}^{(2)})$, resp. $E(a_2^{(1)},b_{-2j}^{(1)})$, a generic expression obtained by taking the (matrix) product of derivatives of order 2, resp. order 1, of a_2 with derivatives of order 2, resp. order 1, of b_{-2j}. Hence for all $j\geq 0$ we may generically write

$$\{a_2,b_{-2j}\}=E(a_2^{(1)},b_{-2j}^{(1)}),\ \text{ and } \{a_2,b_{-2j}\}_{(2)}=E(a_2^{(2)},b_{-2j}^{(2)}).$$

Notice, therefore, that in $\{a_2,b_{-2j}\}_{(2)}$ we have a **constant**-coefficient matrix (given by partial derivatives of order 2 of a_2) times partial derivatives of order 2 of b_{-2j}.

We proceed by induction. We start with the case $j=0$. In this case b_0 is the solution of

$$\begin{cases}\partial_t b_0+a_2b_0=0,\\ \\ b_0\big|_{t=0}=I,\end{cases}\tag{7.20}$$

whence $b_0(t)=O(1)$ as $t\to 0+$.

Next, by induction on ℓ we show that $b_0^{(\ell)}$ has the claimed property.

For $\ell=1$ we take a 1st-order partial derivative with respect to X of (7.20) and have

$$\begin{cases}\partial_t b_0^{(1)}+a_2b_0^{(1)}=-a_2^{(1)}b_0,\\ \\ b_0^{(1)}\big|_{t=0}=0,\end{cases}\tag{7.21}$$

whence

$$b_0^{(1)}(t)=-\int_0^t e^{-(t-t')a_2}a_2^{(1)}b_0(t')dt'=O(t),\ \ t\to 0+.\tag{7.22}$$

For $\ell=2$ we take a 1st-order partial derivative with respect to X of (7.22) (of course, we may equivalently use again (7.20) by taking an extra derivative of (7.21)) and have

$$\begin{aligned}b_0^{(2)}(t)&=-\int_0^t(e^{-(t-t')a_2})^{(1)}a_2^{(1)}b_0(t')dt'-\int_0^t e^{-(t-t')a_2}a_2^{(2)}b_0(t')dt'\\&\quad-\int_0^t e^{-(t-t')a_2}a_2^{(1)}b_0^{(1)}(t')dt'\\&=O(t^2)+O(t)+O(t^2)=O(t),\ t\to 0+.\end{aligned}$$

Next, suppose $b_0^{(2k-1+\ell)}(t)=O(t^k)$ as $t\to 0+$, for $\ell=0,1$ and all $k\geq 0$ We want to prove that $b_0^{(2k+1+\ell)}(t)=O(t^{k+1})$, as $t\to 0+$, for $\ell=0,1$. Using (7.20) and taking a

$2k+1$-st partial derivative with respect to X we get (recall that $a_2^{(\ell)} = 0$ for all $\ell \geq 3$, as a_2 is a polynomial in X of degree 2)

$$\begin{cases} \partial_t b_0^{(2k+1)} + a_2 b_0^{(2k+1)} = -a_2^{(1)} b_0^{(2k)} - a_2^{(2)} b_0^{(2k-1)} = O(t^k) + O(t^k) = O(t^k), \\ b_0^{(2k+1)}\big|_{t=0} = 0, \end{cases}$$

whence $b_0^{(2k+1)}(t) = O(t^{k+1})$ as $t \to 0+$. Then, as before,

$$\begin{aligned} b_0^{(2k+2)}(t) &= -\int_0^t (\partial_X(e^{-(t-t')a_2}))\,(a_2^{(1)} b_0^{(2k)}(t') + a_2^{(2)} b_0^{(2k-1)}(t'))dt' \\ &\quad - \int_0^t e^{-(t-t')a_2}\partial_X\,(a_2^{(1)} b_0^{(2k)}(t') + a_2^{(2)} b_0^{(2k-1)}(t'))dt' \\ &= O(t^{k+2}) + O(t^{k+1}) = O(t^{k+1}),\ t \to 0+. \end{aligned}$$

Hence the result is proved for b_0. So suppose, by induction, that

$$b_{-2j}(t) = O(t^j),\ b_{-2j}^{(2k+1+\ell)}(t) = O(t^{j+k+1}),\ t \to 0+,\ \ell = 0,1.$$

We prove the result for $b_{-2(j+1)}$. To do so, we have to examine r_{-2j} (see (7.7)). In the first place we have from (7.6)

$$r_{-2j}(t) = O(t^j) + O(t^{j-1+1}) + O(t^{j+1}) = O(t^j),\ t \to 0+.$$

Consider next, keeping into account the fact that a_0 is a constant matrix,

$$\begin{aligned} r_{-2j}^{(2k+1)} &= a_0 b_{-2j}^{(2k+1)} + E(a_2^{(2)}, b_{-2(j-1)}^{(2+2k+1)}) + E(a_2^{(2)}, b_{-2j}^{(2k+1)}) + E(a_2^{(1)}, b_{-2j}^{(2k+2)}) \\ &= O(t^{j+k+1}) + O(t^{j-1+k+2}) + O(t^{j+k+1}) + O(t^{j+k+1}) \\ &= O(t^{j+k+1}),\ t \to 0+. \end{aligned}$$

Taking an extra derivative, one immediately sees also that

$$r_{-2j}^{(2k+2)} = O(t^{j+k+1}),\ \ t \to 0+.$$

Hence

$$r_{-2j}^{(2k+1+\ell)} = O(t^{j+k+1}),\ \ t \to 0+,\ \ell = 0,1,\ \forall k \geq -1 \tag{7.23}$$

(when $k = -1$ we take $\ell = 1$). Since

$$\begin{cases} \partial_t b_{-2(j+1)} + a_2 b_{-2(j+1)} = -r_{-2j}, \\ b_{-2(j+1)}\big|_{t=0} = 0, \end{cases} \tag{7.24}$$

we obtain $b_{-2(j+1)}(t) = O(t^{j+1})$ as $t \longrightarrow 0+$. As before, taking one partial derivative with respect to X yields

$$\begin{cases} \partial_t b^{(1)}_{-2(j+1)} + a_2 b^{(1)}_{-2(j+1)} = -E(a_2^{(1)}, b^{(1)}_{-2(j+1)}) - r^{(1)}_{-2j} = O(t^{j+1}) + O(t^{j+1}), \\ b^{(1)}_{-2(j+1)}|_{t=0} = 0, \end{cases}$$

whence it follows that $b^{(1)}_{-2(j+1)}(t) = O(t^{j+2})$, and, taking an extra derivative, also that, as $t \longrightarrow 0+$,

$$b^{(2)}_{-2(j+1)}(t) = -\partial_X \left(\int_0^t e^{-(t-t')a_2} \left(E(a_2^{(1)}, b^{(1)}_{-2(j+1)}) + r^{(1)}_{-2j} \right) dt' \right) = O(t^{j+2}).$$

Supposing then by induction the estimates up to order $2k-1$, using

$$\begin{cases} \partial_t b^{(2k+1)}_{-2(j+1)} + a_2 b^{(2k+1)}_{-2(j+1)} = -E(a_2^{(1)}, b^{(2k)}_{-2(j+1)}) - E(a_2^{(2)}, b^{(2k-1)}_{-2(j+1)}) - r^{(2k+1)}_{-2j} \\ \qquad\qquad\qquad\qquad\qquad = O(t^{j+1+k}) + O(t^{j+1+k}) + O(t^{j+1+k}), \\ b^{(2k+1)}_{-2(j+1)}|_{t=0} = 0, \end{cases}$$

we get $b^{(2k+1)}_{-2(j+1)}(t) = O(t^{j+1+k+1})$, as $t \longrightarrow 0+$, and using

$$\begin{aligned} &b^{(2k+2)}_{-2(j+1)}(t) \\ &= -\partial_X \left(\int_0^t e^{-(t-t')a_2} \left(E(a_2^{(1)}, b^{(2k)}_{-2(j+1)}) + E(a_2^{(2)}, b^{(2k-1)}_{-2(j+1)}) + r^{(2k+1)}_{-2j} \right) dt' \right), \end{aligned}$$

also that

$$b^{(2k+2)}_{-2(j+1)}(t) = O(t^{j+1+k+1}), \quad t \longrightarrow 0+,$$

which proves the proposition. □

7.3 The Ichinose–Wakayama Theorem

In this section we want to outline the proof of Ichinose-Wakayama's theorem on the spectral zeta function ζ_Q of our elliptic NCHO $Q^{\mathrm{w}}_{(\alpha,\beta)}(x,D)$ (for $\alpha, \beta > 0$ and $\alpha\beta > 1$). It is worth noting that, by Corollary 4.1 of Parmeggiani-Wakayama [59, p. 555], when $\alpha = \beta > 1$ one has that $Q^{\mathrm{w}}_{(\alpha,\alpha)}(x,D)$ is **unitarily** equivalent to the scalar harmonic oscillator

$$\sqrt{\alpha^2 - 1}\left(-\frac{\partial_x^2}{2} + \frac{x^2}{2} \right) \begin{bmatrix} 1 & 0 \\ 0 & 1 \end{bmatrix},$$

which has eigenvalues $\sqrt{\alpha^2-1}(j+1/2)$, $j \geq 0$, all of multiplicity 2. We therefore have that when $\alpha = \beta > 1$

$$\zeta_Q(s) = 2\frac{2^s-1}{(\alpha^2-1)^{s/2}}\zeta(s), \tag{7.25}$$

where ζ is the Riemann zeta function. Hence, when $\alpha = \beta > 1$, ζ_Q has a meromorphic extension to the whole complex plane $\mathbb{C}$ with *one single simple pole* at $s = 1$ and it is zero at the non-positive integers $-2k$, $k \in \mathbb{Z}_+$. Notice that the *negative* integers $-2k$, $k \in \mathbb{N}$, are the so-called "trivial zeros" of ζ. We may hence say that ζ_Q has trivial zeros at the non-positive even integers. The fact that all this holds true also when $\alpha \neq \beta$ is a remarkable and elegant theorem due to Ichinose and Wakayama [31].

Theorem 7.3.1. *There exist constants $C_{Q,j}$, $j \in \mathbb{N}$, such that $\zeta_Q(s)$ is represented, for every integer $\nu \in \mathbb{N}$, as*

$$\zeta_Q(s) = \frac{1}{\Gamma(s)}\Big[\frac{\alpha+\beta}{\sqrt{\alpha\beta\,(\alpha\beta-1)}}\frac{1}{s-1} + \sum_{j=1}^{\nu}\frac{C_{Q,j}}{s+2j-1} + H_{Q,\nu}(s)\Big], \tag{7.26}$$

where $\Gamma(s)$ is the Euler gamma function, and $H_{Q,\nu}$ is holomorphic in $\mathrm{Re}\, s > -2\nu$. *Consequently, the spectral zeta function $\zeta_Q(s)$ is meromorphic in the whole complex plane $\mathbb{C}$ with a simple pole at $s = 1$, and has zeros (the so-called "trivial zeros") for $0, -2, -4, \ldots$ (the non-positive even integers $2\mathbb{Z}_-$).*

The non-positive even integers $2\mathbb{Z}_-$, the "trivial zeros" of $\zeta_Q(s)$, are due to the presence of the function $1/\Gamma(s)$, which is holomorphic in $\mathbb{C}$ with zeros at the non-positive integers $\mathbb{Z}_- = \{0, -1, -2, \ldots\}$, in the expression of $\zeta_Q(s)$. As already observed above, the *negative* trivial zeros are the same as those of Riemann's ζ.

As already mentioned in Chapter 6, Section 6.4, the use of Ikehara's theorem gives the following corollary of Theorem 7.3.1 (Weyl's law of the asymptotics of large eigenvalues; see Theorem 6.3.1, Corollary 6.3.2 and Remark 6.3.3; see also Theorem 5.2 in Parmeggiani [52]).

Corollary 7.3.2.

$$N(\lambda) = \sum_{\lambda_j < \lambda} 1 \sim \frac{\alpha+\beta}{\sqrt{\alpha\beta\,(\alpha\beta-1)}}\lambda, \ \text{as } \lambda \to +\infty.$$

Remark 7.3.3. Notice that the fact that ζ_Q has only one simple pole at $s = 1$ is a consequence of Theorem 7.2.1. But (7.26) gives a much more precise result, for it also states that $\zeta_Q(-2k) = 0$ for $k \in \mathbb{Z}_+$. As already remarked, obtaining this information from Theorem 7.2.1 is in principle also possible. We leave it to the reader to explore this direction. △

Proof (sketch of proof of Theorem 7.3.1). Write for short Q in place of $Q^{\mathrm{w}}_{(\alpha,\beta)}(x,D)$. Because of the properties of the semigroup $0 \le t \mapsto e^{-tQ}$ we may use Mellin's transform and write

$$Q^{-s} = \frac{1}{\Gamma(s)} \int_0^{+\infty} t^{s-1} e^{-tQ} dt, \ \operatorname{Re} s > 1, \tag{7.27}$$

so that

$$\zeta_Q(s) = \operatorname{Tr} Q^{-s} = \frac{1}{\Gamma(s)} \int_0^{+\infty} t^{s-1} \operatorname{Tr} e^{-tQ} dt, \ \operatorname{Re} s > 1. \tag{7.28}$$

Write

$$\zeta_Q(s) = \frac{1}{\Gamma(s)} \Big(\int_0^1 + \int_1^{+\infty} \Big) t^{s-1} \operatorname{Tr} e^{-tQ} dt =: Z_0(s) + Z_\infty(s).$$

One then readily sees (as before) that $Z_\infty(s)$ is holomorphic in $\mathbb{C}$, so that the crux of the matter lies, as before, in the study of the function $Z_0(s)$, and for this, Ichinose and Wakayama need an asymptotic expansion for $t > 0$ of the kind

$$\operatorname{Tr} e^{-tQ} \sim c_{-1} t^{-1} + c_0 + c_1 t + c_2 t^2 + \ldots, \ 0 < t \to 0+.$$

Put $\mathsf{A} = \begin{bmatrix} \alpha & 0 \\ 0 & \beta \end{bmatrix}$, and define the pseudodifferential operator $P_1(t)$ (a quantization "to the right") by

$$P_1(t,x,D)u(x) = (2\pi)^{-1} \iint e^{i(x-y)\xi} e^{-t\left(\mathsf{A}(y^2+\xi^2)/2 + iJy\xi\right)} u(y) dy d\xi, \ u \in \mathscr{S}(\mathbb{R};\mathbb{C}^2).$$

Next put $R_2(t) = e^{-tQ} - P_1(t)$. Ichinose and Wakayama then prove that

$$Z_0(s) = \frac{\alpha+\beta}{\sqrt{\alpha\beta(\alpha\beta-1)}} \frac{1}{\Gamma(s)} \frac{1}{s-1} + \frac{1}{\Gamma(s)} \int_0^1 t^{s-1} \operatorname{Tr} R_2(t) dt, \tag{7.29}$$

whence it follows that it suffices to treat the second term on the right-hand side of (7.29). We now use the heat-equation:

$$0 = (\partial_t + Q)e^{-tQ} = (\partial_t + Q)P_1(t) + (\partial_t + Q)R_2(t),$$

so that

$$(\partial_t + Q)R_2(t) = -(\partial_t + Q)P_1(t) =: S(t). \tag{7.30}$$

By the Duhamel principle we may solve (7.30) obtaining

$$R_2(t) = \int_0^t e^{-(t-t')Q} S(t') dt' = \int_0^t \Big(P_1(t-t') + R_2(t-t') \Big) S(t') dt' =: P_2(t) + R_3(t),$$

whence

$$e^{-tQ} = P_1(t) + P_2(t) + R_3(t),$$

and by repeating the process

$$e^{-tQ} = P_1(t) + P_2(t) + \ldots + P_\nu(t) + R_{\nu+1}(t).$$

One has therefore to study $\mathrm{Tr}\,P_k(t)$ and $\mathrm{Tr}\,R_{\nu+1}(t)$ for $0 < t \to 0+$. One has that *for any given (sufficiently small) $\varepsilon > 0$ there exist positive constants C_ε (**independent of** t) and C (**independent of** t and ν) such that*

$$|\mathrm{Tr}\,R_2(t)| \leq C_\varepsilon t^{-\varepsilon}, \text{ and } |\mathrm{Tr}\,R_{\nu+1}(t)| \leq C^\nu \frac{\Gamma(1/2)^\nu}{\Gamma(1+\nu/2)} t^{\nu/2}, \; \nu \geq 1. \tag{7.31}$$

This yields that the Mellin transform of $\mathrm{Tr}\,R_2(t)$ is holomorphic in $\mathrm{Re}\,s > \varepsilon$, and hence in $\mathrm{Re}\,s > 0$, for ε is arbitrary, and that the Mellin tranform of $\mathrm{Tr}\,R_{\nu+1}(t)$ ($\nu \geq 1$) is holomorphic in the region $\mathrm{Re}\,s > -\nu/2$. As for the $P_k(t)$ terms, Ichinose and Wakayama then prove, by highly non-trivial computations, that

$$\mathrm{Tr}\,P_k(t) \sim \sum_{j \geq 0} c_{k,j} t^j, \; \textit{for } k = 2, 3, \ldots, \textit{ and for } t \to 0+, \tag{7.32}$$

with $c_{k,j} = 0$ for $2 \leq j < k-2$ and $j \in 2\mathbb{N}$. It follows that

$$\mathrm{Tr}\,P_1(t) = \frac{\alpha+\beta}{\sqrt{\alpha\beta\,(\alpha\beta-1)}} t^{-1},$$

and that

$$\begin{aligned}
\mathrm{Tr}\,P_2(t) &\sim c_{2,1}t + c_{2,3}t^3 + c_{2,5}t^5 + c_{2,7}t^7 + c_{2,9}t^9 + \ldots \\
\mathrm{Tr}\,P_3(t) &\sim c_{3,1}t + c_{3,3}t^3 + c_{3,5}t^5 + c_{3,7}t^7 + c_{3,9}t^9 + \ldots \\
\mathrm{Tr}\,P_4(t) &\sim c_{4,3}t^3 + c_{4,5}t^5 + c_{4,7}t^7 + c_{4,9}t^9 + \ldots \\
\mathrm{Tr}\,P_5(t) &\sim c_{5,3}t^3 + c_{5,5}t^5 + c_{5,7}t^7 + c_{5,9}t^9 + \ldots \\
\mathrm{Tr}\,P_6(t) &\sim c_{6,5}t^5 + c_{6,7}t^7 + c_{6,9}t^9 + \ldots \\
&\ldots \\
\mathrm{Tr}\,R_{\nu+1}(t) &= O(t^{\nu/2}).
\end{aligned}$$

Via the Mellin transform we thus obtain

$$\begin{aligned}
\zeta_Q(s) = {} & \frac{1}{\Gamma(s)} \frac{\alpha+\beta}{\sqrt{\alpha\beta\,(\alpha\beta-1)}} \frac{1}{s-1} + (c_{2,1}+c_{3,1}) \frac{1}{\Gamma(s)} \frac{1}{s+1} \\
& + (c_{2,3}+c_{3,3}+c_{4,3}+c_{5,3}) \frac{1}{\Gamma(s)} \frac{1}{s+3} \\
& + (c_{2,5}+c_{3,5}+c_{4,5}+c_{5,5}+c_{6,5}) \frac{1}{\Gamma(s)} \frac{1}{s+5} + \ldots + \frac{1}{\Gamma(s)} H_{Q,\nu}(s),
\end{aligned}$$

which concludes the proof. □

Problem 7.3.4. Compute the constants $C_{Q,j}$ appearing in the statement of Theorem 7.3.1 by using the parametrix approximation of e^{-tA}, as in the proof of Theorem 7.2.1 (see also Corollary 7.2.5).

7.4 Notes

In the outline of the proof Theorem 7.3.1 and its consequences we followed Ichinose and Wakayama's paper [33].

We address the reader to the papers by Ichinose and Wakayama [32], by Kimoto and Wakayama [36] (see also the references therein), and by Ochiai [48] for important work on the *special values* of the spectral ζ_Q. This direction of problems is very deep, interesting and promising.

Chapter 8
Some Properties of the Eigenvalues of $Q^{\mathrm{w}}_{(\alpha,\beta)}(x,D)$

In this chapter we shall show some properties of the eigenvalues of a NCHO $Q_{(\alpha,\beta)}$ (when $\alpha\beta > 1$ and $\alpha,\beta > 0$), and in particular give some upper and lower bounds to the lowest eigenvalue.

In the first place, we establish the following simple consequence of Theorem 3.1.12, which is useful when one is willing to study the behavior of the eigenvalues of the NCHO $Q_{(\alpha,\beta)}$ with respect to $\alpha,\beta \in \mathbb{R}$.

Lemma 8.0.1. *We have that, for any given $\alpha,\beta \in \mathbb{R}$, the symbols $Q_{(\alpha,\beta)}(x,\xi)$ and $Q_{(\beta,\alpha)}(x,\xi)$ have the same eigenvalues, for all $(x,\xi) \in \mathbb{R}\times\mathbb{R}$, and, at the operator level, $Q^{\mathrm{w}}_{(\alpha,\beta)}(x,D)$ is* **unitarily equivalent** *to $Q^{\mathrm{w}}_{(\beta,\alpha)}(x,D)$, that is, there exists a unitary transformation $U\colon L^2(\mathbb{R};\mathbb{C}^2) \longrightarrow L^2(\mathbb{R};\mathbb{C}^2)$ such that*

$$U^* Q^{\mathrm{w}}_{(\alpha,\beta)}(x,D) U = Q^{\mathrm{w}}_{(\beta,\alpha)}(x,D).$$

Moreover, U is an automorphism of $\mathscr{S}'(\mathbb{R};\mathbb{C}^2)$ and $\mathscr{S}(\mathbb{R};\mathbb{C}^2)$, so that the operators $Q^{\mathrm{w}}_{(\alpha,\beta)}(x,D)$ and $Q^{\mathrm{w}}_{(\beta,\alpha)}(x,D)$ are also equivalent in $\mathscr{S}'$ and $\mathscr{S}$.

Proof. Let $\mathsf{A} = \begin{bmatrix} \alpha & 0 \\ 0 & \beta \end{bmatrix}$, $J = \begin{bmatrix} 0 & -1 \\ 1 & 0 \end{bmatrix}$, and $K = \begin{bmatrix} 0 & 1 \\ 1 & 0 \end{bmatrix}$. Consider the system

$$KQ_{(\alpha,\beta)}(x,\xi)K = K\mathsf{A}Kp_0(x,\xi) - iJx\xi.$$

As $K\mathsf{A}K = \begin{bmatrix} \beta & 0 \\ 0 & \alpha \end{bmatrix}$, by considering the symplectic transformation $\kappa\colon \mathbb{R}\times\mathbb{R} \ni (x,\xi) \longmapsto (\xi,-x) \in \mathbb{R}\times\mathbb{R}$, we have that

$$Q_{(\beta,\alpha)}(x,\xi) = \big(KQ_{(\alpha,\beta)}K \circ \kappa\big)(x,\xi).$$

It thus follows, since

$$\mathsf{Tr}\, Q_{(\alpha,\beta)}(x,\xi) = \mathsf{Tr}\, Q_{(\alpha,\beta)}(\xi,-x),$$

A. Parmeggiani, *Spectral Theory of Non-Commutative Harmonic Oscillators: An Introduction*, Lecture Notes in Mathematics 1992,
DOI 10.1007/978-3-642-11922-4_8, © Springer-Verlag Berlin Heidelberg 2010

and

$$\det Q_{(\alpha,\beta)}(x,\xi) = \det Q_{(\alpha,\beta)}(\xi,-x),$$

that $Q_{(\alpha,\beta)}(x,\xi)$ and $Q_{(\beta,\alpha)}(x,\xi)$ have **exactly** the same eigenvalues, and that by Theorem 3.1.12, with U_κ the metaplectic operator (the normalized Fourier transform) associated with κ,

$$Q^{\mathrm{w}}_{(\beta,\alpha)}(x,D) = U_\kappa^{-1} K Q^{\mathrm{w}}_{(\alpha,\beta)}(x,D) K U_\kappa = (KQ_{(\alpha,\beta)}K \circ \kappa)^{\mathrm{w}}(x,D)$$

are **unitarily equivalent**, for $U := KU_\kappa \colon L^2(\mathbb{R};\mathbb{C}^2) \longrightarrow L^2(\mathbb{R};\mathbb{C}^2)$ is unitary. Since clearly U is also an automorphism of $\mathscr{S}'(\mathbb{R};\mathbb{C}^2)$ and $\mathscr{S}(\mathbb{R};\mathbb{C}^2)$, this concludes the proof of the lemma. □

From now on, unless otherwise specified, we shall assume $\alpha,\beta > 0$ and $\alpha\beta > 1$, so that $Q = Q^{\mathrm{w}}_{(\alpha,\beta)}$ is elliptic and positive (as a differential operator), and hence has a discrete spectrum contained in $\mathbb{R}_+$.

8.1 The Ichinose and Wakayama Bounds

In [33] Ichinose and Wakayama proved the following theorem about estimates on upper and lower bounds for the eigenvalues of Q.

Theorem 8.1.1. *Let $\lambda_{2j-1}, \lambda_{2j}$, $j = 1,2,\dots$, be the $(2j-1)$-st and $2j$-th eigenvalues of Q. Then*

$$(j-\frac{1}{2})\min\{\alpha,\beta\}\sqrt{\frac{\alpha\beta-1}{\alpha\beta}} \le \lambda_{2j-1} \le \lambda_{2j} \le (j-\frac{1}{2})\max\{\alpha,\beta\}\sqrt{\frac{\alpha\beta-1}{\alpha\beta}}. \tag{8.1}$$

Proof. Put $K(t) = K(t,x,y) = e^{-tQ}(x,y)$, $t > 0$. Then

$$(\partial_t + Q)K(t) = 0,\ \ K\big|_{t=0} = \delta(x-y).$$

Now, following Parmeggiani-Wakayama [58, 59], define

$$\begin{aligned} Q' &:= \mathsf{A}^{-1/2} Q \mathsf{A}^{-1/2} = I(\frac{-\partial_x^2 + x^2}{2}) + \gamma J(x\partial_x + \frac{1}{2}) \\ &= \frac{1}{2}(-i\partial_x + i\gamma Jx)^2 + \frac{1-\gamma^2}{2}x^2, \end{aligned}$$

where $\gamma := 1/\sqrt{\alpha\beta}$. One then has that $K'(t) = e^{-tQ'}(x,y)$ solves

$$(\partial_t + Q')K'(t) = 0,\ \ K'\big|_{t=0} = \delta(x-y),$$

and

$$K'(t,x,y) = (1-\gamma^2)^{1/4} e^{\gamma(x^2-y^2)J/2} p_H(\sqrt{1-\gamma^2}t, \sqrt{1-\gamma^2}x, \sqrt{1-\gamma^2}y),$$

where

$$p_H(t,x,y) = e^{-tH}(x,y)$$

is the heat-kernel of the harmonic oscillator $H = p_0^{\mathrm{w}}$. This follows from [58] (see also [59]), for Q' is unitarily equivalent to

$$Q_0 := \frac{1}{2}\gamma I\Big(-\partial_x^2 + (\frac{1}{\gamma^2} - 1)x^2\Big),$$

so that

$$\mathrm{Spec}(Q') = \Big\{\sqrt{1-\gamma^2}(j - \frac{1}{2});\ j \geq 1\Big\},$$

with each eigenvalue of multiplicity 2. Therefore, it also follows that

$$\zeta_{Q'}(s) = \mathrm{Tr}\, Q'^{-s} = (1-\gamma^2)^{-s/2}\zeta_{Q_0}(s) = 2\frac{2^s - 1}{(1-\gamma^2)^{s/2}}\zeta(s).$$

Now, since each eigenvalue λ_j' of Q' has multiplicity 2, we have that the $(2j-1)$-st eigenvalue λ_{2j-1}' and the $2j$-th eigenvalue λ_{2j}' coincide and

$$\lambda_{2j-1}' = \lambda_{2j}' = (j - \frac{1}{2})\sqrt{1-\gamma^2} = (j - \frac{1}{2})\sqrt{\frac{\alpha\beta - 1}{\alpha\beta}}.$$

We now use the fact that

$$Q = \mathsf{A}^{1/2} Q' \mathsf{A}^{1/2},$$

and the Minimax Principle (4.1). For $n = 2j-1$ or $n = 2j$, with

$$[u_1, \dots, u_{n-1}]^{\perp} := \mathrm{Span}(u_1, \dots, u_{n-1})^{\perp} \cap D(Q),$$

$D(Q)$ being the domain of Q, we thus have that

$$\begin{aligned}
\lambda_n &= \sup_{u_1,\dots,u_{n-1}\,\mathrm{lin.\,ind.}} \left(\inf_{0 \neq u \in [u_1,\dots,u_{n-1}]^{\perp}} \frac{(Qu,u)}{\|u\|_0^2} \right) \\
&= \sup_{u_1,\dots,u_{n-1}\,\mathrm{lin.\,ind.}} \left(\inf_{0 \neq u \in [u_1,\dots,u_{n-1}]^{\perp}} \frac{(Q'\mathsf{A}^{1/2}u, \mathsf{A}^{1/2}u)}{\|u\|_0^2} \right)
\end{aligned}$$

(note that the vectors $u_1, \dots, u_{n-1}$ are linearly independent if and only if the vectors $\mathsf{A}^{-1/2}u_1, \dots, \mathsf{A}^{-1/2}u_{n-1}$ are linearly independent, for $\mathsf{A}^{1/2}\colon L^2(\mathbb{R};\mathbb{C}^2) \longrightarrow L^2(\mathbb{R};\mathbb{C}^2)$ is an isomorphism, i.e. bounded with bounded inverse, and that $u \in D(Q)$ or $u \in D(Q')$ if and only if $\mathsf{A}^{-1/2}u$ does)

$$= \sup_{u_1,\dots,u_{n-1}\,\text{lin. ind.}} \left(\inf_{0\neq u\in[\mathsf{A}^{-1/2}u_1,\dots,\mathsf{A}^{-1/2}u_{n-1}]^\perp} \frac{(Q'\mathsf{A}^{1/2}u,\mathsf{A}^{1/2}u)}{\|u\|_0^2} \right)$$

$$= \sup_{u_1,\dots,u_{n-1}\,\text{lin. ind.}} \left(\inf_{0\neq \mathsf{A}^{1/2}u\in[u_1,\dots,u_{n-1}]^\perp} \frac{(Q'\mathsf{A}^{1/2}u,\mathsf{A}^{1/2}u)}{\|u\|_0^2} \right)$$

(putting $v = \mathsf{A}^{1/2}u$)

$$= \sup_{u_1,\dots,u_{n-1}\,\text{lin. ind.}} \left(\inf_{0\neq v\in[u_1,\dots,u_{n-1}]^\perp} \frac{(Q'v,v)}{\|\mathsf{A}^{-1/2}v\|_0^2} \right)$$

$$= \sup_{u_1,\dots,u_{n-1}\,\text{lin. ind.}} \left(\inf_{0\neq v\in[u_1,\dots,u_{n-1}]^\perp} \frac{(Q'v,v)}{\|v\|_0^2}\frac{\|v\|_0^2}{\|\mathsf{A}^{-1/2}v\|_0^2} \right).$$

Put $m = \min\{\alpha,\beta\}$ and $M = \max\{\alpha,\beta\}$. Then

$$M^{-1}\|v\|_0^2 \leq (\mathsf{A}^{-1}v,v) = \|\mathsf{A}^{-1/2}v\|_0^2 \leq m^{-1}\|v\|_0^2,\ \ \forall v\in L^2(\mathbb{R};\mathbb{C}^2),$$

and

$$m \leq \frac{\|v\|_0^2}{\|\mathsf{A}^{-1/2}v\|_0^2} \leq M,\ \ \forall v\neq 0.$$

We thus have

$$\lambda_{2j-1} \leq \lambda_{2j} \leq M \sup_{u_1,\dots,u_{2j-1}\,\text{lin. ind.}} \left(\inf_{0\neq v\in[u_1,\dots,u_{2j-1}]^\perp} \frac{(Q'v,v)}{\|v\|_0^2} \right)$$

$$= M\lambda'_{2j} = M\lambda'_{2j-1},$$

and

$$\lambda_{2j} \geq \lambda_{2j-1} \geq m \sup_{u_1,\dots,u_{2j-2}\,\text{lin. ind.}} \left(\inf_{0\neq v\in[u_1,\dots,u_{2j-2}]^\perp} \frac{(Q'v,v)}{\|v\|_0^2} \right) = m\lambda'_{2j-1},$$

which concludes the proof. □

The bounds in (8.1) make up an interval

$$I_j := \left[(j-\frac{1}{2})\min\{\alpha,\beta\}\sqrt{\frac{\alpha\beta-1}{\alpha\beta}}, (j-\frac{1}{2})\max\{\alpha,\beta\}\sqrt{\frac{\alpha\beta-1}{\alpha\beta}} \right].$$

When $j<k$ we have

$$I_j \cap I_k = \emptyset \Longleftrightarrow \frac{2k-1}{2j-1} > \frac{\max\{\alpha,\beta\}}{\min\{\alpha,\beta\}}.$$

It follows that the eigenvalue λ_{2j-1} or λ_{2j} has a multiplicity less than or equal to 2 if

$$j < \frac{\max\{\alpha,\beta\}+\min\{\alpha,\beta\}}{2(\max\{\alpha,\beta\}-\min\{\alpha,\beta\})} = \frac{|\alpha+\beta|}{2|\alpha-\beta|}.$$

Otherwise, the eigenvalue will possibly happen to have a multiplicity greater than 2.

However, we will see in Chapter 12, Section 12.4, that, as a consequence of a more general argument (see Section 12.3), under certain assumptions on the periods of the Hamilton-trajectories associated with the eigenvalues of the symbol, the spectrum of $Q^{\mathrm{w}}_{(\alpha,\beta)}(x,D)$ "clusters" and is simple (see also Parmeggiani [52,55]).

As regards the lowest eigenvalue, it was shown by Nakao, Nagatou and Wakayama in [45], and by Parmeggiani in [51], that it is always simple for $\alpha\beta$ large. Moreover, in [51], using perturbation theory in the limit $\alpha\beta \to +\infty$ **and** α/β a fixed constant $\neq 1$, it is seen that the lowest eigenvalue is always smaller, for $\alpha\beta$ sufficiently large (α/β =constant$\neq 1$), than the lowest eigenvalue of the operator

$$\begin{bmatrix} \alpha & 0 \\ 0 & \beta \end{bmatrix} \left(\frac{-\partial_x^2}{2} + \frac{x^2}{2}\right).$$

Perturbation theory may be used, for one writes

$$Q^{\mathrm{w}}_{(\alpha,\beta)}(x,D) = \sqrt{\alpha\beta}\left(\begin{bmatrix} \sqrt{\frac{\alpha}{\beta}} & 0 \\ 0 & \sqrt{\frac{\beta}{\alpha}} \end{bmatrix}\left(\frac{-\partial_x^2+x^2}{2}\right) + \frac{1}{\sqrt{\alpha\beta}} J\left(x\partial_x + \frac{1}{2}\right)\right).$$

Put then $\varepsilon = 1/\sqrt{\alpha\beta} \to 0+$ and $\omega_0 = \sqrt{\alpha/\beta}$ with ω_0 fixed with $\omega_0 \neq 1$. Since

$$Q^{\mathrm{w}}_{\omega_0}(x,D) = \begin{bmatrix} \omega_0 & 0 \\ 0 & \omega_0^{-1} \end{bmatrix}\left(\frac{-\partial_x^2+x^2}{2}\right),$$

is an elliptic system of GPDOs, it therefore possesses a parametrix. It is then easy to see that $E^{\mathrm{w}}(x,D) = J\left(x\partial_x + \frac{1}{2}\right)$ is bounded with respect to $Q^{\mathrm{w}}_{\omega_0}(x,D)$ (see Kato's book [35]). Hence Rellich's theory can be applied as $\varepsilon \to 0+$ to study the spectrum of

$$\varepsilon Q^{\mathrm{w}}_{(\alpha,\beta)}(x,D) = Q^{\mathrm{w}}_{\omega_0}(x,D) + \varepsilon E^{\mathrm{w}}(x,D),$$

in terms of the spectrum of $Q^{\mathrm{w}}_{\omega_0}(x,D)$. See Parmeggiani [51] for more on this.

8.2 A Better Upper-Bound for the Lowest Eigenvalue

We now show, by using some elementary symplectic linear algebra, how to obtain an upper bound for the lowest eigenvalue of Q which is more accurate than that of Theorem 8.1.1.

Theorem 8.2.1. *For the lowest eigenvalue λ_1 of Q we have*

$$\frac{1}{2}\min\{\alpha,\beta\}\sqrt{\frac{\alpha\beta-1}{\alpha\beta}} \le \lambda_1 \le \frac{\sqrt{\alpha\beta}\sqrt{\alpha\beta-1}}{\alpha+\beta+|\alpha-\beta|\frac{(\alpha\beta-1)^{1/4}}{\sqrt{\alpha\beta}}\operatorname{Re}\omega}, \tag{8.2}$$

where $\omega \in \mathbb{C}$ is the solution of $\omega^2 = \sqrt{\alpha\beta-1} - i$ with $\operatorname{Re}\omega > 0$.

Remark 8.2.2. Since, as is readily seen,

$$\frac{\sqrt{\alpha\beta}\sqrt{\alpha\beta-1}}{\alpha+\beta+|\alpha-\beta|\frac{(\alpha\beta-1)^{1/4}}{\sqrt{\alpha\beta}}\operatorname{Re}\omega} \le \frac{1}{2}\max\{\alpha,\beta\}\frac{\sqrt{\alpha\beta-1}}{\sqrt{\alpha\beta}},$$

the upper bound in (8.2) is better then the one by Ichinose and Wakayama given in (8.1).
When $\alpha = \beta$ all the bounds reduce to the actual value of $\lambda_1 = \sqrt{\alpha^2-1}/2$. △

Proof (of Theorem 8.2.1). The lower bound in (8.2) has already been proved in Theorem 8.1.1.

Let $e(x,\xi) := x\xi$, and $\delta := \sqrt{\alpha\beta}$. We may therefore write

$$Q = \frac{1}{\delta}\mathsf{A}^{1/2}\Big(\delta\, p_0^{\mathrm{w}}(x,D)I + iJe^{\mathrm{w}}(x,D)\Big)\mathsf{A}^{1/2}.$$

Let next $v_\pm = \frac{1}{\sqrt{2}}\begin{bmatrix} 1 \\ \mp i \end{bmatrix}$ be the normalized eigenvectors of J, so that $Jv_\pm = \pm i v_\pm$.
Notice that $KJv_\pm = v_\mp$.

Set

$$L_\delta(x,\xi) := \frac{1}{2}(\xi^2 + (\delta^2-1)x^2).$$

Consider the following symplectic transformations of $\mathbb{R}\times\mathbb{R}$, and the corresponding metaplectic operators,

$$\kappa_\delta(x,\xi) = (\delta^{1/2}x, \frac{1}{\delta^{1/2}}\xi),\ (U_\delta f)(x) = \frac{1}{\delta^{1/4}}f(\frac{x}{\delta^{1/2}}),$$

$$\kappa_\pm(x,\xi) = (x, \xi \pm x),\ (U_\pm f)(x) = e^{\pm ix^2/2}f(x).$$

Define also

$$a_\pm^{\mathrm{w}}(x,D) := \delta\, p_0^{\mathrm{w}}(x,D) \mp e^{\mathrm{w}}(x,D).$$

With $\mu_\pm := (\sqrt{\alpha} \pm \sqrt{\beta})/2$, we may write

$$\mathsf{A}^{-1/2} = \frac{1}{\delta}\Big(\mu_+ I - \mu_- KJ\Big).$$

We also write

$$f = f_+ v_+ + f_- v_-, \ \forall f \in \mathscr{S}(\mathbb{R};\mathbb{C}^2).$$

Let hence

$$\begin{aligned}\lambda_1 &= \inf_{0\neq f\in\mathscr{S}(\mathbb{R};\mathbb{C}^2)} \frac{(Qf,f)}{\|f\|_0^2} \\ &= \inf_{0\neq f_+v_++f_-v_-\in\mathscr{S}(\mathbb{R};\mathbb{C}^2)} \frac{\left(\left(\delta\, p_0^{\mathrm{w}}(x,D)I + iJe^{\mathrm{w}}(x,D)\right)(f_+v_+ + f_-v_-), f_+v_+ + f_-v_-\right)}{\delta\|\mathsf{A}^{-1/2}(f_+v_+ + f_-v_-)\|_0^2}.\end{aligned}$$

In the basis $\{v_+, v_-\}$ of $\mathbb{C}^2$,

$$\left(\delta\, p_0^{\mathrm{w}}(x,D)I + iJe^{\mathrm{w}}(x,D)\right)(f_+v_+ + f_-v_-)$$

is represented by

$$\begin{bmatrix} a_+^{\mathrm{w}}(x,D) & 0 \\ 0 & a_-^{\mathrm{w}}(x,D)\end{bmatrix}\begin{bmatrix} f_+ \\ f_-\end{bmatrix},$$

for we have that

$$\begin{aligned}\left(\delta\, p_0^{\mathrm{w}}(x,D)I + iJe^{\mathrm{w}}(x,D)\right)(f_+v_+ + f_-v_-) &= \left((\delta\, p_0^{\mathrm{w}}(x,D) - e^{\mathrm{w}}(x,D))f_+\right)v_+ \\ &\quad + \left((\delta\, p_0^{\mathrm{w}}(x,D) + e^{\mathrm{w}}(x,D))f_-\right)v_- \\ &= (a_+^{\mathrm{w}}(x,D)f_+)v_+ + (a_-^{\mathrm{w}}(x,D)f_-)v_-.\end{aligned}$$

Hence, $\{v_+, v_-\}$ being a unitary basis of $\mathbb{C}^2$,

$$\lambda_1 = \frac{1}{\delta}\inf_{0\neq f_+v_++f_-v_-\in\mathscr{S}(\mathbb{R};\mathbb{C}^2)} \frac{(a_+^{\mathrm{w}}(x,D)f_+, f_+) + (a_-^{\mathrm{w}}(x,D)f_-, f_-)}{\|\mathsf{A}^{-1/2}(f_+v_+ + f_-v_-)\|_0^2}.$$

Now, it is readily seen that

$$\delta\, p_0(x,\xi) \mp x\xi = (L_\delta \circ \kappa_\mp \circ \kappa_\delta^{-1})(x,\xi).$$

Hence, by Theorem 3.1.12, we have

$$\begin{aligned}a_\pm^{\mathrm{w}}(x,D) &= (L_\delta \circ \kappa_\mp \circ \kappa_\delta^{-1})^{\mathrm{w}}(x,D) = U_\delta (L_\delta \circ \kappa_\mp)^{\mathrm{w}}(x,D) U_\delta^{-1} \\ &= U_\delta U_\mp^{-1} L_\delta^{\mathrm{w}}(x,D) U_\mp U_\delta^{-1}.\end{aligned}$$

One next computes

$$\mathsf{A}^{-1/2}(f_+v_+ + f_-v_-) = \frac{1}{\delta}\left((\mu_+ f_+ - \mu_- f_-)v_+ + (\mu_+ f_- - \mu_- f_+)v_-\right),$$

whence

$$\|\mathsf{A}^{-1/2}(f_+v_+ + f_-v_-)\|_0^2 = \frac{1}{\delta^2}\Big(\|\mu_+f_+ - \mu_-f_-\|_0^2 + \|\mu_+f_- - \mu_-f_+\|_0^2\Big).$$

We thus have

$$\lambda_1 = \delta \inf_{(0,0)\neq(f_+,f_-)\in\mathscr{S}(\mathbb{R},\mathbb{C}^2)} \Bigg[\frac{(L^{\mathrm{w}}_\delta(x,D)U_-U_\delta^{-1}f_+, U_-U_\delta^{-1}f_+)}{\|\mu_+f_+ - \mu_-f_-\|_0^2 + \|\mu_+f_- - \mu_-f_+\|_0^2}$$
$$+\frac{(L^{\mathrm{w}}_\delta(x,D)U_+U_\delta^{-1}f_-, U_+U_\delta^{-1}f_-)}{\|\mu_+f_+ - \mu_-f_-\|_0^2 + \|\mu_+f_- - \mu_-f_+\|_0^2}\Bigg].$$

Let $\varphi_0 = \varphi_0(x) = ce^{-(\alpha\beta-1)^{1/2}x^2/2}$ be the ground state of $L^{\mathrm{w}}_\delta(x,D)$, with c so chosen that $\|\varphi_0\|_0 = 1$. Thus

$$L^{\mathrm{w}}_\delta(x,D)\varphi_0 = \frac{\sqrt{\alpha\beta-1}}{2}\varphi_0, \text{ and } c = \left(\frac{(\alpha\beta-1)^{1/4}}{\sqrt{\pi}}\right)^{1/2}.$$

We now choose $f_\pm$ to be

$$f_\pm = U_\delta U_\mp^{-1}\varphi_0,$$

that is

$$f_\pm(x) = \frac{1}{\delta^{1/4}}e^{\pm ix^2/2\delta}\varphi_0\Big(\frac{x}{\delta^{1/2}}\Big).$$

It follows that

$$(a^{\mathrm{w}}_\pm(x,D)f_\pm, f_\pm) = \frac{1}{2}\sqrt{\alpha\beta-1}.$$

Now, for $r,s\in\mathbb{R}$,

$$\begin{aligned}\|rf_+ + sf_-\|_0^2 &= \|(re^{ix^2/2} + se^{-ix^2/2})\varphi_0\|_0^2\\ &= (r^2+s^2)\|\varphi_0\|_0^2 + 2rs\int\cos(x^2)|\varphi_0(x)|^2dx,\end{aligned}$$

so that

$$\begin{aligned}\|\mathsf{A}^{-1/2}f\|_0^2 &= \frac{2}{\delta^2}\Big((\mu_+^2+\mu_-^2) - 2\mu_+\mu_-\int\cos(x^2)|\varphi_0(x)|^2dx\Big)\\ &= \frac{1}{\delta^2}\Big(\alpha+\beta-(\alpha-\beta)\int\cos(x^2)|\varphi_0(x)|^2dx\Big).\end{aligned}$$

We next compute

$$\begin{aligned}\int \cos(x^2)|\varphi_0(x)|^2 dx &= \frac{(\alpha\beta-1)^{1/4}}{\sqrt{\pi}} \int \cos(x^2) e^{-\sqrt{\alpha\beta-1}x^2} dx \\ &= \frac{(\alpha\beta-1)^{1/4}}{\sqrt{\pi}} \mathrm{Re} \int e^{-(\sqrt{\alpha\beta-1}-i)x^2} dx.\end{aligned}$$

Let $\omega \in \mathbb{C}$ be the unique solution to $\omega^2 = \sqrt{\alpha\beta-1} - i$ with $\mathrm{Re}\,\omega > 0$. Then, as is well-known,

$$\int e^{-(\sqrt{\alpha\beta-1}-i)x^2} dx = \frac{\sqrt{\pi}}{\omega},$$

whence

$$\int \cos(x^2)|\varphi_0(x)|^2 dx = \frac{(\alpha\beta-1)^{1/4}}{\sqrt{\pi}} \sqrt{\pi}\, \mathrm{Re}\, \frac{1}{\omega} = \frac{(\alpha\beta-1)^{1/4}}{\sqrt{\alpha\beta}} \mathrm{Re}\,\omega,$$

and

$$\|\mathsf{A}^{-1/2} f\|_0^2 = \frac{1}{\alpha\beta}\Big(\alpha+\beta-(\alpha-\beta)\frac{(\alpha\beta-1)^{1/4}}{\sqrt{\alpha\beta}} \mathrm{Re}\,\omega\Big) > 0.$$

We therefore have that

$$\begin{aligned}\lambda_1 &\leq \frac{1}{\sqrt{\alpha\beta}} \frac{\frac{\sqrt{\alpha\beta-1}}{2} + \frac{\sqrt{\alpha\beta-1}}{2}}{\frac{1}{\alpha\beta}\Big(\alpha+\beta-(\alpha-\beta)\frac{(\alpha\beta-1)^{1/4}}{\sqrt{\alpha\beta}} \mathrm{Re}\,\omega\Big)} \\ &= \frac{\sqrt{\alpha\beta}\sqrt{\alpha\beta-1}}{\alpha+\beta-(\alpha-\beta)\frac{(\alpha\beta-1)^{1/4}}{\sqrt{\alpha\beta}} \mathrm{Re}\,\omega}.\end{aligned}$$

Since, by Lemma 8.0.1, $Q^{\mathrm{w}}_{(\alpha,\beta)}(x,D)$ and $Q^{\mathrm{w}}_{(\beta,\alpha)}(x,D)$ are unitarily equivalent, they have the same lowest eigenvalue. Thus (with the **same** ω)

$$\lambda_1 \leq \min\left\{ \frac{\sqrt{\alpha\beta}\sqrt{\alpha\beta-1}}{\alpha+\beta-(\alpha-\beta)\frac{(\alpha\beta-1)^{1/4}}{\sqrt{\alpha\beta}} \mathrm{Re}\,\omega}, \frac{\sqrt{\alpha\beta}\sqrt{\alpha\beta-1}}{\alpha+\beta-(\beta-\alpha)\frac{(\alpha\beta-1)^{1/4}}{\sqrt{\alpha\beta}} \mathrm{Re}\,\omega} \right\},$$

that is

$$\lambda_1 \leq \frac{\sqrt{\alpha\beta}\sqrt{\alpha\beta-1}}{\alpha+\beta+|\alpha-\beta|\frac{(\alpha\beta-1)^{1/4}}{\sqrt{\alpha\beta}} \mathrm{Re}\,\omega}.$$

This shows that inequality (8.2) holds and concludes the proof of the theorem. □

8.3 Notes

The problem of determining the lowest eigenvalue, and its multiplicity, of a generic elliptic positive NCHO is an open, and important, problem, which should be explored in depth.

Chapter 9
Some Tools from the Semiclassical Calculus

In this chapter we shall describe some tools from Semiclassical Analysis, that will be used in the subsequent chapters to obtain localization properties of the large eigenvalues of an elliptic positive NCHO $Q^{\mathrm{w}}_{(\alpha,\beta)}(x,D)$, and more generally of an elliptic positive NCHO (with no sub-principal term), relating them to properties of the *periods of the bicharacteristics of the (principal) symbol.*

9.1 The Semiclassical Calculus

We recall here some basic properties of semiclassical the Weyl-quantization and the semiclassical calculus. (In general, for Semiclassical Analysis, we address the reader to Dimassi-Sjöstrand [7], Evans-Zworski [15], Ivrii [34], Martinez [40], Robert [65], Shubin [67] and Voros [71].)

Definition 9.1.1 (Semiclassical symbols). We shall say that a function $a(X;h) = a(\cdot;h) \in C^\infty(\mathbb{R}^{2n}_X)$, possibly depending on a parameter $h \in (0,h_0]$, $h_0 \in (0,1]$, belongs to the symbol class $S^k_\delta(m^\mu,g)$, $k,\mu \in \mathbb{R}$ and $\delta \in [0,1/2]$, if for all $\alpha \in \mathbb{Z}^{2n}_+$ there exists $C_\alpha > 0$ such that

$$|\partial^\alpha_X a(X;h)| \le C_\alpha m(X)^{\mu-|\alpha|} h^{-k-|\alpha|\delta}, \ \forall X \in \mathbb{R}^{2n}, \ \forall h \in (0,h_0]. \qquad (9.1)$$

As before, $S^k_\delta(m^\mu,g;\mathsf{M}_N) = S^k_\delta(m^\mu,g) \otimes \mathsf{M}_N$, and $S^k_\delta(m^\mu,g;\mathsf{V}) = S^k_\delta(m^\mu,g) \otimes \mathsf{V}$, for any given finite-dimensional complex vector space V.

Given $a \in S^k_\delta(m^\mu,g;\mathsf{V})$, we shall consider its h-Weyl quantization

$$a^{\mathrm{w}}(x,hD)u(x) = (2\pi h)^{-n} \iint e^{ih^{-1}\langle x-y,\xi\rangle} a(\frac{x+y}{2},\xi;h)u(y)dyd\xi, \ u \in \mathscr{S}.$$

It is easy to see that $a^{\mathrm{w}}(x,hD)$ is in fact the Weyl-quantization of the symbol $a(x,h\xi;h)$.

We shall also need more general classes, defined as follows (see Dimassi-Sjöstrand [7]). We say that a smooth function $a(\cdot;h) \in C^\infty(\mathbb{R}^{2n}_X)$, possibly depending

A. Parmeggiani, *Spectral Theory of Non-Commutative Harmonic Oscillators: An Introduction*, Lecture Notes in Mathematics 1992,
DOI 10.1007/978-3-642-11922-4_9,

on a parameter $h \in (0, h_0]$, $h_0 \in (0,1]$, belongs to the symbol class $S^k_\delta(m^\mu)$, $k, \mu \in \mathbb{R}$ and $\delta \in [0, 1/2]$, if for all $\alpha \in \mathbb{Z}^n_+$ there exists $C_\alpha > 0$ such that

$$|\partial_X^\alpha a(X;h)| \leq C_\alpha m(X)^\mu h^{-k-|\alpha|\delta}, \ \forall X \in \mathbb{R}^{2n}, \ \forall h \in (0, h_0]. \tag{9.2}$$

Notice that, compared to (9.1), no further decay in m is obtained by taking derivatives of the symbol. The h-Weyl quantization of a symbol $a \in S^k_\delta(m^\mu)$ is defined as before. The vector and matrix-valued symbol classes are defined as before.

Notice that

$$S^k_\delta(m^\mu) = S^k_\delta(m^\mu, |dX|^2),$$

where $|dX|^2$ is the Euclidean metric in $\mathbb{R}^{2n}_X$, and that

$$S(m^\mu, g) \subset S^0_0(m^\mu, g) \subset S^k_\delta(m^\mu, g) \subset S^k_\delta(m^\mu). \tag{9.3}$$

Remark 9.1.2. Let $E > 0$. Define the L^2-isometry (also automorphism of $\mathscr{S}'$ and $\mathscr{S}$),

$$U_E \colon u(x) \longmapsto E^{-n/4} u(x/\sqrt{E}). \tag{9.4}$$

Then, given any symbol $a \in S^k_0(m^\mu; \mathsf{M}_N)$ one has from Theorem 3.1.12

$$U_E^{-1} a^{\mathrm{w}}(x, hD) U_E = a^{\mathrm{w}}(\sqrt{E}x, \sqrt{E}\tilde{h}D), \ \textit{where} \ \tilde{h} = \frac{h}{E}. \tag{9.5}$$

In particular

$$U_h^{-1} a^{\mathrm{w}}(x, hD) U_h = a^{\mathrm{w}}(\sqrt{h}x, \sqrt{h}D). \tag{9.6}$$

Notice that since U_h corresponds to the symplectic transformation

$$\kappa_h \colon (x, \xi) \longmapsto (h^{1/2}x, h^{-1/2}\xi),$$

by using $h^{1/2} m(X) \leq m(h^{1/2}X) \leq m(X)$, we have that if $a \in S(m^\mu, g)$, then

$$(a \circ \kappa_h)(\cdot, h\cdot) = a(\sqrt{h}\cdot, \sqrt{h}\cdot) \in S_0^{-\min\{\mu, 0\}/2}(m^\mu, g).$$

△

By Proposition 3.2.15 we have the following result, that we shall frequently use when dealing with "classical semiclassical symbols" in the class $S^k_{0,\mathrm{cl}}(m^\mu, g)$ (see Definition 9.1.9 below).

Proposition 9.1.3. *Let $\mu \in \mathbb{R}$. Let $a_j \in S(m^{\mu-2j}, g)$, $j \in \mathbb{Z}_+$ (in particular, the a_j are* **independent** *of h). Then there exists $a \in S^0_0(m^\mu, g)$ such that*

$$a \sim \sum_{j \geq 0} h^j a_j,$$

that is, for all $r \in \mathbb{Z}_+$ *we have*

$$a - \sum_{j=0}^{r} h^j a_j \in S_0^{-(r+1)}(m^{\mu-2(r+1)}, g).$$

If another symbol a' *has the same property, then*

$$a - a' \in S^{-\infty}(m^{-\infty}, g) := \bigcap_{k \in \mathbb{Z}_+} S_0^{-k}(m^{\mu-2k}, g). \tag{9.7}$$

Proof. One takes an excision function $\chi \in C^\infty(\mathbb{R}_t; [0,1])$, such that $\chi(t) = 0$ for $|t| \leq 2$ and $\chi(t) = 1$ for $|t| \geq 4$, and defines

$$a(X;h) := \sum_{j=0}^{\infty} h^j \chi\left(\frac{m(X)}{hR_j}\right) a_j(X),$$

where $\{R_j\}_{j\in\mathbb{Z}_+} \subset [1, +\infty)$ is an increasing sequence with $R_j \nearrow +\infty$ as $j \to +\infty$ sufficiently fast. Following the steps of the proof of Proposition 3.2.15, using the fact that $h \in (0,1]$ and that on the support of $1 - \chi(m(X)/hR_j)$ we have $|X| \leq 4R_j$ and $1 \leq h^{-1} \leq 4R_j$ (so that we may divide and multiply by h^r, for any given $r \in \mathbb{N}$), one sees that the sequence R_j can indeed be so chosen that $a(X;h)$ has the required properties. □

Using a Borel-summation argument one also has the following proposition (see Evans-Zworski [15]).

Proposition 9.1.4. *Let* $k_j \searrow -\infty$, $k_j > k_{j+1}$, $j \in \mathbb{Z}_+$, *be a monotone decreasing sequence of real numbers. Let* $a_j \in S_\delta^{k_j}(m^\mu)$. *Then there exists* $a \in S_\delta^{k_0}(m^\mu)$ *such that* $a \sim \sum_{j \geq 0} a_j$, *that is, for all* $r \in \mathbb{Z}_+$

$$a - \sum_{j=0}^{r} a_j \in S_\delta^{k_{r+1}}(m^\mu).$$

If another symbol a' *has the same property, then*

$$a - a' \in S^{-\infty}(m^\mu) := \bigcap_{k \in \mathbb{R}} S_\delta^k(m^\mu). \tag{9.8}$$

Proof. One chooses $\chi \in C^\infty([0,+\infty); \mathbb{R})$ such that

$$0 \leq \chi \leq 1,\ \chi\big|_{[0,1]} = 1,\ \chi\big|_{[2,+\infty)} \equiv 0,$$

and then defines

$$a(X) := \sum_{j \geq 0} \chi(\lambda_j h) a_j(X),\ X \in \mathbb{R}^{2n},$$

where the sequence $1 \leq \lambda_1 \leq \ldots \leq \lambda_j \leq \lambda_{j+1} \leq \ldots \to +\infty$ must be picked. The choice of the λ_j is therefore made as follows: for each multiindex α with $|\alpha| \leq j$, we have

$$\begin{aligned}
|\partial_X^\alpha(\chi(\lambda_j h)a_j(X))| &= |\chi(\lambda_j h)\partial_X^\alpha a_j(X)| \leq C_{j,\alpha} h^{-k_j-\delta|\alpha|}\chi(\lambda_j h)m(X)^\mu \\
&= C_{j,\alpha} h^{-k_j-\delta|\alpha|}\frac{\lambda_j h}{\lambda_j h}\chi(\lambda_j h)m(X)^\mu \\
&\qquad \text{(since } \lambda_j h \leq 2 \text{ in the support of } \chi) \\
&\leq 2C_{j,\alpha}\frac{1}{\lambda_j}h^{-k_j-1-\delta|\alpha|}m(X)^\mu \\
&\leq h^{-k_j-1-\delta|\alpha|}2^{-j}m(X)^\mu, \ \forall X \in \mathbb{R}^{2n},
\end{aligned}$$

if λ_j is picked sufficiently large. Since we can accomplish this for all j and all α with $|\alpha| \leq j$, and trivially make it possible to have $\lambda_{j+1} \geq \lambda_j$, the sequence $\{\lambda_j\}_j$ is therefore determined. One then concludes in a way similar to that used in the proof of Proposition 3.2.15. □

We leave it as an exercise for the reader to fill in the details of the proofs of Proposition 9.1.3 and of Proposition 9.1.4.

From Evans-Zworski [15] (see also Dimassi-Sjöstrand [7]) we have the following important result.

Theorem 9.1.5. *We have that for $a \in S^0_\delta(m^\mu; \mathsf{M}_N)$, the operator $a^{\mathrm{w}}(x,hD)$ as a linear map*

$$a^{\mathrm{w}}(x,hD) \colon \mathscr{S}(\mathbb{R}^n;\mathbb{C}^N) \longrightarrow \mathscr{S}(\mathbb{R}^n;\mathbb{C}^N)$$

and as a linear map

$$a^{\mathrm{w}}(x,hD) \colon \mathscr{S}'(\mathbb{R}^n;\mathbb{C}^N) \longrightarrow \mathscr{S}'(\mathbb{R}^n;\mathbb{C}^N)$$

is **continuous**. *In particular, by virtue of (9.3), this is the case also for $a \in S^0_\delta(m^\mu, g; \mathsf{M}_N)$.*

As regards the L^2-continuity we have the following theorem.

Theorem 9.1.6. *Let $a \in S^0_\delta(1; \mathsf{M}_N)$, $0 \leq \delta \leq 1/2$. Then*

$$a^{\mathrm{w}}(x,hD) \colon L^2(\mathbb{R}^n;\mathbb{C}^N) \longrightarrow L^2(\mathbb{R}^n;\mathbb{C}^N)$$

is bounded and there is a constant $C > 0$, **independent of** *h, such that*

$$\|a^{\mathrm{w}}(x,hD)\|_{L^2 \to L^2} \leq C, \ \ \forall h \in (0,1].$$

As regards the composition one has the following theorems.

For the classes $S^k_\delta(m^\mu; \mathsf{M}_N)$ we have the following result (see Dimassi-Sjöstrand [7] and Evans-Zworski [15]).

Theorem 9.1.7. *Given $a \in S^0_\delta(m^{\mu_1};\mathsf{M}_N)$ and $b \in S^0_\delta(m^{\mu_2};\mathsf{M}_N)$, one has*

$$a^{\mathrm{w}}(x,hD)b^{\mathrm{w}}(x,hD) = (a\sharp_h b)^{\mathrm{w}}(x,hD),$$

where

$$a\sharp_h b = e^{ih\sigma(D_X;D_Y)/2}(a(X)b(Y))\big|_{X=Y} \in S^0_\delta(m^{\mu_1+\mu_2};\mathsf{M}_N).$$

Here, recall, $\sigma(D_X;D_Y) = \sigma(D_x,D_\xi;D_y,D_\eta)$. Furthermore, when $\delta \in [0,1/2)$ one has

$$a\sharp_h b \sim \sum_{k=0}^{\infty} \frac{1}{k!}\Big(\frac{ih}{2}\sigma(D_X;D_Y)\Big)^k a(X)b(Y)\big|_{X=Y}. \tag{9.9}$$

When $\delta = 1/2$ one has (see [15]) that if $a,b \in S^0_{1/2}(1;\mathsf{M}_N)$ with

$$\operatorname{supp} a \subset K, \quad \textit{and} \quad \operatorname{dist}(\operatorname{supp} a, \operatorname{supp} b) \geq \gamma > 0,$$

where the **compact** *K and the constant γ are* **independent of** *h, then*

$$\|a^{\mathrm{w}}(x,hD)b^{\mathrm{w}}(x,hD)\|_{L^2\to L^2} = O(h^\infty). \tag{9.10}$$

Recall that $f(h) = O(h^{N_0})$ for some $N_0 \in \mathbb{Z}_+$ if there exists $C_{N_0} > 0$ such that $|f(h)| \leq C_{N_0} h^{N_0}$. We say that $f(h) = O(h^\infty)$ if for any given $N_0 \in \mathbb{Z}_+$ one has $f(h) = O(h^{N_0})$.

For the classes $S^k_\delta(m^\mu, g;\mathsf{M}_N)$ we have the following result (see Shubin [67, p. 245]).

Theorem 9.1.8. *Let $\delta \in [0,1/2)$. For any given symbols $a \in S^{k_1}_\delta(m^{\mu_1},g;\mathsf{M}_N)$ and $b \in S^{k_2}_\delta(m^{\mu_2},g;\mathsf{M}_N)$ we have*

$$a^{\mathrm{w}}(x,hD)b^{\mathrm{w}}(x,hD) = (a\sharp_h b)^{\mathrm{w}}(x,hD)$$

where

$$a\sharp_h b = e^{ih\sigma(D_X;D_Y)/2}(a(X)b(Y))\big|_{X=Y} \in S^{k_1+k_2}_\delta(m^{\mu_1+\mu_2},g;\mathsf{M}_N),$$

and for all $N_0 \in \mathbb{Z}_+$

$$(a\sharp_h b) = \sum_{k=0}^{N_0} \frac{1}{k!}\Big(\frac{ih}{2}\sigma(D_X;D_Y)\Big)^k a(X)b(Y)\big|_{X=Y} + h^{N_0+1} r_{N_0+1}, \tag{9.11}$$

where $r_{N_0+1} \in S^{k_1+k_2+2(N_0+1)\delta}_\delta(m^{\mu_1+\mu_2-2(N_0+1)},g;\mathsf{M}_N)$.

We next define the classical symbols in the semiclassical setting.

Definition 9.1.9 (Classical semiclassical symbols).

1. We shall say that a semiclassical symbol $a \in S_0^k(m^\mu;\mathsf{M}_N)$ is **classical** and write $a \in S_{\rm cl}^k(m^\mu;\mathsf{M}_N)$ if

$$a(X;h) \sim h^{-k} \sum_{j\geq 0} h^j a_j(X) \text{ in } S_0^k(m^\mu;\mathsf{M}_N),$$

   where the $a_j \in S_0^0(m^\mu;\mathsf{M}_N)$ are **independent of** h, $j \geq 0$.
2. We shall say that a semiclassical symbol $a \in S_0^k(m^\mu,g;\mathsf{M}_N)$ is **classical** and write $a \in S_{0,\rm cl}^k(m^\mu,g;\mathsf{M}_N)$ if there exists a sequence $\{a_{\mu-2j}\}_{j\geq 0}$ of symbols $a_{\mu-2j} \in S(m^{\mu-2j},g;\mathsf{M}_N)$, $j \geq 0$, with the $a_{\mu-2j}$ **independent of** h, such that for any given $N_0 \in \mathbb{Z}_+$

$$a(X;h) - h^{-k} \sum_{j=0}^{N_0} h^j a_{\mu-2j}(X) \in S_0^{k-(N_0+1)}(m^{\mu-2(N_0+1)},g;\mathsf{M}_N).$$

   We shall write

$$a(X;h) \sim h^{-k} \sum_{j\geq 0} h^j a_{\mu-2j}(X) \text{ in } S_0^k(m^\mu,g;\mathsf{M}_N).$$

   For a classical semiclassical symbol $a \sim a_\mu + h a_{\mu-2} + \ldots$, one calls a_μ the **principal** symbol of a, and $a_{\mu-2}$ the **subprincipal** symbol of a.
3. We shall say that a classical semiclassical symbol $a \in S_{0,\rm cl}^0(m^\mu,g;\mathsf{M}_N)$ is **elliptic** if its principal symbol a_μ belongs to $S(m^\mu,g;\mathsf{M}_N)$ and $a_\mu(X)^{-1}$ exists for all $X \in \mathbb{R}^{2n}$ and belongs to $S(m^{-\mu},g;\mathsf{M}_N)$. Equivalently, one requires that $\det a_\mu \in S(m^{N\mu},g)$ be such that $|\det a_\mu(X)| \gtrsim m(X)^{N\mu}$ for all $X \in \mathbb{R}^{2n}$.
4. We shall say that a classical semiclassical symbol $a = a^* \in S_{0,\rm cl}^0(m^\mu,g;\mathsf{M}_N)$ is **positive elliptic** if its principal symbol $a_\mu = a_\mu^*$ belongs to $S(m^\mu,g;\mathsf{M}_N)$ and there are $0 < c_1 < c_2$ such that

$$c_1 m(X)^\mu |v|_{\mathbb{C}^N}^2 \leq \langle a_\mu(X)v,v\rangle_{\mathbb{C}^N} \leq c_2 m(X)^\mu |v|_{\mathbb{C}^N}^2,$$

   for all $v \in \mathbb{C}^N$ and all $X \in \mathbb{R}^{2n}$.

Remark 9.1.10. From Theorem 9.1.7 and Theorem 9.1.8, respectively, we have that also the classes $S_{\rm cl}^k(m^\mu;\mathsf{M}_N)$ and $S_{0,\rm cl}^k(m^\mu,g;\mathsf{M}_N)$, respectively, are well-behaved under composition. △

We now wish to consider *inverses* in the semiclassical calculus.

We need the following result, due to R. Beals, which characterizes pseudodifferential operators in the semiclassical setting (see Dimassi-Sjöstrand [7] or Evans-Zworski [15]; we give a statement in the scalar case for simplicity).

Theorem 9.1.11. *Let $A = A_h \colon \mathscr{S}(\mathbb{R}^n) \longrightarrow \mathscr{S}'(\mathbb{R}^n)$ be a continuous linear operator, $0 < h \leq 1$. Then the following statements are equivalent:*

1. *$A = a^{\mathrm{w}}(x,hD)$ for some $a(x,\xi;h) = a \in S_0^0(1)$;*
2. *For every $N \in \mathbb{N}$ and for every family $\ell_1(x,\xi),\ldots,\ell_N(x,\xi)$ of linear forms on $\mathbb{R}^{2n}$, the operator $\mathrm{ad}_{\ell_1(x,hD)} \circ \ldots \circ \mathrm{ad}_{\ell_N(x,hD)} A_h$ is* **continuous** *in $L^2(\mathbb{R}^n)$, i.e. it belongs to $\mathscr{L}(L^2,L^2)$, with norm $O(h^N)$ in that space. (Recall that $\mathrm{ad}_A B = [A,B]$.)*

One may then prove the following proposition about "elliptic" elements in $S_0^0(m^\mu;\mathsf{M}_N)$.

Theorem 9.1.12. *Let $a(X;h) = a \in S_0^0(m^\mu;\mathsf{M}_N)$ be elliptic, that is, by definition, $a(X;h)^{-1} = a^{-1} \in S_0^0(m^{-\mu};\mathsf{M}_N)$, for $0 < h \leq h_0$. Then, by possibly shrinking h_0, there exists a symbol $b \in S_0^0(m^{-\mu};\mathsf{M}_N)$ such that*

$$b^{\mathrm{w}}(x,hD)a^{\mathrm{w}}(x,hD) = I = a^{\mathrm{w}}(x,hD)b^{\mathrm{w}}(x,hD), \quad 0 < h \leq h_0,$$

and, furthermore, $b^{\mathrm{w}}(x,hD)$ possesses an asymptotic expansion, that is,

$$b \sim a^{-1} + h(a^{-1}\sharp_h r) + h^2(a^{-1}\sharp_h r\sharp_h r) + \ldots,$$

where $r \in S_0^0(1;\mathsf{M}_N)$ is such that $(a^{-1})^{\mathrm{w}}(x,hD)a^{\mathrm{w}}(x,hD) = I - hr^{\mathrm{w}}(x,hD)$.

Proof. Let $\tilde{b} = a^{-1} \in S_0^0(m^{-\mu};\mathsf{M}_N)$. Then

$$\tilde{b}^{\mathrm{w}}(x,hD)a^{\mathrm{w}}(x,hD) = I - hr_L^{\mathrm{w}}(x,hD), \ \ r_L \in S_0^0(1;\mathsf{M}_N),$$

and

$$a^{\mathrm{w}}(x,hD)\tilde{b}^{\mathrm{w}}(x,hD) = I - hr_R^{\mathrm{w}}(x,hD), \ \ r_R \in S_0^0(1;\mathsf{M}_N).$$

Then by Theorem 9.1.6

$$I - hr_L^{\mathrm{w}}(x,hD) \colon L^2 \longrightarrow L^2 \text{ and } I - hr_R^{\mathrm{w}}(x,hD) \colon L^2 \longrightarrow L^2$$

are both invertible, provided $h \in (0,h_0]$, if $h_0 \in (0,1]$ is taken sufficiently small. By the Beals Theorem 9.1.11 it follows that there exists $c_L, c_R \in S_0^0(1;\mathsf{M}_N)$ such that

$$\Big(I - hr_L^{\mathrm{w}}(x,hD)\Big)^{-1} = c_L(x,hD), \ \Big(I - hr_R^{\mathrm{w}}(x,hD)\Big)^{-1} = c_R(x,hD),$$

so that

$$\Big(I - hr_L^{\mathrm{w}}(x,hD)\Big)^{-1}\tilde{b}^{\mathrm{w}}(x,hD) =: b_L^{\mathrm{w}}(x,hD), \ \ b_L \in S_0^0(m^{-\mu};\mathsf{M}_N),$$

and

$$\tilde{b}^{\mathrm{w}}(x,hD)\Big(I - hr_R^{\mathrm{w}}(x,hD)\Big)^{-1} =: b_R^{\mathrm{w}}(x,hD), \ \ b_R \in S_0^0(m^{-\mu};\mathsf{M}_N),$$

whence we get the existence of a left-inverse $b_L(x,\xi;h) = b_L$ and of a right-inverse $b_R(x,\xi;h) = b_R$, both belonging to the symbol class $S_0^0(m^{-\mu};\mathsf{M}_N)$, such that

$$b_L^{\mathrm{w}}(x,hD)a^{\mathrm{w}}(x,hD) = I = a^{\mathrm{w}}(x,hD)b_R^{\mathrm{w}}(x,hD), \quad 0 < h \le h_0,$$

and, finally,

$$b_L^{\mathrm{w}}(x,hD) = b_L^{\mathrm{w}}(x,hD)a^{\mathrm{w}}(x,hD)b_R^{\mathrm{w}}(x,hD) = b_R^{\mathrm{w}}(x,hD).$$

Put therefore $b = b_L = b_R$.

In addition, we have that b possesses an asymptotic expansion. In fact, for any given $N_0 \in \mathbb{Z}_+$, let

$$b_{N_0}^{\mathrm{w}}(x,hD)$$
$$:= \tilde{b}^{\mathrm{w}}(x,hD)\Big(I + hr_R^{\mathrm{w}}(x,hD) + h^2 r_R^{\mathrm{w}}(x,hD)^2 + \ldots + h^{N_0} r_R^{\mathrm{w}}(x,hD)^{N_0}\Big).$$

Then

$$a^{\mathrm{w}}(x,hD)b_{N_0}^{\mathrm{w}}(x,hD) = I - h^{N_0+1} r_R(x,hD)^{N_0+1},$$

and it follows that

$$\begin{aligned} b_{N_0}^{\mathrm{w}}(x,hD) &= b^{\mathrm{w}}(x,hD)a^{\mathrm{w}}(x,hD)b_{N_0}^{\mathrm{w}}(x,hD) \\ &= b^{\mathrm{w}}(x,hD) - h^{N_0+1} b^{\mathrm{w}}(x,hD) r_R^{\mathrm{w}}(x,hD)^{N_0+1} \\ &= b^{\mathrm{w}}(x,hD) + h^{N_0+1} r_{N_0+1}^{\mathrm{w}}(x,hD), \end{aligned}$$

where we have put $r_{N_0+1}^{\mathrm{w}}(x,hD) := -b^{\mathrm{w}}(x,hD)r_R^{\mathrm{w}}(x,hD)^{N_0+1}$. By Theorem 9.1.7 we have $r_{N_0+1} \in S_0^0(m^{-\mu};\mathsf{M}_N)$, and this concludes the proof. □

It will be also useful to have the following variation of Theorem 9.1.12 for elliptic classical semiclassical symbols $a \in S_{0,\mathrm{cl}}^0(m^{\mu},g;\mathsf{M}_N)$. The usual parametrix construction (using h as "parameter of homogeneity") gives the following theorem.

Theorem 9.1.13. *Let $a \in S_{0,\mathrm{cl}}^0(m^{\mu},g;\mathsf{M}_N)$ be elliptic, that is, with $a \sim \sum_{j\ge 0} h^j a_{\mu-2j}$, let $a_{\mu}^{-1} \in S(m^{-\mu},g;\mathsf{M}_N)$. Then there exists a classical semiclassical symbol $b \in S_{0,\mathrm{cl}}^0(m^{-\mu},g;\mathsf{M}_N)$ such that*

$$b^{\mathrm{w}}(x,hD)a^{\mathrm{w}}(x,hD) = I + r_L^{\mathrm{w}}(x,hD),$$

$$a^{\mathrm{w}}(x,hD)b^{\mathrm{w}}(x,hD) = I + r_R^{\mathrm{w}}(x,hD),$$

where $r_L, r_R \in S^{-\infty}(m^{-\infty},g;\mathsf{M}_N)$ (see (9.7)).

9.2 Decoupling a System

In this section, using an adaptation taken from Parenti-Parmeggiani [49] of the classical decoupling argument of Taylor [69], we prove the following result. (See also Helffer-Sjöstrand [21].)

Theorem 9.2.1. *Let $\mu > 0$ and let $a = a^* \sim \sum_{j\geq 0} h^j a_{\mu-2j} \in S^0_{0,\mathrm{cl}}(m^\mu, g; \mathsf{M}_N)$. Suppose there exists $e_0 \in S(1, g; \mathsf{M}_N)$ such that $e_0^* e_0 = e_0 e_0^* = I$ and*

$$e_0^* a_\mu e_0 = b_\mu = \begin{bmatrix} \lambda_{1,\mu} & 0 \\ 0 & \lambda_{2,\mu} \end{bmatrix}, \tag{9.12}$$

where $\lambda_{j,\mu} = \lambda_{j,\mu}^* \in S(m^\mu, g; \mathsf{M}_{N_j})$, $j = 1, 2$, $N_1 + N_2 = N$, and

$$d_{\lambda_1,\lambda_2}(X) \gtrsim m(X)^\mu, \quad \forall X \in \mathbb{R}^{2n}, \tag{9.13}$$

where, for each $X \in \mathbb{R}^{2n}$,

$$d_{\lambda_1,\lambda_2}(X) = \inf\Big\{ |\mu_1 - \mu_2|;\ \mu_1 \in \mathrm{Spec}(\lambda_{1,\mu}(X)), \mu_2 \in \mathrm{Spec}(\lambda_{2,\mu}(X)) \Big\}. \tag{9.14}$$

Then there exists $e \in S^0_{0,\mathrm{cl}}(1, g; \mathsf{M}_N)$ with principal symbol e_0 such that:

1. *One has*

$$e^{\mathrm{w}}(x,hD)^* e^{\mathrm{w}}(x,hD) - I,\ e^{\mathrm{w}}(x,hD) e^{\mathrm{w}}(x,hD)^* - I \in S^{-\infty}(m^{-\infty}, g; \mathsf{M}_N);$$

2. $e^{\mathrm{w}}(x,hD)^* a^{\mathrm{w}}(x,hD) e^{\mathrm{w}}(x,hD) - b^{\mathrm{w}}(x,hD) \in S^{-\infty}(m^{-\infty}, g; \mathsf{M}_N)$, *where the symbol $b \sim \sum_{j\geq 0} h^j b_{\mu-2j} \in S^0_{0,\mathrm{cl}}(m^\mu, g; \mathsf{M}_N)$ is* **blockwise diagonal**, *with*

$$b_{\mu-2j}(X) = \begin{bmatrix} b_{1,\mu-2j}(X) & 0 \\ 0 & b_{2,\mu-2j}(X) \end{bmatrix}, \quad \forall X \in \mathbb{R}^{2n}, \ \forall j \geq 0,$$

with blocks $b_{j,\mu}$ of sizes N_j, $j = 1, 2$, respectively, and with principal symbol

$$b_\mu(X) = \begin{bmatrix} \lambda_{1,\mu}(X) & 0 \\ 0 & \lambda_{2,\mu}(X) \end{bmatrix}, \quad \forall X \in \mathbb{R}^{2n}.$$

We shall call b **an h^∞-(block)-diagonalization** *of a. Notice that b depends on a and e_0.*

Proof. We immediately observe that once $e^{\mathrm{w}}(x,hD)$ has been found with the property that its principal symbol is e_0 and

$$e^{\mathrm{w}}(x,hD) e^{\mathrm{w}}(x,hD)^* = I + r^{\mathrm{w}}(x,hD), \quad \text{with } r \in S^{-\infty}(m^{-\infty}, g; \mathsf{M}_N),$$

then by the ellipticity (using Theorem 9.1.13) we also get

$$e^{\mathrm{w}}(x,hD)^* e^{\mathrm{w}}(x,hD) = I + s^{\mathrm{w}}(x,hD), \text{ with } s \in S^{-\infty}(m^{-\infty}, g; \mathsf{M}_N).$$

Hence it suffices to prove the existence of $e^{\mathrm{w}}(x,hD)$ and b with the required properties. We show that for every integer $k \in \mathbb{Z}_+$ there exist

$$(i) \qquad e_{-2k} \in S(m^{-2k}, g; \mathsf{M}_N),$$

and

$$(ii) \qquad b_{j,\mu-2k} \in S(m^{\mu-2k}, g; \mathsf{M}_{N_j}), \quad j = 1,2,$$

such that, with $E_{N_0}(X) := \sum_{k=0}^{N_0} h^k e_{-2k}(X)$,

$$E_{N_0} \sharp_h E_{N_0}^* = I + h^{N_0+1} S^0_{0,\mathrm{cl}}(m^{-2(N_0+1)}, g; \mathsf{M}_N),$$

and

$$E_{N_0}^* \sharp_h a \sharp_h E_{N_0} = \sum_{k=0}^{N_0} b_{\mu-2k} + h^{N_0+1} S^0_{0,\mathrm{cl}}(m^{\mu-2(N_0+1)}, g; \mathsf{M}_N),$$

where the $b_{\mu-2k} = \begin{bmatrix} b_{1,\mu-2k} & 0 \\ 0 & b_{2,\mu-2k} \end{bmatrix}$ are in block-diagonal form. We shall then take $e \sim \sum_{k\geq 0} h^k e_{-2k}$.

Hence we proceed by induction. So, suppose we have already constructed symbols $e_0, e_{-2}, \ldots, e_{-2N_0}$, and $b_\mu, b_{\mu-2}, \ldots, b_{\mu-2N_0}$, *independent of* h, with the required properties. Put hence

$$S^{\mathrm{w}}_{N_0}(x,hD) := E^{\mathrm{w}}_{N_0}(x,hD) E^{\mathrm{w}}_{N_0}(x,hD)^* - I, \tag{9.15}$$

where $S_{N_0} \in h^{N_0+1} S^0_{0,\mathrm{cl}}(m^{-2(N_0+1)}, g; \mathsf{M}_N)$. We shall write, for short, $E^{\mathrm{w}}_{N_0}$, $S^{\mathrm{w}}_{N_0}$ and e^{w}_{-2k} in place of $E^{\mathrm{w}}_{N_0}(x,hD)$ etc.

We look for a matrix-symbol $e_{-2(N_0+1)} \in S(m^{-2(N_0+1)}, g; \mathsf{M}_N)$ such that

$$\Big(E^{\mathrm{w}}_{N_0} + h^{N_0+1} e^{\mathrm{w}}_{-2(N_0+1)}\Big)\Big((E^{\mathrm{w}}_{N_0})^* + h^{N_0+1}(e^{\mathrm{w}}_{-2(N_0+1)})^*\Big) - I = h^{N_0+2} r^{\mathrm{w}}(x,hD),$$

where $r \in S^0_{0,\mathrm{cl}}(m^{-2(N_0+2)}, g; \mathsf{M}_N)$, that is, we look for $e_{-2(N_0+1)}$ such that

$$S^{\mathrm{w}}_{N_0} + h^{N_0+1}\Big(e^{\mathrm{w}}_0 (e^{\mathrm{w}}_{-2(N_0+1)})^* + e^{\mathrm{w}}_{-2(N_0+1)} (e^{\mathrm{w}}_0)^*\Big) = h^{N_0+2} \tilde{r}^{\mathrm{w}}(x,hD),$$

with $\tilde{r} \in S^0_{0,\mathrm{cl}}(m^{-2(N_0+2)}, g; \mathsf{M}_N)$. Using the composition formula (9.11) we thus look at the coefficient of h^{N_0+1} and require that it be zero, obtaining the equation

$$s_{-2(N_0+1)} + e_0 (e_{-2(N_0+1)})^* + e_{-2(N_0+1)} e_0^* = 0. \tag{9.16}$$

Notice that $s^*_{-2(N_0+1)} = s_{-2(N_0+1)}$ (this follows from (9.15)). Equation (9.16) has general solution

$$e_{-2(N_0+1)} = -\frac{1}{2}s_{-2(N_0+1)}e_0 + \phi_{-2(N_0+1)}, \tag{9.17}$$

where $\phi_{-2(N_0+1)}$ solves

$$e_0\phi^*_{-2(N_0+1)} + \phi_{-2(N_0+1)}e_0^* = 0,$$

which in turn gives that $\phi_{-2(N_0+1)}$ has the form

$$\phi_{-2(N_0+1)} = \alpha_{-2(N_0+1)}e_0, \tag{9.18}$$

where $\alpha_{-2(N_0+1)} \in S(m^{-2(N_0+1)}, g; \mathsf{M}_N)$ and

$$\alpha^*_{-2(N_0+1)} + \alpha_{-2(N_0+1)} = 0.$$

We next perform a choice of $\alpha_{-2(N_0+1)}$ for obtaining the blocks $b_{j,\mu-2(N_0+1)}$, $j = 1,2$. Since on the one hand

$$(E^{\mathrm{w}}_{N_0+1})a^{\mathrm{w}}E^{\mathrm{w}}_{N_0+1} = \sum_{k=0}^{N_0+1} h^k b^{\mathrm{w}}_{\mu-2k} + h^{N_0+2}r_1^{\mathrm{w}},$$

with $r_1 \in S^0_{0,\mathrm{cl}}(m^{\mu-2(N_0+2)}, g; \mathsf{M}_N)$, and on the other

$$\begin{aligned}(E^{\mathrm{w}}_{N_0+1})a^{\mathrm{w}}E^{\mathrm{w}}_{N_0+1} &= (E^{\mathrm{w}}_{N_0})^*a^{\mathrm{w}}E^{\mathrm{w}}_{N_0} + h^{N_0+1}\Big((e^{\mathrm{w}}_{-2(N_0+1)})^*a^{\mathrm{w}}e_0^{\mathrm{w}} \\ &\quad + (e_0^{\mathrm{w}})^*a^{\mathrm{w}}e^{\mathrm{w}}_{-2(N_0+1)}\Big) + h^{N_0+2}r_2^{\mathrm{w}} \\ &=: T^{\mathrm{w}}_{N_0} + h^{N_0+2}r_2^{\mathrm{w}},\end{aligned} \tag{9.19}$$

with $r_2 \in S^0_{0,\mathrm{cl}}(m^{\mu-2(N_0+2)}, g; \mathsf{M}_N)$, the conditions for the blocks $b_{j,\mu-2k}$ are already satisfied for $0 \le k \le N_0$, *independently of* $e_{-2(N_0+1)}$. Let $q_{\mu-2(N_0+1)}$ be the coefficient of h^{N_0+1} in $E^*_{N_0}\sharp_h a\sharp_h E_{N_0}$. Then the coefficient of h^{N_0+1} in T_{N_0} is

$$\begin{aligned}&q_{\mu-2(N_0+1)} + e^*_{-2(N_0+1)}a_\mu e_0 + e_0^*a_\mu e_{-2(N_0+1)} \\ &\quad = q_{\mu-2(N_0+1)} + (e^*_{-2(N_0+1)}e_0)e_0^*a_\mu e_0 + e_0^*a_\mu e_0(e_0^*e_{-2(N_0+1)}).\end{aligned} \tag{9.20}$$

Using (9.17) and (9.18), we write

$$\begin{cases} e_0^*e_{-2(N_0+1)} = -\dfrac{1}{2}e_0^*s_{-2(N_0+1)}e_0 + e_0^*\alpha_{-2(N_0+1)}e_0 =: \tau + \beta, \\[2ex] e^*_{-2(N_0+1)}e_0 = \tau - \beta, \end{cases} \tag{9.21}$$

where

$$\tau = -\frac{1}{2} e_0^* s_{-2(N_0+1)} e_0 = \tau^*, \text{ and } \beta = e_0^* \alpha_{-2(N_0+1)} e_0 = -\beta^*.$$

By (9.21), the term (9.20) goes over to

$$q_{\mu-2(N_0+1)} + (e_0^* a_\mu e_0)\tau + \tau(e_0^* a_\mu e_0) + (e_0^* a_\mu e_0)\beta - \beta(e_0^* a_\mu e_0). \tag{9.22}$$

We now show that β, hence in turn

$$\alpha_{-2(N_0+1)} = e_0 \beta e_0^*, \tag{9.23}$$

can be so chosen as to kill the off-diagonal terms in (9.22). In fact, upon writing

$$q_{\mu-2(N_0+1)} + (e_0^* a_\mu e_0)\tau + \tau(e_0^* a_\mu e_0) = \begin{bmatrix} u_1 & \gamma \\ \gamma^* & u_2 \end{bmatrix},$$

where the $u_j = u_j^*$ are $N_j \times N_j$ blocks, $j = 1, 2$, we look for β in the form

$$\beta = \begin{bmatrix} 0 & \delta \\ -\delta^* & 0 \end{bmatrix},$$

and, using (9.12), we are therefore led to the matrix equation

$$\lambda_{1,\mu}\delta - \delta\lambda_{2,\mu} = -\gamma. \tag{9.24}$$

By Lemma 9.2.2 below, hypothesis (9.13) yields that equation (9.24) has a unique smooth $N_1 \times N_2$ matrix-valued solution $\delta \in S(m^{-2(N_0+1)}, g; \mathrm{Mat}_{N_1 \times N_2}(\mathbb{C}))$. Since this fixes β, and hence $\alpha_{-2(N_0+1)}$, the terms $b_{j,\mu-2(N_0+1)}$ are then the block-diagonal terms in (9.22). This concludes the inductive step and the proof of the theorem. □

Lemma 9.2.2. *Let $E = E^* \in S(m^\mu, g; \mathsf{M}_{N_1})$ and $F = F^* \in S(m^\mu, g; \mathsf{M}_{N_2})$ be such that (recall (9.14))*

$$d_{E,F}(X) \geq c_0\, m(X)^\mu, \ \forall X \in \mathbb{R}^{2n}.$$

Then for each $X \in \mathbb{R}^{2n}$ the map

$$\begin{cases} \Phi_{E,F}(X)\colon \mathrm{Mat}_{N_1 \times N_2}(\mathbb{C}) \longrightarrow \mathrm{Mat}_{N_1 \times N_2}(\mathbb{C}), \\ \Phi_{E,F}(X)T = E(X)T - TF(X), \end{cases} \tag{9.25}$$

is an isomorphism. Moreover,

$$\|\Phi_{E,F}(X)^{-1}\| \leq \frac{C}{m(X)^\mu}, \ \forall X \in \mathbb{R}^{2n}, \tag{9.26}$$

for a **universal** *constant* $C>0$. *Hence, if* $S\in S(m^{\mu-2k},g;\mathrm{Mat}_{N_1\times N_2}(\mathbb{C}))$, *for some* $k\in\mathbb{Z}_+$, *we have that*

$$X\longmapsto T(X):=\Phi_{E,F}(X)^{-1}\big(S(X)\big)\in S(m^{-2k},g;\mathrm{Mat}_{N_1\times N_2}(\mathbb{C})). \tag{9.27}$$

Proof. For each fixed $X\in\mathbb{R}^{2n}$, let $\mathrm{Spec}(E(X))=\{e_1(X),\dots,e_{\nu_X}(X)\}$. Hence $e_j(X)\neq e_{j'}(X)$ if $j\neq j'$. Consider the contour in $\mathbb{C}$

$$\gamma(X)=\bigcup_{j=1}^{\nu_X}\gamma_j(X),\ \text{ with }\ \gamma_j(X)\cap\gamma_{j'}(X)=\emptyset\ \text{ for }\ j\neq j',$$

counter-clockwise oriented, where each $\gamma_j(X)=\partial D_j(X)$ is a small circle that encloses only the eigenvalue $e_j(X)$ of $E(X)$, $1\le j\le\nu_X$, and

$$\mathrm{Spec}(F(X))\subset\mathbb{C}\setminus\bigcup_{j=1}^{\nu_X}D_j(X).$$

Then, on considering the equation $E(X)T-TF(X)=S(X)$ we have that the solution can be written as

$$\begin{aligned}T=T(X)&=\frac{1}{2\pi i}\int_{\gamma(X)}(\zeta-E(X))^{-1}S(X)(\zeta-F(X))^{-1}d\zeta\\&=\sum_{j=1}^{\nu_X}P_{E,j}(X)S(X)\big(e_j(X)-F(X)\big)^{-1},\end{aligned} \tag{9.28}$$

where $P_{E,j}(X)\colon\mathbb{C}^{N_1}\longrightarrow\mathbb{C}^{N_1}$ is the orthogonal projection associated with the eigenvalue $e_j(X)$. This shows that for each fixed X the map $\Phi_{E,F}(X)$ is an isomorphism (for it is injective and linear). Moreover, using the fact that the norm of the resolvent of a normal operator equals the spectral radius, we also obtain from (9.28) that

$$|T(X)|=|\Phi_{E,F}(X)^{-1}\big(S(X)\big)|\le\frac{C}{m(X)^{\mu}}|S(X)|,\ \ \forall X\in\mathbb{R}^{2n},$$

where $C=N_1/c_0$. This shows (9.26). Now, (9.26) gives that $T(X)$ is continuous and differentiable to all orders, for we have that for any given $X_1,X_2\in\mathbb{R}^{2n}$

$$\begin{aligned}&\Phi_{E,F}(X_1)\Big(T(X_1)-T(X_2)\Big)\\&\quad=\Big(E(X_2)-E(X_1)\Big)T(X_2)+T(X_2)\Big(F(X_1)-F(X_2)\Big)+S(X_1)-S(X_2),\end{aligned} \tag{9.29}$$

from which, since (9.26) yields $|T(X_2)|\lesssim|S(X_2)|$ for all X_2 (recall that $m(X)\ge 1$ for all X), we obtain the continuity of T by considering the map $\Phi_{E,F}(X_1)^{-1}$, taking the limit as $X_2 \to X_1$, and using the continuity of E, F and S. The differentiability claim follows from (9.29) by induction.

It remains to control the growth of $|\partial_X^\alpha T(X)|$. For any given $\alpha \in \mathbb{Z}_+^{2n}$, the matrix $\partial_X^\alpha T$ is a solution to

$$E\partial_X^\alpha T - \partial_X^\alpha T F + S_\alpha = \partial_X^\alpha S,$$

where, by the Leibniz rule,

$$S_\alpha := \sum_{\substack{\beta \leq \alpha \\ \beta \neq \alpha}} \binom{\alpha}{\beta} \left(\partial_X^{\alpha-\beta} E \partial_X^\beta T - \partial_X^\beta T \partial_X^{\alpha-\beta} F \right).$$

Hence

$$\partial_X^\alpha T(X) = \Phi_{E,F}(X)^{-1} \Big(\partial_X^\alpha S(X) - S_\alpha(X) \Big),$$

so that assuming by induction on α that $|\partial_X^\beta T| \lesssim m^{-2k-|\beta|}$, gives by (9.26)

$$|\partial_X^\alpha T(X)| \leq C_\alpha m(X)^{-2k-|\alpha|}, \ \forall X \in \mathbb{R}^{2n},$$

which concludes the proof. □

As a consequence of the proof of Theorem 9.2.1 we have the following analogue for classical symbols (which is used in Parmeggiani [52]).

Theorem 9.2.3. *Let $\mu > 0$ and let $a = a^* \sim \sum_{j \geq 0} a_{\mu-2j} \in S_{\mathrm{cl}}(m^\mu, g; \mathsf{M}_N)$. Suppose there exists $e_0 \in C^\infty(\mathbb{R}^{2n} \setminus \{0\}; \mathsf{M}_N)$, positively homogeneous of degree* 0, *such that*

$$e_0^* e_0 = e_0 e_0^* = I, \ \text{and} \ e_0^* a_\mu e_0 = b_\mu = \begin{bmatrix} \lambda_{1,\mu} & 0 \\ 0 & \lambda_{2,\mu} \end{bmatrix}, \ X \neq 0,$$

where $\lambda_{j,\mu} = \lambda_{j,\mu}^* \in C^\infty(\mathbb{R}^{2n} \setminus \{0\}; \mathsf{M}_{N_j})$ are **positively homogeneous** *of degree μ, $j = 1,2$, and*

$$\mathrm{Spec}(\lambda_{1,\mu}(X)) \cap \mathrm{Spec}(\lambda_{2,\mu}(X)) = \emptyset, \ \forall X \in \mathbb{R}^{2n}, \ |X| = 1. \tag{9.30}$$

Then there exists $e \in S_{\mathrm{cl}}(1, g; \mathsf{M}_N)$ with principal symbol e_0 such that

1. *$e^{\mathrm{w}}(x,D)^* e^{\mathrm{w}}(x,D) - I$, $e^{\mathrm{w}}(x,D) e^{\mathrm{w}}(x,D)^* - I \in S(m^{-\infty}, g; \mathsf{M}_N)$;*
2. *$e^{\mathrm{w}}(x,D)^* a^{\mathrm{w}}(x,D) e^{\mathrm{w}}(x,D) - b^{\mathrm{w}}(x,D) \in S(m^{-\infty}, g; \mathsf{M}_N)$, where the symbol $b \sim \sum_{j \geq 0} b_{\mu-2j} \in S_{\mathrm{cl}}(m^\mu, g; \mathsf{M}_N)$ is* **blockwise diagonal**, *with*

$$b_{\mu-2j}(X) = \begin{bmatrix} b_{1,\mu-2j}(X) & 0 \\ 0 & b_{2,\mu-2j}(X) \end{bmatrix}, \ \forall X \in \mathbb{R}^{2n} \setminus \{0\}, \forall j \geq 0,$$

with blocks $b_{j,\mu}$ of sizes N_j, $j = 1,2$, respectively, and with principal symbol

$$b_\mu(X) = \begin{bmatrix} \lambda_{1,\mu}(X) & 0 \\ 0 & \lambda_{2,\mu}(X) \end{bmatrix}, \quad \forall X \in \mathbb{R}^{2n} \setminus \{0\}.$$

We shall call such a symbol b **a (block)-diagonalization** *of a. Notice that b depends on a and e_0.*

Remark 9.2.4. Given $N \times N$ matrices a and b, a straightforward computation gives the following very useful formula:

$$\{a,b\}^* = -\{b^*,a^*\} \tag{9.31}$$

which yields in particular that

$$\frac{i}{2}\{a,a^*\} \text{ and } \frac{i}{2}\{a^*,a\} \text{ are } \textbf{Hermitian} \text{ matrices.}$$

△

It will be useful to compute the subprincipal part $b_{\mu-2}$ (that is, the coefficient of the h term) of the block-diagonal system obtained in Theorem 9.2.1. The same formula holds for the version given in Theorem 9.2.3 for classical operators. We have the following proposition.

Proposition 9.2.5. *For the subprincipal part $b_{\mu-2}$ of the h^∞-diagonalization given in Theorem 9.2.1 one has, by (9.20), the formula*

$$b_{\mu-2} = e_{-2}^* e_0 b_\mu + b_\mu e_0^* e_{-2} + e_0^* a_{\mu-2} e_0 - \frac{i}{2}\Big(e_0^*\{a_\mu, e_0\} + \{e_0^*, a_\mu e_0\}\Big), \tag{9.32}$$

where

$$e_{-2} = \frac{i}{4}\{e_0, e_0^*\}e_0 + \alpha_{-2}e_0,$$

with $\alpha_{-2}^* = -\alpha_{-2}$ determined by equation (9.24) through $\beta_{-2} = e_0^*\alpha_{-2}e_0$.

In the case $N = 2$, supposing that $a_\mu = a_\mu^* > 0$ and that there exist positive smooth functions $\lambda_1, \lambda_2 \in S(m^\mu, g)$ (where we now write λ_j, $j = 1,2$, for the eigenvalues of a_μ) such that

$$|\lambda_1(X) - \lambda_2(X)| \gtrsim m(X)^\mu, \ \forall X \in \mathbb{R}^{2n}, \tag{9.33}$$

whence the existence of a smooth unitary matrix e_0 such that

$$e_0(X)^* a_\mu(X) e_0(X) = \begin{bmatrix} \lambda_1(X) & 0 \\ 0 & \lambda_2(X) \end{bmatrix}, \forall X \in \mathbb{R}^{2n}, \tag{9.34}$$

we have the following corollary.

Corollary 9.2.6. *Suppose* $a_\mu = a_\mu^* > 0$ *possesses smooth eigenvalues* λ_1, λ_2 *satisfying (9.33). Let* $\{w_1, w_2\}$ *be the* **canonical** *basis of* $\mathbb{C}^2$, $w_1 = \begin{bmatrix} 1 \\ 0 \end{bmatrix}$, $w_2 = \begin{bmatrix} 0 \\ 1 \end{bmatrix}$, *so that* $b_\mu(X)w_j = \lambda_j(X)w_j$, $j = 1,2$, *for all* $X \in \mathbb{R}^{2n}$. *Then the symbol of the* h^∞*-diagonalization* b *is a* **diagonal** 2×2 *matrix. Moreover, for the subprincipal symbol* $b_{\mu-2} = \begin{bmatrix} b_{\mu-2}^{(11)} & 0 \\ 0 & b_{\mu-2}^{(22)} \end{bmatrix}$ *we have*

$$\begin{aligned} b_{\mu-2}^{(jj)} &= \langle b_{\mu-2}w_j, w_j\rangle \\ &= \langle e_0^* a_{\mu-2} e_0 w_j, w_j\rangle + \frac{1}{2}\mathrm{Im}\Big(\langle\{e_0^*, \lambda_j\}e_0 w_j, w_j\rangle\Big) \\ &\quad + \frac{1}{2}\mathrm{Im}\Big(\langle e_0^*\{a_\mu, e_0\}w_j, w_j\rangle\Big), \quad j = 1,2. \end{aligned} \tag{9.35}$$

Proof. From the proof of Theorem 9.2.1 we have that the matrix $\beta = \beta_{-2}$ is skew-adjoint, whence

$$\langle \beta_{-2}^* b_\mu w_j, w_j\rangle + \langle b_\mu \beta_{-2} w_j, w_j\rangle = -\lambda_j\langle \beta_{-2}w_j, w_j\rangle + \lambda_j\langle \beta_{-2}w_j, w_j\rangle = 0. \tag{9.36}$$

Then, from (9.32), recalling that

$$e_0^* e_{-2} = \frac{i}{4} e_0^*\{e_0, e_0^*\}e_0 + \beta_{-2},$$

using $a_\mu e_0 = e_0 b_\mu$ and (9.36), and since $(\partial b_\mu)w_j = (\partial\lambda_j)w_j$, we obtain

$$\begin{aligned} \langle b_{\mu-2}w_j, w_j\rangle &= \mathrm{Re}\,\langle b_{\mu-2}w_j, w_j\rangle \\ &= \langle e_0^* a_{\mu-2} e_0 w_j, w_j\rangle + 2\,\mathrm{Re}\Big(\frac{i}{4}\langle e_0^*\{e_0, e_0^*\}e_0 b_\mu w_j, w_j\rangle\Big) \\ &\quad -\mathrm{Re}\Big(\frac{i}{2}\langle e_0^*\{a_\mu, e_0\}w_j, w_j\rangle\Big) - \mathrm{Re}\Big(\frac{i}{2}\langle\{e_0^*, e_0 b_\mu\}w_j, w_j\rangle\Big) \\ &= \langle e_0^* a_{\mu-2} e_0 w_j, w_j\rangle \\ &\quad -\frac{1}{2}\mathrm{Im}\Big(\langle e_0^*\{e_0, e_0^*\}e_0 b_\mu w_j, w_j\rangle\Big) + \frac{1}{2}\mathrm{Im}\Big(\langle e_0^*\{a_\mu, e_0\}w_j, w_j\rangle\Big) \\ &\quad +\frac{1}{2}\mathrm{Im}\Big(\langle\{e_0^*, e_0\}b_\mu w_j, w_j\rangle\Big) \\ &\quad +\frac{1}{2}\sum_{\ell=1}^{n}\mathrm{Im}\Big(\langle(\frac{\partial e_0^*}{\partial \xi_\ell}e_0\frac{\partial \lambda_j}{\partial x_\ell} - \frac{\partial e_0^*}{\partial x_\ell}e_0\frac{\partial \lambda_j}{\partial \xi_\ell})w_j, w_j\rangle\Big). \end{aligned}$$

Now, since $e_0 e_0^* = I = e_0^* e_0$, we have that

$$(\partial e_0)e_0^* + e_0(\partial e_0^*) = 0 \tag{9.37}$$

whence

$$\{e_0, e_0^* e_0\} = 0 = \{e_0, e_0^*\}e_0 + \sum_{\ell=1}^{n}\Big(\frac{\partial e_0}{\partial \xi_\ell} e_0^* \frac{\partial e_0}{\partial x_\ell} - \frac{\partial e_0}{\partial x_\ell} e_0^* \frac{\partial e_0}{\partial \xi_\ell}\Big)$$

(using (9.37))

$$= \{e_0, e_0^*\}e_0 - e_0\{e_0^*, e_0\},$$

that is

$$e_0^*\{e_0, e_0^*\}e_0 = \{e_0^*, e_0\}. \tag{9.38}$$

Using (9.38) in the above expression for $\langle b_{\mu-2}w_j, w_j\rangle$ and noting that

$$\frac{\partial e_0^*}{\partial \xi_\ell} e_0 \frac{\partial \lambda_j}{\partial x_\ell} - \frac{\partial e_0^*}{\partial x_\ell} e_0 \frac{\partial \lambda_j}{\partial \xi_\ell} = \frac{\partial e_0^*}{\partial \xi_\ell}\frac{\partial \lambda_j}{\partial x_\ell} e_0 - \frac{\partial e_0^*}{\partial x_\ell}\frac{\partial \lambda_j}{\partial \xi_\ell} e_0 = \{e_0^*, \lambda_j\}e_0$$

($\partial \lambda_j$ being a scalar), proves (9.35). □

We must now study the "transformation properties" (we are interested just in the 2×2 case) of the subprincipal terms depending on the choice of e_0. More precisely, we have the following proposition.

Proposition 9.2.7. *Let $0 < a_\mu = a_\mu^* \in S(m^\mu, g; \mathsf{M}_2)$ satisfy (9.33). Let e_0 and $\tilde{e}_0$ be smooth, unitary 2×2 matrices in $S(1, g; \mathsf{M}_2)$ such that*

$$e_0^* a_\mu e_0 = \tilde{e}_0^* a_\mu \tilde{e}_0 = b_\mu = \begin{bmatrix} \lambda_1 & 0 \\ 0 & \lambda_2 \end{bmatrix}.$$

Denote by $b_{\mu-2}$ and $\tilde{b}_{\mu-2}$ the subprincipal terms given in Corollary 9.2.6, associated respectively with e_0 and $\tilde{e}_0$. Let hence $f \in S(1, g; \mathsf{M}_2)$ be the unitary matrix

$$f = \begin{bmatrix} f_1 & 0 \\ 0 & f_2 \end{bmatrix}, \quad \text{such that } f^* f = f f^* = I \text{ and } e_0 = \tilde{e}_0 f,$$

so that the $f_j \in C^\infty(\mathbb{R}^{2n}; \mathbb{C})$ belong to $S(1, g)$ and $|f_j(X)| = 1$, for all $X \in \mathbb{R}^{2n}$, $j = 1, 2$. Then, with $\{w_1, w_2\}$ the canonical basis of $\mathbb{C}^2$ as before,

$$b_{\mu-2}^{(jj)} = \langle b_{\mu-2} w_j, w_j\rangle = \langle \tilde{b}_{\mu-2} w_j, w_j\rangle + \mathrm{Im}\big(f_j\{\bar{f}_j, \lambda_j\}\big), \; j = 1, 2. \tag{9.39}$$

Proof. In the first place we have, since $f w_j = f_j w_j$ and $|f_j| = 1$, $j = 1, 2$,

$$\begin{aligned}\langle e_0^* a_{\mu-2} e_0 w_j, w_j\rangle &= \langle f^* \tilde{e}_0^* a_{\mu-2} \tilde{e}_0 f w_j, w_j\rangle \\ &= \langle f_j a_{\mu-2} \tilde{e}_0 w_j, f_j \tilde{e}_0 w_j\rangle = \langle a_{\mu-2} \tilde{e}_0 w_j, \tilde{e}_0 w_j\rangle.\end{aligned} \tag{9.40}$$

Next, using $(\partial f)w_j = (\partial f_j)w_j$, we compute

$$\begin{aligned}\langle\{e_0^*,\lambda_j\}e_0w_j,w_j\rangle &= \langle\{f^*\tilde{e}_0^*,\lambda_j\}\tilde{e}_0f_jw_j,w_j\rangle\\ &= \langle f_j\{\tilde{e}_0^*,\lambda_j\}\tilde{e}_0w_j,f_jw_j\rangle\\ &\quad+\langle f_j\sum_{\ell=1}^n\Big(\frac{\partial f^*}{\partial\xi_\ell}\tilde{e}_0^*\frac{\partial\lambda_j}{\partial x_\ell}-\frac{\partial f^*}{\partial x_\ell}\tilde{e}_0^*\frac{\partial\lambda_j}{\partial\xi_\ell}\Big)\tilde{e}_0w_j,w_j\rangle\\ &= \langle\{\tilde{e}_0^*,\lambda_j\}\tilde{e}_0w_j,w_j\rangle\\ &\quad+\sum_{\ell=1}^n\Big(\langle f_j\tilde{e}_0^*\frac{\partial\lambda_j}{\partial x_\ell}\tilde{e}_0w_j,\frac{\partial f_j}{\partial\xi_\ell}w_j\rangle-\langle f_j\tilde{e}_0^*\frac{\partial\lambda_j}{\partial\xi_\ell}\tilde{e}_0w_j,\frac{\partial f_j}{\partial x_\ell}w_j\rangle\Big)\\ &= \langle\{\tilde{e}_0^*,\lambda_j\}\tilde{e}_0w_j,w_j\rangle+\langle f_j\{\bar{f}_j,\lambda_j\}w_j,w_j\rangle\\ &= \langle\{\tilde{e}_0^*,\lambda_j\}\tilde{e}_0w_j,w_j\rangle+f_j\{\bar{f}_j,\lambda_j\},\quad j=1,2,\end{aligned}$$

that is

$$\langle\{e_0^*,\lambda_j\}e_0w_j,w_j\rangle=\langle\{\tilde{e}_0^*,\lambda_j\}\tilde{e}_0w_j,w_j\rangle+f_j\{\bar{f}_j,\lambda_j\},\ j=1,2.$$

Hence, for $j=1,2$,

$$\frac{1}{2}\mathrm{Im}\Big(\langle\{e_0^*,\lambda_j\}e_0w_j,w_j\rangle\Big)=\frac{1}{2}\mathrm{Im}\Big(\langle\{\tilde{e}_0^*,\lambda_j\}\tilde{e}_0w_j,w_j\rangle\Big)+\frac{1}{2}\mathrm{Im}\big(f_j\{\bar{f}_j,\lambda_j\}\big).\tag{9.41}$$

We now consider

$$\begin{aligned}\langle e_0^*\{a_\mu,e_0\}w_j,w_j\rangle &= \bar{f}_j\langle\{a_\mu,\tilde{e}_0f\}w_j,\tilde{e}_0w_j\rangle\\ &= \langle\{a_\mu,\tilde{e}_0\}w_j,\tilde{e}_0w_j\rangle+\bar{f}_j\langle\{a_\mu,f_j\}\tilde{e}_0w_j,\tilde{e}_0w_j\rangle\\ &= \langle\tilde{e}_0^*\{a_\mu,\tilde{e}_0\}w_j,w_j\rangle+\bar{f}_j\langle\{a_\mu,f_j\}\tilde{e}_0w_j,\tilde{e}_0w_j\rangle.\end{aligned}$$

Since

$$(\partial a_\mu)\tilde{e}_0+a_\mu(\partial\tilde{e}_0)=\tilde{e}_0(\partial b_\mu)+(\partial\tilde{e}_0)b_\mu,$$

we get

$$\partial(a_\mu-\lambda_j)\tilde{e}_0w_j=-(a_\mu-\lambda_j)(\partial\tilde{e}_0)w_j,\ j=1,2.\tag{9.42}$$

It hence follows, with ∂ and ∂' generic first-order derivatives,

$$(\partial a_\mu)(\partial' f_j)\tilde{e}_0w_j=(\partial' f_j)(\partial a_\mu)\tilde{e}_0w_j=(\partial' f_j)\Big((\partial\lambda_j)\tilde{e}_0w_j-(a_\mu-\lambda_j)(\partial\tilde{e}_0)w_j\Big),$$

whence

$$\{a_\mu,f_j\}\tilde{e}_0w_j=\{\lambda_j,f_j\}\tilde{e}_0w_j-(a_\mu-\lambda_j)\{\tilde{e}_0,f_j\}w_j,\ j=1,2,$$

so that, by using

$$\langle (a_\mu - \lambda_j)\{\tilde{e}_0, f_j\}w_j, \tilde{e}_0 w_j\rangle = \langle \{\tilde{e}_0, f_j\}w_j, (a_\mu - \lambda_j)\tilde{e}_0 w_j\rangle = 0,$$

we obtain

$$\bar{f}_j\langle \{a_\mu, f_j\}\tilde{e}_0 w_j, \tilde{e}_0 w_j\rangle = \bar{f}_j\langle \{\lambda_j, f_j\}\tilde{e}_0 w_j, \tilde{e}_0 w_j\rangle = \bar{f}_j\{\lambda_j, f_j\},\ j = 1,2.$$

Hence

$$\langle \tilde{e}_0^*\{a_\mu, \tilde{e}_0\}w_j, w_j\rangle + \bar{f}_j\langle \{a_\mu, f_j\}\tilde{e}_0 w_j, \tilde{e}_0 w_j\rangle = \langle \tilde{e}_0^*\{a_\mu, \tilde{e}_0\}w_j, w_j\rangle + \bar{f}_j\{\lambda_j, f_j\},$$

for $j = 1,2$, and thus

$$\frac{1}{2}\mathrm{Im}\Big(\langle e_0^*\{a_\mu, e_0\}w_j, w_j\rangle\Big) = \frac{1}{2}\mathrm{Im}\Big(\langle \tilde{e}_0^*\{a_\mu, \tilde{e}_0\}w_j, w_j\rangle\Big) + \frac{1}{2}\mathrm{Im}\big(\bar{f}_j\{\lambda_j, f_j\}\big), \tag{9.43}$$

for $j = 1,2$. We now observe that

$$0 = \{1, \lambda_j\} = \{|f_j|^2, \lambda_j\} = f_j\{\bar{f}_j, \lambda_j\} + \bar{f}_j\{f_j, \lambda_j\},$$

so that

$$\bar{f}_j\{\lambda_j, f_j\} = f_j\{\bar{f}_j, \lambda_j\},\ \ j = 1,2. \tag{9.44}$$

Plugging (9.40), (9.41), (9.43) and (9.44) in (9.35) gives

$$\begin{aligned}\langle b_{\mu-2}w_j, w_j\rangle &= \langle \tilde{b}_{\mu-2}w_j, w_j\rangle + \frac{1}{2}\mathrm{Im}\big(f_j\{\bar{f}_j, \lambda_j\}\big) + \frac{1}{2}\mathrm{Im}\big(\bar{f}_j\{\lambda_j, f_j\}\big)\\ &= \langle \tilde{b}_{\mu-2}w_j, w_j\rangle + \mathrm{Im}\big(f_j\{\bar{f}_j, \lambda_j\}\big),\ j = 1,2,\end{aligned}$$

which proves the proposition. □

Remark 9.2.8. It is useful to remark that Proposition 9.2.5, Corollary 9.2.6 and Proposition 9.2.7 all hold true in the case of **classical** symbols (with the usual Weyl-quantization). △

Remark 9.2.9. Notice that when $a = a^*$ is an $N \times N$ **globally positive elliptic differential system** of order μ (hence μ is even), by virtue of the homogeneity

$$\lambda_{j,\mu}(X) = |X|^\mu \lambda_{j,\mu}(\frac{X}{|X|}),\ \ \forall X \in \mathbb{R}^{2n} \setminus \{0\},\ \ j = 1,\dots,N,$$

we have that condition (9.33) on the eigenvalues of a_μ becomes

$$|\lambda_{j,\mu}(\omega) - \lambda_{j',\mu}(\omega)| \gtrsim 1,\ \ \forall \omega \in \mathbb{S}^{2n-1},\ j \neq j'. \tag{9.45}$$

△

9.3 Some Estimates for Semiclassical Operators

Recall from Definition 3.2.25 and Proposition 3.2.26 that for $s \in \mathbb{Z}_+$,

$$B^s(\mathbb{R}^n) = \{u \in L^2;\ x^\alpha \partial_x^\beta u \in L^2,\ |\alpha| + |\beta| \le s\},$$

and $B^{-s} = (B^s)^*$. It is also useful to recall that

$$B^2(\mathbb{R}^n) = D(p_0^{\mathrm{w}}(x,D)),$$

where p_0^{w} is the usual harmonic oscillator $(|x|^2 + |D|^2)/2 = (|x|^2 - \Delta)/2$ in $\mathbb{R}^n$. Recall also that

$$B^s(\mathbb{R}^n;\mathbb{C}^N) = B^s(\mathbb{R}^n) \otimes \mathbb{C}^N,$$

and that, from (5.7), on B^{2s} we have the equivalent norms

$$\|u\|^2_{B^{2s},1} := \sum_{|\alpha|+|\beta|\le 2s} \|x^\alpha \partial_x^\beta u\|_0^2, \quad \text{and} \quad \|u\|^2_{B^{2s}} := \|u\|_0^2 + \|p_0^{\mathrm{w}}(x,D)^s u\|_0^2.$$

We next wish to introduce the semiclassical parameter $h \in (0,1]$ into the game.

Consider the L^2-isometry, also automorphism of $\mathscr{S}'$ and $\mathscr{S}$,

$$U_h \colon u \longmapsto (U_h u)(x) = h^{-n/4} u(x/\sqrt{h}).$$

From (9.6) we have

$$U_h^{-1} p_0^{\mathrm{w}}(x,hD) U_h = p_0^{\mathrm{w}}(\sqrt{h}x, \sqrt{h}D) = h\, p_0^{\mathrm{w}}(x,D). \tag{9.46}$$

Since

$$\partial_x^\beta \left((U_h^{-1}u)(x) \right) = h^{n/4} h^{|\beta|/2} (\partial_x^\beta u)(\sqrt{h}x) = h^{|\beta|/2} (U_h^{-1}(\partial_x^\beta u))(x),$$

we get

$$\|x^\alpha \partial_x^\beta (U_h^{-1}u)\|_0^2 = h^{|\beta|-|\alpha|} \|U_h^{-1}(x^\alpha \partial_x^\beta u)\|_0^2 = h^{|\beta|-|\alpha|} \|x^\alpha \partial_x^\beta u\|_0^2, \tag{9.47}$$

since U_h^{-1} is an L^2-isometry as well.

Consider now the h-dependent norm

$$\|u\|^2_{B^{2s},h} := \sum_{|\alpha|+|\beta|\le 2s} h^{2|\beta|} \|x^\alpha \partial_x^\beta u\|_0^2,$$

which is an equivalent norm of B^{2s}, for one readily has

$$h^{4s} \|u\|^2_{B^{2s},1} \le \|u\|^2_{B^{2s},h} \le \|u\|^2_{B^{2s},1}, \quad \forall u \in B^{2s}.$$

Using (9.47) we get

$$\|u\|^2_{B^{2s},h} = \sum_{|\alpha|+|\beta|\le 2s} h^{2|\beta|}\|U_h^{-1}(x^\alpha \partial_x^\beta u)\|_0^2 = \sum_{|\alpha|+|\beta|\le 2s} h^{|\alpha|+|\beta|}\|x^\alpha \partial_x^\beta (U_h^{-1}u)\|_0^2,$$

from which it follows, with constants independent of h, that

$$h^{2s}\Big(\|U_h^{-1}u\|_0^2 + \|p_0^{\mathrm{w}}(x,D)^s U_h^{-1}u\|_0^2\Big) \lesssim \|u\|^2_{B^{2s},h} \lesssim \|U_h^{-1}u\|_0^2 + \|p_0^{\mathrm{w}}(x,D)^s U_h^{-1}u\|_0^2,$$

and since

$$\|U_h^{-1}u\|_0^2 + \|p_0^{\mathrm{w}}(x,D)^s U_h^{-1}u\|_0^2 = \|U_h^{-1}u\|_0^2 + \|U_h p_0^{\mathrm{w}}(x,D)^s U_h^{-1}u\|_0^2 =$$

(by (9.46))

$$= \|u\|_0^2 + h^{-2s}\|p_0^{\mathrm{w}}(x,hD)^s u\|_0^2 = h^{-2s}\Big(h^{2s}\|u\|_0^2 + \|p_0^{\mathrm{w}}(x,hD)^s u\|_0^2\Big),$$

we get, with constants independent of h,

$$\begin{cases} h^{2s}\Big(\|u\|_0^2 + \|p_0^{\mathrm{w}}(x,hD)^s u\|_0^2\Big) \lesssim \|u\|^2_{B^{2s},h} \lesssim h^{-2s}\Big(\|u\|_0^2 + \|p_0^{\mathrm{w}}(x,hD)^s u\|_0^2\Big), \\ h^{4s}\|u\|^2_{B^{2s},1} \le \|u\|^2_{B^{2s},h} \le \|u\|^2_{B^{2s},1}, \end{cases} \tag{9.48}$$

for all $u \in B^{2s}$. We may hence prove the following useful fact.

Proposition 9.3.1. *Let $r \in S_0^0(m^{-2N_0}; \mathsf{M}_N)$, for some $N_0 \in \mathbb{N}$. Then, given any integer k_0 with $0 \le k_0 \le N_0$,*

$$r^{\mathrm{w}}(x,hD)\colon L^2(\mathbb{R}^n;\mathbb{C}^N) \longrightarrow B^{2k_0}(\mathbb{R}^n;\mathbb{C}^N)$$

is continuous, satisfying the following estimates: there exists $C > 0$ (dependent on k_0 but **independent of** *h) such that*

$$\|r^{\mathrm{w}}(x,hD)u\|_{B^{2k_0},1} \le Ch^{-3k_0}\|u\|_0, \ \forall u \in L^2(\mathbb{R}^n;\mathbb{C}^N),$$

and

$$\|r^{\mathrm{w}}(x,hD)u\|_{B^{2k_0},h} \le Ch^{-k_0}\|u\|_0, \ \forall u \in L^2(\mathbb{R}^n;\mathbb{C}^N),$$

for all $h \in (0,1]$. In particular, when $0 < k_0 \le N_0$,

$$r^{\mathrm{w}}(x,hD)\colon L^2(\mathbb{R}^n;\mathbb{C}^N) \longrightarrow B^{2k_0}(\mathbb{R}^n;\mathbb{C}^N) \hookrightarrow\hookrightarrow L^2(\mathbb{R}^n;\mathbb{C}^N)$$

is compact, for all $h \in (0,1]$.

Proof. We want to estimate $\|r^{\mathrm{w}}(x,hD)u\|_{B^{2k_0},h}$ (of course, we may suppose $k_0>0$, otherwise there is nothing to prove). Now, $r^{\mathrm{w}}(x,hD)$ is also a *bounded* operator $L^2 \longrightarrow L^2$ with norm bounded independently of $h\in(0,1]$. Take a sequence $\{u_j\}_{j\in\mathbb{Z}_+}\subset\mathscr{S}(\mathbb{R}^n;\mathbb{C}^N)$ such that $u_j\to u$ in L^2 as $j\to+\infty$. Since $(\underbrace{p_0\sharp_h\cdots\sharp_h p_0}_{k_0})\sharp_h r\in S^0_0(1;\mathsf{M}_N)$, we have, by the L^2-boundedness, with constants independent of h,

$$\|p_0^{\mathrm{w}}(x,hD)^{k_0}r^{\mathrm{w}}(x,hD)(u_j-u_{j'})\|_0\lesssim\|u_j-u_{j'}\|_0\longrightarrow 0,\ \text{as } j,j'\to+\infty.$$

Hence, by (9.48), with constants independent of h,

$$\begin{aligned}
h^{4k_0}\|r^{\mathrm{w}}(x,hD)(u_j-u_{j'})\|^2_{B^{2k_0},1} &\le \|r^{\mathrm{w}}(x,hD)(u_j-u_{j'})\|^2_{B^{2k_0},h}\\
&\lesssim h^{-2k_0}\big(\|r^{\mathrm{w}}(x,hD)(u_j-u_{j'})\|_0^2+\|p_0^{\mathrm{w}}(x,hD)^{k_0}r^{\mathrm{w}}(x,hD)(u_j-u_{j'})\|_0^2\big)\\
&\lesssim h^{-2k_0}\|u_j-u_{j'}\|_0^2\longrightarrow 0,\ \text{as } j,j'\to+\infty.
\end{aligned}$$

Hence $r^{\mathrm{w}}(x,hD)u_j\xrightarrow{B^{2k_0}}v\in B^{2k_0}$ as $j\to+\infty$. But on the other hand we also have $r^{\mathrm{w}}(x,hD)u_j\xrightarrow{\mathscr{S}'}r^{\mathrm{w}}(x,hD)u$ as $j\to+\infty$. It therefore follows that

$$r^{\mathrm{w}}(x,hD)u\in B^{2k_0},\quad \forall u\in L^2,$$

and

$$\|r^{\mathrm{w}}(x,hD)u\|_{B^{2k_0},1}\le Ch^{-3k_0}\|u\|_0,\ \ \forall u\in L^2,$$

$$\|r^{\mathrm{w}}(x,hD)u\|_{B^{2k_0},h}\le Ch^{-k_0}\|u\|_0,\ \ \forall u\in L^2,$$

which concludes the proof. □

Corollary 9.3.2. *Let $r\in S^0_0(m^{-2N_0};\mathsf{M}_N)$, for some $N_0\in\mathbb{N}$. Then, given any integers k_0,k_1 with $0\le k_0\le k_1\le N_0$,*

$$r^{\mathrm{w}}(x,hD)\colon B^{2k_0}(\mathbb{R}^n;\mathbb{C}^N)\longrightarrow B^{2k_1}(\mathbb{R}^n;\mathbb{C}^N)$$

is continuous, satisfying the following estimates: there exists $C>0$ (independent of h) such that

$$\|r^{\mathrm{w}}(x,hD)u\|_{B^{2k_1},1}\le Ch^{-3k_1}\|u\|_{B^{2k_0},1},\ \ \forall u\in B^{2k_0}(\mathbb{R}^n;\mathbb{C}^N),$$

and

$$\|r^{\mathrm{w}}(x,hD)u\|_{B^{2k_1},h}\le Ch^{-k_1}\|u\|_{B^{2k_0},h},\ \ \forall u\in B^{2k_0}(\mathbb{R}^n;\mathbb{C}^N),$$

for all $h\in(0,1]$. In particular, when $0\le k_0<k_1\le N_0$,

$$r^{\mathrm{w}}(x,hD)\colon B^{2k_0}(\mathbb{R}^n;\mathbb{C}^N)\longrightarrow B^{2k_1}(\mathbb{R}^n;\mathbb{C}^N)\hookrightarrow\!\!\!\hookrightarrow B^{2k_0}(\mathbb{R}^n;\mathbb{C}^N)$$

is compact, for all $h\in(0,1]$.

Proof. Given any $u \in \mathscr{S}(\mathbb{R}^n;\mathbb{C}^N)$, we have that $r^{\mathrm{w}}(x,hD)u \in \mathscr{S}(\mathbb{R}^n;\mathbb{C}^N)$. Hence it suffices to prove the inequalities for Schwartz functions u, with constants independent of u and h. Since $B^{2k_0}(\mathbb{R}^n;\mathbb{C}^N) \subset L^2(\mathbb{R}^n;\mathbb{C}^N)$ with

$$\|u\|_0^2 \leq \|u\|_{B^{2k_0},h}^2, \quad \text{and} \quad \|u\|_0^2 \leq \|u\|_{B^{2k_0},1}^2,$$

from Proposition 9.3.1 we have

$$\|r^{\mathrm{w}}(x,hD)u\|_{B^{2k_1},h}^2 \leq Ch^{-2k_1}\|u\|_0^2 \leq Ch^{-2k_1}\|u\|_{B^{2k_0},h}^2,$$

and

$$\|r^{\mathrm{w}}(x,hD)u\|_{B^{2k_1},1}^2 \leq Ch^{-6k_1}\|u\|_0^2 \leq h^{-6k_1}\|u\|_{B^{2k_0},1}^2,$$

for a constant $C>0$ independent of u and of h, which concludes the proof. □

9.4 Some Spectral Properties of Semiclassical GPDOs

We establish in this section a few useful results about spectral properties of h-Weyl quantizations of $N \times N$ *semiclassical* GPD systems which are positive elliptic. We start by giving the definition of semiclassical global polynomial differential system.

Definition 9.4.1 (Semiclassical GPD). We shall say that a classical semiclassical symbol $a \in S^0_{0,\mathrm{cl}}(m^\mu,g;\mathsf{M}_N)$ is a **semiclassical GPD system of order** μ if $\mu \in \mathbb{N}$ and

$$a = \sum_{j=0}^{[\mu/2]} h^j a_{\mu-2j}, \quad a_{\mu-2j} \in S(m^{\mu-2j},g;\mathsf{M}_N)$$

(with $[\mu/2]$ denoting, as usual, the integer part of $\mu/2$), where the entries of the $a_{\mu-2j}$ are **homogeneous polynomials** in $X \in \mathbb{R}^{2n}$ of degree $\mu-2j$.

We say that a semiclassical GPD system $a \in S^0_{0,\mathrm{cl}}(m^\mu,g;\mathsf{M}_N)$ of order μ is **elliptic** (resp. **positive elliptic**, when $a=a^*$) if the principal part a_μ is a homogeneous globally elliptic (resp. globally positive elliptic) symbol.

The h-Weyl quantization of a semiclassical GPD will be called an h-GPDO.

Suppose

$$a = \sum_{j=0}^{[\mu/2]} h^j a_{\mu-2j}$$

is an $N \times N$ semiclassical GPD symbol of order $\mu \in \mathbb{N}$ in $\mathbb{R}^n$. Since (9.5) holds also on $\mathscr{S}'(\mathbb{R}^n;\mathbb{C}^N)$ and $\mathscr{S}(\mathbb{R}^n;\mathbb{C}^N)$, we have the following lemma.

Lemma 9.4.2. *Let $E>0$ and let $a\in S^0_{0,\mathrm{cl}}(m^\mu,g;\mathsf{M}_N)$ be an $N\times N$* **semiclassical** GPD *system of order $\mu\in\mathbb{N}$. Let U_E be the isometry introduced in Remark 9.1.2. Then*

$$U_E^{-1}a^{\mathrm{w}}(x,hD)U_E=E^{\mu/2}a^{\mathrm{w}}(x,\tilde{h}D), \tag{9.49}$$

where $\tilde{h}=h/E$ and

$$a^{\mathrm{w}}(x,\tilde{h}D)=\sum_{j=0}^{[\mu/2]}\tilde{h}^j a^{\mathrm{w}}_{\mu-2j}(x,\tilde{h}D).$$

In particular, when $E=h$, we have

$$U_h^{-1}a^{\mathrm{w}}(x,hD)U_h=h^{\mu/2}a^{\mathrm{w}}(x,D), \tag{9.50}$$

where

$$a^{\mathrm{w}}(x,D)=\sum_{j=0}^{[\mu/2]}a^{\mathrm{w}}_{\mu-2j}(x,D).$$

Proof. The proof follows immediately from Remark 9.1.2. In fact, it suffices to observe that for $\tilde{h}=h/E$ we have

$$\begin{aligned}
a^{\mathrm{w}}(\sqrt{E}x,\sqrt{E}\tilde{h}D)&=U_E^{-1}a^{\mathrm{w}}(x,hD)U_E\\
&=\sum_{j=0}^{[\mu/2]}h^jU_E^{-1}a^{\mathrm{w}}_{\mu-2j}(x,hD)U_E=\sum_{j=0}^{[\mu/2]}\tilde{h}^jE^ja^{\mathrm{w}}_{\mu-2j}(\sqrt{E}x,\sqrt{E}\tilde{h}D)\\
&=\sum_{j=0}^{[\mu/2]}\tilde{h}^jE^jE^{\mu/2-j}a^{\mathrm{w}}_{\mu-2j}(x,\tilde{h}D)=E^{\mu/2}a^{\mathrm{w}}(x,\tilde{h}D),
\end{aligned} \tag{9.51}$$

in view of the fact that $h=E\tilde{h}$ and that the $a_{\mu-2j}$ have entries which are all homogeneous of degree $\mu-2j$. □

We therefore get the following scaling properties of eigenvalues.

Lemma 9.4.3. *Let $E>0$ and let $a\in S^0_{0,\mathrm{cl}}(m^\mu,g;\mathsf{M}_N)$ be an $N\times N$* **semiclassical** GPD *system of order $\mu\in\mathbb{N}$. Let U_E be the isometry introduced in Remark 9.1.2. Then*

$$a^{\mathrm{w}}(x,hD)u(h)=\lambda(h)u(h),\quad u(h)\in L^2(\mathbb{R}^n;\mathbb{C}^N),\ u(h)\neq 0,$$

that is $u(h)$ is an eigenfunction of $a^{\mathrm{w}}(x,hD)$ belonging to the eigenvalue $\lambda(h)$, iff, with $\tilde{u}(\tilde{h}):=U_E^{-1}u(h)\in L^2(\mathbb{R}^n;\mathbb{C}^N)$,

$$a^{\mathrm{w}}(x,\tilde{h}D)\tilde{u}(\tilde{h})=\frac{\lambda(E\tilde{h})}{E^{\mu/2}}\tilde{u}(\tilde{h}),\ \text{where}\ \tilde{h}=\frac{h}{E},$$

that is, $\tilde{u}(\tilde{h})$ is an eigenfunction of $a^{\mathrm{w}}(x,\tilde{h}D)$ belonging to the eigenvalue $\dfrac{\lambda(E\tilde{h})}{E^{\mu/2}}$. *Hence,*

$$\lambda(h)\in\operatorname{Spec}(a^{\mathrm{w}}(x,hD))\Longleftrightarrow\frac{\lambda(E\tilde{h})}{E^{\mu/2}}\in\operatorname{Spec}(a^{\mathrm{w}}(x,\tilde{h}D)),\ \tilde{h}=\frac{h}{E}.$$

In particular, when $E=h$, one obtains that

$$\lambda(h)\in\operatorname{Spec}(a^{\mathrm{w}}(x,hD))\Longleftrightarrow\frac{\lambda(h)}{h^{\mu/2}}\in\operatorname{Spec}(a^{\mathrm{w}}(x,D)).$$

Proof. The proof follows immediately from Lemma 9.4.2. □

As a (by now elementary) consequence of the results of Section 3.2, namely Remark 9.1.2, Lemma 9.4.2 and Theorem 3.3.13 we have the following fact concerning the spectrum of semiclassical GPD systems.

Proposition 9.4.4. *Let $A=A^*\in S^0_{0,\mathrm{cl}}(m^\mu,g;\mathsf{M}_N)$ be an $N\times N$ positive elliptic semiclassical* GPD *system of order $\mu\in 2\mathbb{N}$. Consider the unbounded operator $A(h)$ defined by*

$$A(h)\colon B^\mu(\mathbb{R}^n;\mathbb{C}^N)\subset L^2(\mathbb{R}^n;\mathbb{C}^N)\longrightarrow L^2(\mathbb{R}^n;\mathbb{C}^N),$$
$$A(h)u=A^{\mathrm{w}}(x,hD)u,\ \forall u\in B^\mu(\mathbb{R}^n;\mathbb{C}^N).$$

Then $A(h)$ is semi-bounded from below for all $h\in(0,1]$. Hence, $\operatorname{Spec}(A(h))$ *is made of a sequence of eigenvalues $\{\lambda_j(h)\}_{j\geq 1}\subset\mathbb{R}$ with finite multiplicities, such that*

$$-\infty<\lambda_1(h)\leq\lambda_2(h)\leq\ldots\leq\lambda_j(h)\leq\ldots\longrightarrow+\infty,$$

with repetitions according to the multiplicity. As before, the eigenfunctions of $A(h)$ all belong to the Schwartz space and form, possibly after an orthonormalization procedure, a basis of $L^2(\mathbb{R}^n;\mathbb{C}^N)$.

Proof. In fact, by Lemma 9.4.2 we have

$$U_h^{-1}A(h)U_h=h^{\mu/2}A(1),$$

so that

$$\begin{aligned}(A(h)u,u)&=(U_h^{-1}A(h)U_hU_h^{-1}u,U_h^{-1}u)=h^{\mu/2}(A(1)U_h^{-1}u,U_h^{-1}u)\\&\geq -h^{\mu/2}C\|U_h^{-1}u\|_0^2=-h^{\mu/2}C\|u\|_0^2,\ \forall u\in\mathscr{S}(\mathbb{R}^n;\mathbb{C}^N).\end{aligned}$$

This concludes the proof of the lemma. □

Hence, when A is an $N\times N$ *positive elliptic semiclassical* GPD system of order μ we immediately obtain from the Minimax Principle, Lemma 9.4.3 and Proposition 9.4.4 the following corollary.

Corollary 9.4.5. *Let $A = A^* \in S^0_{0,\mathrm{cl}}(m^\mu, g; \mathsf{M}_N)$ be an $N \times N$ positive elliptic semiclassical* GPD *system of order $\mu \in 2\mathbb{N}$. Let $\phi_j \in \mathscr{S}(\mathbb{R}^n; \mathbb{C}^N)$, $j \in \mathbb{N}$, be an eigenfunction of $A^{\mathrm{w}}(x,D)$ (i.e. with $h = 1$) belonging to the eigenvalue λ_j. Then*

$$\varphi_j(h;x) := (U_h\phi_j)(x) = h^{-n/4}\phi_j(\frac{x}{\sqrt{h}}) \tag{9.52}$$

belongs to the eigenvalue $\lambda_j(h) := h^{\mu/2}\lambda_j$ of $A^{\mathrm{w}}(x,hD)$. In particular

$$\varphi_j(\frac{h}{E};x) = E^{n/4}\varphi_j(h;\sqrt{E}x), \quad j \geq 1, \tag{9.53}$$

belongs to the eigenvalue $\lambda_j(\frac{h}{E}) = \left(\frac{h}{E}\right)^{\mu/2}\lambda_j$ of $A^{\mathrm{w}}(x, \frac{h}{E}D)$. Hence, using the Minimax, for every $j \in \mathbb{N}$

$$\lambda_j \in \mathrm{Spec}(A^{\mathrm{w}}(x,D)) \Longleftrightarrow h^{\mu/2}\lambda_j = \lambda_j(h) \in \mathrm{Spec}(A^{\mathrm{w}}(x,hD)).$$

We finally consider the following situation, that will be very useful later on. Let $A = A^* \in S^0_{0,\mathrm{cl}}(m^\mu, g; \mathsf{M}_N)$ be an $N \times N$ semiclassical positive elliptic GPD system. Let $R = R^* \in S^0_0(1, g; \mathsf{M}_N)$ with $\mathrm{supp}(R) \subset K$, where K is a compact set *independent of* h. Then

$$R^{\mathrm{w}}(x,hD)\colon \mathscr{S}'(\mathbb{R}^n;\mathbb{C}^N) \longrightarrow \mathscr{S}(\mathbb{R}^n;\mathbb{C}^N)$$

is **continuous** and, as an operator in L^2, $\|R^{\mathrm{w}}(x,hD)\|_{L^2\to L^2} = O(1)$ for all $h \in (0,1]$ (see Dimassi-Sjöstrand [7]). We have the following important result.

Proposition 9.4.6. *Let $A = A^* \in S^0_{0,\mathrm{cl}}(m^\mu, g; \mathsf{M}_N)$ be an $N \times N$ semiclassical positive elliptic* GPD *system, and let $R = R^* \in S^0_0(1, g; \mathsf{M}_N)$ with* $\mathrm{supp}(R) \subset K$, *where K is a compact set* **independent of** *h. Let us consider $A_0^{\mathrm{w}}(x,hD) = A^{\mathrm{w}}(x,hD) + R^{\mathrm{w}}(x,hD)$. Then for all $h \in (0,1]$ the unbounded operator $A_0(h)$ defined by*

$$A_0(h)\colon B^\mu(\mathbb{R}^n;\mathbb{C}^N) \subset L^2(\mathbb{R}^n;\mathbb{C}^N) \longrightarrow L^2(\mathbb{R}^n;\mathbb{C}^N),$$

$$A_0(h)u = A_0^{\mathrm{w}}(x,hD)u, \ \forall u \in B^\mu(\mathbb{R}^n;\mathbb{C}^N),$$

is self-adjoint with a discrete spectrum bounded from below, made of a sequence of eigenvalues $\{\lambda_j(h)\}_{j\geq 1} \subset \mathbb{R}$ with finite multiplicities, such that

$$-\infty < \lambda_1(h) \leq \lambda_2(h) \leq \ldots \leq \lambda_j(h) \leq \ldots \longrightarrow +\infty,$$

with repetitions according to the multiplicity.

Proof. That $A_0(h) = A_0(h)^*$ with the same domain B^μ of $A(h)$ is trivial, for $R^{\mathrm{w}}(x,hD)$ is bounded in L^2 and symmetric. By Theorem 10.1.1 below the resolvent set of $A_0(h)$ is non-empty, and since B^μ is compactly embedded into L^2, $A_0(h)$

also has a discrete spectrum, that must be bounded from below, for the spectrum of $A(h)$ is bounded from below and $R^{\mathrm{w}}(x,hD)$ is bounded in L^2. □

Remark 9.4.7. One may prove the discreteness of $\mathrm{Spec}(A_0(h))$ also as follows. Since the Schwartz-kernel

$$\begin{aligned} \mathsf{K}_R(x,y;h) &= (2\pi)^{-n}\int e^{i\langle x-y,\xi\rangle}R(\frac{x+y}{2},h\xi)d\xi \\ &= (2\pi h)^{-n}\int e^{ih^{-1}\langle x-y,\xi\rangle}R(\frac{x+y}{2},\xi)d\xi \\ &= h^{-n}(\mathscr{F}_{\xi\to t}R)(\frac{x+y}{2},t)\big|_{t=h^{-1}(x-y)} \end{aligned}$$

of $R^{\mathrm{w}}(x,hD)$ belongs to $\mathscr{S}(\mathbb{R}^n\times\mathbb{R}^n;\mathsf{M}_N)$, the operator $R^{\mathrm{w}}(x,hD)$ is actually Hilbert-Schmidt. Hence $A_0(h)$ and $A(h)$ have the same essential spectrum. Since the essential spectrum of $A(h)$ is empty, the same is true for that of $A_0(h)$. △

Remark 9.4.8. More generally, one may prove that if a classical semiclassical symbol $A=A^*\in S^0_{0,\mathrm{cl}}(m^\mu,g;\mathsf{M}_N)$, with $\mu\geq 1$, has **globally positive elliptic** principal symbol, then there exists $h_0\in(0,1]$ such that the unbounded operator $A(h)$ defined by

$$\begin{aligned} &A(h)\colon B^\mu(\mathbb{R}^n;\mathbb{C}^N)\subset L^2(\mathbb{R}^n;\mathbb{C}^N)\longrightarrow L^2(\mathbb{R}^n;\mathbb{C}^N), \\ &A(h)u=A^{\mathrm{w}}(x,hD)u,\ \forall u\in B^\mu(\mathbb{R}^n;\mathbb{C}^N), \end{aligned}$$

is **self-adjoint**, **semi-bounded** from below and with a **discrete spectrum**, for all $h\in(0,h_0]$. In addition, from the elliptic regularity we have that its eigenfunctions are in $\mathscr{S}(\mathbb{R}^n;\mathbb{C}^N)$, and, possibly after an orthonormalization procedure, they form a basis of $L^2(\mathbb{R}^n;\mathbb{C}^N)$.

In fact, one uses (9.5) of Remark 9.1.2 (with $E=h$), namely

$$U_h^{-1}A^{\mathrm{w}}(x,hD)U_h=A^{\mathrm{w}}(\sqrt{h}x,\sqrt{h}D)$$

to reduce matters to Theorem 3.3.13. △

Chapter 10
On Operators Induced by General Finite-Rank Orthogonal Projections

In this chapter we further prepare the ground for the eigenvalue localization of elliptic global systems. Namely, in the following chapters we shall have to control the sandwitch $(I-\Pi)B(I-\Pi)$ of an operator B, semi-bounded from below, by the orthogonal projectors $(I-\Pi)$ relative to **another** operator A, semi-bounded from below. We present things in an abstract setting, for this is a useful machinery. Throughout this chapter H will always stand for a separable (infinite-dimensional) Hilbert space endowed with the scalar product $(\cdot,\cdot)=(\cdot,\cdot)_H$.

10.1 Reductions by a Finite-Rank Orthogonal Projection

The first result we need is the following theorem (see Kato's book [35, Theorem 4.10, p. 291]).

Theorem 10.1.1. *Let $T\colon D(T)\subset H\longrightarrow H$ be self-adjoint (hence it is densely defined), and let $B=B^*\in\mathscr{L}(H,H)$ (where, recall, $\mathscr{L}(H,H)$ denotes the space of bounded linear operators on H). Then $S=T+B\colon D(T)\subset H\longrightarrow H$ is self-adjoint and*

$$\mathrm{dist}\Big(\mathrm{Spec}(T),\mathrm{Spec}(S)\Big)\leq\|B\|,$$

that is,

$$\sup_{\zeta\in\mathrm{Spec}(S)}\mathrm{dist}(\zeta,\mathrm{Spec}(T))\leq\|B\|,\quad \sup_{\zeta\in\mathrm{Spec}(T)}\mathrm{dist}(\zeta,\mathrm{Spec}(S))\leq\|B\|,\tag{10.1}$$

where $\|B\|=\|B\|_{H\to H}$.

Let next $T\colon D(T)\subset H\longrightarrow H$ be a densely defined semi-bounded from below self-adjoint operator with compact resolvent. Hence T has a discrete spectrum made of a sequence $\{\lambda_j\}_{j\geq1}$ of eigenvalues $-\infty<\lambda_1\leq\lambda_2\leq\ldots\leq\lambda_j\leq\ldots\to+\infty$, *repeated according to the multiplicity*. Let $\{u_j\}_{j\in\mathbb{Z}_+}\subset D(T)$ be *any* fixed orthonormal basis of H.

A. Parmeggiani, *Spectral Theory of Non-Commutative Harmonic Oscillators: An Introduction*, Lecture Notes in Mathematics 1992,
DOI 10.1007/978-3-642-11922-4_10, © Springer-Verlag Berlin Heidelberg 2010

Remark 10.1.2. It is important to note that we **do not** assume that the u_j be eigenfunctions of T (otherwise the result we wish to prove, namely Theorem 10.1.3 below, is a trivial consequence of the Spectral Theorem). △

Fix $N_0 \in \mathbb{N}$ and let $\Pi = \Pi_{N_0} : H \longrightarrow H$ be the finite-rank orthogonal projector onto $\mathrm{Span}\{u_1, \dots, u_{N_0}\}$,

$$\Pi = \sum_{j=1}^{N_0} u_j^* \otimes u_j \colon u \longmapsto \sum_{j=1}^{N_0} (u, u_j) u_j.$$

Note that

$$\Pi u \in D(T), \ \forall u \in H,$$

and that

$$(I - \Pi)u \in D(T) \Longleftrightarrow u \in D(T).$$

We have the following consequence of Theorem 10.1.1.

Theorem 10.1.3. *The operator*

$$T_+ := (I - \Pi)T(I - \Pi) = T - \big(\Pi T + T\Pi\big) + \Pi T \Pi \colon D(T) \subset H \longrightarrow H$$

is self-adjoint, semi-bounded from below, with compact resolvent. Hence its spectrum is discrete and real.

Proof. Consider the operators $\Pi T, T\Pi, \Pi T \Pi \colon D(T) \subset H \longrightarrow H$. It is immediately seen that $T_- := \Pi T \Pi \in \mathscr{L}(H,H)$ is symmetric and compact. Now, since the u_j belong to $D(T)$ and T is self-adjoint, we have

$$D(T) \ni u \longmapsto \Pi T u = \sum_{j=1}^{N_0} (Tu, u_j) u_j = \sum_{j=1}^{N_0} (u, Tu_j) u_j,$$

which shows that ΠT extends to a bounded operator $T_1 \in \mathscr{L}(H,H)$ such that $T_1\big|_{D(T)} = \Pi T$. Similarly,

$$D(T) \ni u \longmapsto T\Pi u = \sum_{j=1}^{N_0} (u, u_j) Tu_j,$$

shows that $T\Pi = T_2 \in \mathscr{L}(H,H)$. We now have that $T_1 + T_2$ is symmetric. In fact,

$$(T_1 u, v) = (\Pi T u, v) = (u, T\Pi v) = (u, T_2 v), \ \forall u, v \in D(T),$$

from which, T_1 and T_2 being bounded operators, the claim follows. Hence, the boundedness of T_1, T_2, T_- and Theorem 10.1.1 give that

$$T_+ = T - (T_1 + T_2) + T_- \ \textit{is self-adjoint and semi-bounded from below.}$$

Pick now $\zeta_0 \in \mathbb{C}$ such that

$$\operatorname{dist}(\zeta_0, \operatorname{Spec}(T)) > \|B\|, \tag{10.2}$$

where $B = -T_1 - T_2 + T_-$. Then, again by Theorem 10.1.1, $\zeta_0 \notin \operatorname{Spec}(T_+)$. Let $R(\zeta) = (\zeta - T)^{-1}$ be the resolvent operator of T and $R_+(\zeta)$ the one of T_+. Then, since

$$\|R(\zeta)\| = \frac{1}{\operatorname{dist}(\zeta, \operatorname{Spec}(T))},$$

(10.2) gives that the Neumann series relative to $(I - BR(\zeta_0))^{-1}$ converges in $\mathscr{L}(H,H)$, whence

$$R_+(\zeta_0) = R(\zeta_0)\big(I - BR(\zeta_0)\big)^{-1}.$$

Since $R(\zeta_0)$ is a compact operator, this completes the proof of the theorem. □

Corollary 10.1.4. *The operator*

$$\tilde{T} := T_- + T_+ \colon D(T) \subset H \longrightarrow H \tag{10.3}$$

is self-adjoint, semi-bounded from below, with compact resolvent. Hence its spectrum is discrete and real.

Proof. It is immediate from Theorem 10.1.3 and Theorem 10.1.1. □

Remark 10.1.5. It is important to notice that $T_- u \in D(T) = D(T_+)$ for all $u \in H$, so that it makes always sense to consider $T_+ T_-$, and that

$$T_+ T_- u = 0,\ \forall u \in H,\ \textit{and}\ T_- T_+ u = 0,\ \forall u \in D(T). \tag{10.4}$$

In particular

$$[T_+, T_-] = 0\ \textit{on}\ D(T). \tag{10.5}$$

△

Define now the operators

$$T_{+-} := (I - \Pi) T \Pi \colon H \longrightarrow H,\ T_{+-} \in \mathscr{L}(H,H),$$

$$T_{-+} := \Pi T (I - \Pi) \colon D(T) \subset H \longrightarrow H.$$

It is readily seen that $T_{-+} = \Pi T - T_-$ can be extended to a bounded operator in H, that we keep denoting by T_{-+}, so that we also have

$$T_{-+}^* = T_{+-}.$$

Moreover, on setting

$$H_+ := (I - \Pi) H,\ H_- := \Pi H,$$

we have that the $H_\pm$ are closed subspaces of H with $H_\pm^\perp = H_\mp$, so that

$$H = H_- \oplus H_+$$

with orthogonal direct-sum. Notice that, since $H_- \subset D(T)$, it makes sense to consider $T_+\big|_{H_-}$. Notice also that

$$T_+\big|_{H_-} = 0, \quad \text{and} \quad T_-\big|_{H_+} = 0. \tag{10.6}$$

Take now an ON basis $\{e_j\}_{j\geq 1} \subset D(T)$ of H which diagonalizes T_+, such that

$$\mathrm{Span}\{e_j\}_{1\leq j\leq N_0} = H_-, \quad \overline{\mathrm{Span}\{e_j\}_{j\geq N_0+1}} = H_+.$$

(This is possible, for $[\Pi, T_+] = 0$ on $D(T)$.) Rotate next the basis $\{e_j\}_{1\leq j\leq N_0}$ into a basis that diagonalizes the $N_0 \times N_0$ Hermitian matrix $T_-\big|_{H_-}$. Call $\{\tilde{e}_j\}_{1\leq j\leq N_0}$ the resulting basis, and notice that $T_+\tilde{e}_j = 0$, $j = 1,\ldots,N_0$. Set $\tilde{e}_j = e_j$ for $j \geq N_0+1$. We thus have that $\{\tilde{e}_j\}_{j\geq 1} \subset D(T)$ is an ON basis of H which diagonalizes $\tilde{T}$ (see (10.3)). This proves that

$$\mathrm{Spec}(\tilde{T}) = \mathrm{Spec}(T_-\big|_{H_-}) \bigcup \mathrm{Spec}(T_+\big|_{H_+}),$$

where

$$T_+\big|_{H_+} : D(T)\cap H_+ \subset H_+ \longrightarrow H_+$$

is the *part of* T_+ *in* H_+ (i.e. T_+ is *reduced by* H_+; see Kato [35, pp. 172, 178 and 278]). Notice that

$$D(T)\cap H_+ = (I-\Pi)D(T),$$

for on the one hand

$$(I-\Pi)D(T) \subset D(T) \quad \text{and} \quad (I-\Pi)D(T) \subset (I-\Pi)H = H_+,$$

and on the other

$$u \in D(T)\cap H_+ \Longrightarrow u = \Pi u + (I-\Pi)u = (I-\Pi)u \in (I-\Pi)D(T).$$

Since

$$T = \tilde{T} + T_0, \quad \text{where} \quad T_0 = T_{+-} + T_{-+} = T_0^* \in \mathscr{L}(H,H),$$

once more from Theorem 10.1.1 we have

$$\mathrm{dist}\Big(\mathrm{Spec}(T), \mathrm{Spec}(\tilde{T})\Big) = \mathrm{dist}\Big(\mathrm{Spec}(T), \bigcup_\pm \mathrm{Spec}(T_\pm\big|_{H_\pm})\Big) \leq \|T_0\|. \tag{10.7}$$

Since (T_- being symmetric and compact)

$$\operatorname{Spec}(T_-) = \operatorname{Spec}(T_-|_{H_-}) \cup \{0\} \subset [-\|T_-\|, \|T_-\|],$$

and since T_+ has a discrete spectrum which is positively diverging to $+\infty$ (T_+ is semi-bounded from below, for T is so, and it has a discrete spectrum by Theorem 10.1.3) one obtains the following theorem.

Theorem 10.1.6. *In the hypotheses of Theorem 10.1.3, upon denoting $\delta := \|T_0\|$ and choosing $E > \|T_-\| + 3\delta$, we have that*

$$\operatorname{Spec}(T) \cap (E, +\infty) \subset \bigcup_{\lambda_+ \in \operatorname{Spec}(T_+) \cap (E-\delta, +\infty)} [\lambda_+ - \delta, \lambda_+ + \delta]. \tag{10.8}$$

Remark 10.1.7. Notice that when the u_j are eigenfunctions of T, $T_{+-} = 0$ and $T_{-+} = 0$ then, so that $\delta = \|T_0\| = 0$. △

Notice that one could have obtained (10.7) and the result of Theorem 10.1.6 by writing T as the (unbounded) system

$$\begin{bmatrix} T_-|_{H_-} & T_{-+}|_{H_+} \\ T_{+-}|_{H_-} & T_+|_{H_+} \end{bmatrix} : H_- \oplus (D(T) \cap H_+) \subset H_- \oplus H_+ \longrightarrow H_- \oplus H_+,$$

by using Proposition 10.1.8 below that shows that $T_+|_{H_+}$ is self-adjoint in H_+ (semi-bounded from below) with compact resolvent, and finally by using once more Theorem 10.1.1.

Proposition 10.1.8. *The operator $T_+|_{H_+} : D(T) \cap H_+ \subset H_+ \longrightarrow H_+$ is self-adjoint in H_+ with compact resolvent (and is obviously semi-bounded from below).*

Proof. Put, for short, $T_{(+)} = T_+|_{H_+}$. Recall that

$$(u, v)_{H_+} := ((I - \Pi)u, (I - \Pi)v)_H = (u, v)_H, \ \forall u, v \in H_+.$$

Then

$$D(T^*_{(+)}) = \{u \in H_+;\ \exists v \in H_+,\ (T_{(+)}u', u)_{H_+} = (u', v)_{H_+}, \forall u' \in D(T_{(+)})\}.$$

Since for all $u, v \in D(T_{(+)}) = D(T) \cap H_+$

$$(T_{(+)}u, v)_{H_+} = (T_+u, v)_H = (u, T_+v)_H = (u, T_{(+)}v)_{H_+},$$

we always have

$$D(T_{(+)}) \subset D(T^*_{(+)}).$$

Take $u \in D(T_{(+)}^*)$. Then, for some $v \in H_+$,

$$(T_{(+)}u', u)_{H_+} = (u', v)_{H_+}, \ \forall u' \in D(T_{(+)}),$$

that is,

$$(T_+u', u)_H = (u', v)_H, \ \forall u' \in D(T) \cap H_+.$$

Since $T_+\big|_{H_-} = 0$ and H_- is orthogonal to H_+, it also follows that

$$(T_+u', u)_H = (u', v)_H, \ \forall u' \in D(T),$$

for, given any $u' \in D(T)$,

$$\begin{aligned}(T_+u', u)_H &= (T_+[\Pi u' + (I-\Pi)u'], u)_H = (T_+(I-\Pi)u', u)_H \\ &= ((I-\Pi)u', v)_H = (\Pi u' + (I-\Pi)u', v)_H = (u', v)_H.\end{aligned}$$

Hence $u \in D(T^*) = D(T)$, and since $u \in H_+$, we thus obtain $u \in D(T) \cap H_+ = D(T_{(+)})$. It follows that

$$D(T_{(+)}^*) \subset D(T_{(+)})$$

and this proves the self-adjointness claim.

To prove the compactness of the resolvent we observe, in the first place, that if ζ belongs to the resolvent set of T_+ (which is non-empty by the semi-boundedness assumption) we have that the operator

$$H_+ \ni u \longmapsto (I-\Pi)(\zeta - T_+)^{-1}(I-\Pi)u \in D(T) \cap H_+ \subset H_+$$

is continuous and compact, for $(\zeta - T_+)^{-1}$ is a compact operator and $(I-\Pi)$ is bounded. On the other hand, for all $u \in H_+$,

$$\begin{aligned}(\zeta - T_{(+)})(I-\Pi)(\zeta - T_+)^{-1}(I-\Pi)u &= (\zeta - T_+)(I-\Pi)(\zeta - T_+)^{-1}(I-\Pi)u \\ &= (I-\Pi)(\zeta - T_+)(\zeta - T_+)^{-1}(I-\Pi)u \\ &= (I-\Pi)u = u,\end{aligned}$$

and for all $u \in D(T) \cap H_+$,

$$\begin{aligned}(I-\Pi)(\zeta - T_+)^{-1}(I-\Pi)(\zeta - T_{(+)})u &= (I-\Pi)(\zeta - T_+)^{-1}(I-\Pi)(\zeta - T_+)u \\ &= (I-\Pi)(\zeta - T_+)^{-1}(\zeta - T_+)(I-\Pi)u \\ &= (I-\Pi)u = u.\end{aligned}$$

This proves that $(\zeta - T_{(+)})^{-1} = (I-\Pi)(\zeta - T_+)^{-1}(I-\Pi)$ for all ζ in the resolvent set of T_+, and concludes the proof. □

10.2 Semiclassical Reduction by a Finite-Rank Orthogonal Projection

We now make things more precise, and introduce also the dependence on the semiclassical parameter $h \in (0,1]$. Let $D \subset H$ be **independent of** h, dense and compactly embedded in H, and let $A = A(h)\colon D(A) = D \subset H \longrightarrow H$, be a self-adjoint operator with $A \geq -CI$ on D. Let therefore $\{\lambda_j(h)\}_{j\geq 1}$ be the sequence $-\infty < \lambda_1(h) \leq \lambda_2(h) \leq \ldots \leq \lambda_j(h) \leq \ldots \to +\infty$ of its eigenvalues, repeated according to their multiplicities. Let $\{u_j(h)\}_{j\geq 1} \subset D$ be a corresponding ON basis of eigenfunctions. Let $E_0 \geq 10$ be fixed (independent of h), and let us consider the orthogonal projector

$$\Pi = \Pi(h) = \sum_{\lambda_j(h)\leq E_0} u_j(h)^* \otimes u_j(h) = \sum_{j=1}^{N_0} u_j^* \otimes u_j\colon H \longrightarrow \mathrm{Span}\{u_1,\ldots,u_{N_0}\} \subset D,$$

where $N_0 = N_0(h)$ is the number of eigenvalues $\lambda_j(h)$ of A, *repeated* according to multiplicity, which satisfy $\lambda_j(h) \leq E_0$ (in the sequel we will often drop the explicit dependence on h). Hence

$$[\Pi, A] = 0 \text{ on } D,$$

and, by the Spectral Theorem,

$$\mathrm{Spec}((I-\Pi)A(I-\Pi)) \cap (E_0,+\infty) = \mathrm{Spec}(A) \cap (E_0,+\infty),$$

or, equivalently,

$$\mathrm{Spec}((I-\Pi)A(I-\Pi)) \setminus \{0\} = \mathrm{Spec}(A) \cap (E_0,+\infty).$$

Let $R = R(h) = R^*\colon H \longrightarrow H$ be in $\mathscr{L}(H,H)$ with

$$\|R\|_{H\to H} \leq c_R, \ \forall h \in (0,1], \tag{10.9}$$

where $c_R > 0$ is **independent of** h, and suppose that

$$\|(I-\Pi)R\|_{H\to H} = \|R(I-\Pi)\|_{H\to H} = O(h^\infty) \tag{10.10}$$

(we shall write $(I-\Pi)R = O(h^\infty)$ and $R(I-\Pi) = O(h^\infty)$, respectively). Consider

$$B = B(h) = A + R\colon D \subset H \longrightarrow H. \tag{10.11}$$

By Theorem 10.1.1 (and the proof of Theorem 10.1.3), we have that $B = B^*$ with compact resolvent for all $h \in (0,1]$. Moreover, $B \geq -C'I$ on D. Let hence $\{\mu_j(h)\}_{j\geq 1}$ be the sequence $-\infty < \mu_1 \leq \mu_2 \leq \ldots \to +\infty$ of its eigenvalues, repeated according to multiplicity. Define, as before,

$$B_- := \Pi B \Pi, \ B_+ = (I-\Pi)B(I-\Pi), \text{ and } \tilde{B} = B_- + B_+.$$

The operator B_- is bounded and symmetric (in fact, a finite-rank operator), and it follows from Theorems 10.1.1 and 10.1.3 that B_+ and $\tilde{B}$ are self-adjoint with the same domain D, semi-bounded from below, with a discrete spectrum. Let

$$\mathrm{Spec}(\tilde{B}) = \{\tilde{\mu}_j\}_{j\geq 1}, \;\; -\infty < \tilde{\mu}_1 \leq \tilde{\mu}_2 \leq \ldots \to +\infty,$$

$$\mathrm{Spec}(B_+) = \{\mu_j^+\}_{j\geq 1}, \;\; -\infty < \mu_1^+ \leq \mu_2^+ \leq \ldots \to +\infty,$$

with repetitions according to multiplicity. Notice, moreover, that since $B_- u \in D$ for all $u \in H$, it always makes sense to consider B_+B_-. Let, as before,

$$H_- := \Pi H, \;\; H_+ := (I - \Pi)H,$$

so that $H_- \subset D$, $H_+ = H_-^{\perp}$, and

$$H = H_- \oplus H_+.$$

We have

$$B_+\big|_{H_-} = 0, \;\; B_-\big|_{H_+} = 0, \;\; \text{and } B_+B_- = B_-B_+ = 0 \text{ on } D \tag{10.12}$$

(hence, in particular, $[B_+, B_-] = 0$ on D). It is readily seen, by (10.10), that

$$(I - \Pi)A(I - \Pi) = B_+ + R_1, \;\; \text{where } R_1 = O(h^\infty) \tag{10.13}$$

and that (since $(I - \Pi)A\Pi = \Pi A(I - \Pi) = 0$ by the Spectral Theorem)

$$B = \tilde{B} + R_2, \;\; \text{where } R_2 = O(h^\infty). \tag{10.14}$$

Notice that there exists $h_0 \in (0, 1]$ such that for any given $h \in (0, h_0]$ we have

$$B_+\big|_{H_+} > 0. \tag{10.15}$$

This follows from (10.13). In fact, let us fix an arbitrary integer $N \geq 1$. Then we have from (10.13) that there exists $C_{N+1} > 0$ such that

$$\|R_j\|_{H\to H} \leq C_{N+1}h^{N+1}, \;\; j = 1, 2, \;\; \forall h \in (0, 1].$$

We may therefore fix $h_0 \in (0, 1]$, such that $C_{N+1}h < 1/10^2$ for all $h \in (0, h_0]$, whence

$$\|R_j\|_{H\to H} < 10^{-2}h^N, \;\; j = 1, 2, \;\; \forall h \in (0, h_0]. \tag{10.16}$$

Then, for all $u \in D \cap H_+$,

$$(B_+u, u) = ((I - \Pi)A(I - \Pi)u, u) - (R_1u, u) \geq$$

$$\geq (E_0 - 10^{-2}h^N)\|u\|^2 > (E_0 - 1)\|u\|^2, \;\; \forall h \in (0, h_0],$$

which proves the claim.

Hence, from (10.15) we have that

$$0 \in \mathrm{Spec}(B_+) \textit{ with multiplicity } N_0 \tag{10.17}$$

(notice that $N_0 = \dim H_-$).

From (10.13) and (10.16) we have that for all $u \in D$ with $\|u\| = 1$

$$-10^{-2}h^N + (B_+u,u) \le ((I-\Pi)A(I-\Pi)u,u) \le (B_+u,u) + 10^{-2}h^N, \quad \forall h \in (0,h_0]. \tag{10.18}$$

Since for each $j \ge N_0 + 1$

$$\begin{aligned} E_0 < \lambda_j &= (Au_j,u_j) = ((I-\Pi)A(I-\Pi)u_j,u_j) \\ &= \sup_{v_1,\ldots,v_{j-1}\,\mathrm{lin.\,ind.}} \left[\inf_{\substack{v \in \mathrm{Span}\{v_1,\ldots,v_{j-1}\}^\perp \cap D \\ \|v\|=1}} ((I-\Pi)A(I-\Pi)v,v) \right], \end{aligned}$$

and since

$$\mu_j^+ = \sup_{v_1,\ldots,v_{j-1}\,\mathrm{lin.\,ind.}} \left[\inf_{\substack{v \in \mathrm{Span}\{v_1,\ldots,v_{j-1}\}^\perp \cap D \\ \|v\|=1}} (B_+v,v) \right],$$

it thus follows from (10.18) and the Minimax Principle that for any given $h \in (0,h_0]$

$$|\lambda_j(h) - \mu_j^+(h)| < 10^{-2}h^N, \text{ and } 0 < E_0 - 1 < \mu_j^+(h), \ \forall j \ge N_0(h) + 1. \tag{10.19}$$

Hence (10.19) says that *for any given $h \in (0,h_0]$, for $j \ge N_0(h)+1$ the j-th eigenvalue of $A(h)$ is within distance $10^{-2}h^N$ to the j-th eigenvalue of $B_+(h)$.*

Therefore, for the eigenvalues of B_+ we have

$$\mu_1^+ = \ldots = \mu_{N_0}^+ = 0, \ 0 < \mu_{N_0+1}^+ \le \mu_{N_0+2}^+ \le \cdots. \tag{10.20}$$

Notice also that $B_-\big|_{H_-}$ is represented in the basis $\{u_1,\ldots,u_{N_0}\}$ by the $N_0 \times N_0$ matrix acting on N_0-dimensional vectors

$$u = \sum_{j'=1}^{N_0} (u,u_{j'})u_{j'} \simeq \begin{bmatrix} (u,u_1) \\ \vdots \\ (u,u_{N_0}) \end{bmatrix},$$

as

$$B_-\big|_{H_-} u = \sum_{j,j'=1}^{N_0} (u,u_{j'})(Bu_{j'},u_j)u_j,$$

i.e. by the matrix whose jj'-entry is given by

$$\left(B_-\big|_{H_-}\right)_{jj'} = \overline{(Bu_j, u_{j'})},\ 1 \le j, j' \le N_0,$$

and that

$$\operatorname{Spec}(B_-) = \operatorname{Spec}(B_-\big|_{H_-}) \cup \{0\} \subset [-c_0, c_0], \tag{10.21}$$

where $c_0 := c_R + E_0$. In fact, B_- is a compact self-adjoint operator, so that its spectrum is contained in the interval $[-\|B_-\|_{H\to H}, \|B_-\|_{H\to H}]$. Since

$$\|B_-\|_{H\to H} = \|\Pi(A+R)\Pi\|_{H\to H} \le E_0 + \|\Pi R \Pi\|_{H\to H} \le E_0 + c_R = c_0,$$

(10.21) follows. Put $\operatorname{Spec}(B_-\big|_{H_-}) = \{\mu_1^-, \dots, \mu_{N_0}^-\}$. We next consider an ON basis $\{e_j\}_{j\ge 1} \subset D$ of H, where

$$\operatorname{Span}\{e_1, \dots, e_{N_0}\} = H_-, \quad \text{and} \quad \overline{\operatorname{Span}\{e_j\}_{j\ge N_0+1}} = H_+,$$

which diagonalizes B_+ (this is possible, for $[\Pi, B_+] = 0$ on D):

$$B_+ e_j = 0,\ 1 \le j \le N_0,\ \ B_+ e_j = \mu_j^+ e_j,\ j \ge N_0 + 1. \tag{10.22}$$

Notice then that, from (10.12),

$$B_- e_j = 0,\ \forall j \ge N_0 + 1.$$

Let us then rotate the $e_1, \dots, e_{N_0}$ to obtain a new ON basis $\tilde{e}_1, \dots, \tilde{e}_{N_0}$ of H_- which diagonalizes the Hermitian matrix $B_-\big|_{H_-}$. Defining $\tilde{e}_j = e_j$ for $j \ge N_0 + 1$ gives therefore an ON basis $\{\tilde{e}_j\}_{j\ge 1} \subset D$ of H which simultaneously diagonalizes $\tilde{B}$, B_- and B_+. Hence, in particular,

$$\tilde{B}\tilde{e}_j = \mu_j^- \tilde{e}_j,\ 1 \le j \le N_0,\ \ \tilde{B}\tilde{e}_j = \mu_j^+ \tilde{e}_j,\ j \ge N_0 + 1.$$

Notice that a-priori we do not know whether or not $\mu_{N_0}^- \le \mu_{N_0+1}^+$, so that we do not have, as yet, that the sequence $\{\mu_1^-, \dots, \mu_{N_0}^-, \mu_{N_0+1}^+, \dots\}$ coincides with the non-decreasing sequence $\{\tilde{\mu}_j\}_{j\ge 1}$ of the eigenvalues of $\tilde{B}$. We have therefore to rearrange the $\mu_j^\pm$ in a non-decreasing order. Let

$$S_+ = \{\mu_j^+;\ 0 < \mu_j^+ \le c_0\}.$$

Then

$$\{j \in \mathbb{Z}_+;\ \mu_j^+ \in S_+\} = \{N_0 + 1, \dots, N_0 + \nu_+\},$$

for some $\nu_+ = \nu_+(h) \in \mathbb{Z}_+$ (where $\nu_+ = 0$ if $S_+ = \emptyset$). Since

$$\{\mu_1^-, \dots, \mu_{N_0}^-\} \cup S_+$$

is a finite set, we may arrange, in a non-decreasing order, with repetitions (according to multiplicity among the μ_j^-, resp. the μ_j^+, and equalities between the μ_j^- and $\mu_{j'}^+$), the elements

$$\mu_j^-,\ 1 \le j \le N_0, \text{ and } \mu_j^+,\ N_0+1 \le j \le N_0+\nu_+,$$

the resulting sequence being $\tilde{\mu}_j$, $1 \le j \le N_0+\nu_+$, and call $\tilde{u}_j$ the associated eigenvectors. Put finally $\tilde{u}_j = \tilde{e}_j$ when $j \ge N_0+\nu_+ + 1$. We therefore obtain the ON sequence $\{\tilde{u}_j\}_{j\ge1}$ of eigenvectors of $\tilde{B}$, and have that

$$j \ge N_0+\nu_+ +1 \Longleftrightarrow \mu_j^+ > c_0.$$

Hence, since for $j \ge N+1+\nu_+$

$$\mu_j^+ = (B_+\tilde{u}_j,\tilde{u}_j) = \sup_{v_1,\dots,v_{j-1}\,\text{lin. ind.}} \left[\inf_{\substack{v\in \mathrm{Span}\{v_1,\dots,v_{j-1}\}^\perp\cap D \\ \|v\|=1}} (\tilde{B}v,v) \right],$$

since

$$\mu_j = \sup_{v_1,\dots,v_{j-1}\,\text{lin. ind.}} \left[\inf_{\substack{v\in \mathrm{Span}\{v_1,\dots,v_{j-1}\}^\perp\cap D \\ \|v\|=1}} (Bv,v) \right],$$

and since by (10.14) and (10.16) we have that for all $u \in D$ with $\|u\| = 1$

$$-10^{-2}h^N + (\tilde{B}u,u) \le (Bu,u) \le (\tilde{B}u,u) + 10^{-2}h^N, \quad \forall h \in (0,h_0],$$

it therefore follows that for any given $h \in (0,h_0]$, for $j \ge N+1+\nu_+$

$$|\mu_j^+(h) - \mu_j(h)| < 10^{-2}h^N. \tag{10.23}$$

Hence (10.23) says that *for any given $h \in (0,h_0]$, for $j \ge N_0(h)+\nu_+(h)+1$ the j-th eigenvalue of $B_+(h)$ is the j-th eigenvalue of $\tilde{B}(h)$ and is within distance $10^{-2}h^N$ to the j-th eigenvalue of $B(h)$.*

Therefore, combining (10.19) and (10.23) yields the following theorem.

Theorem 10.2.1. *Let $E, E' > 0$ with $E > E' \ge E_0 + c_R + 10^2$ be fixed. Given any fixed $N \ge 1$, there exists $h_0 \in (0,1]$ such that for any given $h \in (0,h_0]$ the following holds:*

$$\mathrm{Spec}(A) \cap (E,+\infty) \subset \bigcup_{\mu(h)\in \mathrm{Spec}(B)\cap(E',+\infty)} \left[\mu(h)-h^N, \mu(h)+h^N\right], \tag{10.24}$$

and, more precisely, given any $j \ge 1$ so large that $\lambda_j(h) > E$, then, with the **same** *j,*

$$|\lambda_j(h) - \mu_j(h)| \le h^N, \tag{10.25}$$

where, recall, $\mathrm{Spec}(A(h)) = \{-\infty < \lambda_1(h) \le \lambda_2(h) \le \ldots \to +\infty\}$ *and* $\mathrm{Spec}(B(h)) = \{-\infty < \mu_1(h) \le \mu_2(h) \le \ldots \to +\infty\}$.

Chapter 11
Energy-Levels, Dynamics, and the Maslov Index

Our aim here is to study periodicity properties of the integral trajectories of the Hamilton vector field of a given pseudodifferential symbol p, lying in energy-level sets of the kind $p^{-1}(E)$, $E \in [E_1, E_2]$, with which we shall associate the *action integral*. This will be done in the next section. In section 11.2 we shall then give a crash introduction to the Maslov index of a periodic trajectory, with the aim of enabling the reader to compute it in the cases of interest for us. In the notes to the chapter we shall also give a rapid overview of the reason why one needs this symplectic invariant, that will systematically appear in Chapter 12.

11.1 Introducing the Dynamics

We now introduce the Dynamics into our considerations. Recall that, given a smooth real-valued function p defined on phase-space $\mathbb{R}^n_x \times \mathbb{R}^n_\xi = \mathbb{R}^{2n}_X$, $X = (x, \xi)$, we may consider the associated **Hamilton vector-field** defined by

$$H_p = \sum_{j=1}^{n} \Big(\frac{\partial p}{\partial \xi_j} \frac{\partial}{\partial x_j} - \frac{\partial p}{\partial x_j} \frac{\partial}{\partial \xi_j} \Big).$$

Notice that, using the canonical symplectic 2-form σ on $\mathbb{R}^{2n}$ and its non-degeneracy, one has that H_p is *uniquely* defined, at $(x, \xi) \in \mathbb{R}^{2n}$, by the relation

$$dp(v) = \sigma(v, H_p), \ \ \forall v \in \mathbb{R}^{2n}, \tag{11.1}$$

where v is thought of as a tangent vector to $\mathbb{R}^{2n}$ at (x, ξ).

The integral trajectories γ of H_p, called the **bicharacteristic curves of** p (or simply **bicharacteristics of** p), are the solutions of

$$\frac{d\gamma}{dt}(t) = H_p(\gamma(t)).$$

A. Parmeggiani, *Spectral Theory of Non-Commutative Harmonic Oscillators: An Introduction*, Lecture Notes in Mathematics 1992,
DOI 10.1007/978-3-642-11922-4_11, © Springer-Verlag Berlin Heidelberg 2010

We shall write

$$\gamma_{X_0}(t) = \exp(tH_p)(x_0, \xi_0) = \exp(tH_p)(X_0), \quad \gamma_{X_0}(0) = (x_0, \xi_0) = X_0,$$

for the integral trajectory issued from X_0. It is easy to see that if $X_0 \in p^{-1}(E)$ then $\gamma_{X_0}(t) \in p^{-1}(E)$ for all t in the interval of existence of γ_{X_0}. When $p^{-1}(E)$ is compact, γ_{X_0} exists for all times. One calls $p^{-1}(E)$ an **energy surface** (or **energy level**).

Following Dimassi-Sjöstrand [7] and Helffer-Robert [20], we make the following assumptions on p.

Assumption 11.1.1 (Dynamical assumptions (*H*1)-(*H*3)). *Let $E_1 < E_2$, and let $0 < \varepsilon$ be sufficiently small. We assume that:*

- *$(H1)$ $dp \neq 0$ for all $X \in p^{-1}([E_1 - \varepsilon, E_2 + \varepsilon])$;*
- *$(H2)$ $p^{-1}(E)$ is connected for all $E \in [E_1 - \varepsilon, E_2 + \varepsilon]$;*
- *$(H3)$ There exists a smooth function $T = T(X) > 0$ defined in $p^{-1}([E_1 - \varepsilon, E_2 + \varepsilon])$ such that*

$$\exp(T(X)H_p)(X) = X, \quad \forall X \in p^{-1}([E_1 - \varepsilon, E_2 + \varepsilon]). \tag{11.2}$$

We have the following result.

Lemma 11.1.2. *Assume $(H1)$, $(H2)$ and $(H3)$. Let $X \in p^{-1}([E_1 - \varepsilon, E_2 + \varepsilon])$ and let*

$$\gamma_X : [0, T(X)] \ni t \longmapsto \exp(tH_p)(X).$$

Hence γ_X is a closed curve in $p^{-1}([E_1 - \varepsilon, E_2 + \varepsilon])$ (a **periodic bicharacteristic** *of* **period** *$T(X)$). The functions*

$$X \longmapsto T(X), \text{ and } X \longmapsto J(X) := \int_{\gamma_X} \xi dx$$

are smooth and depend only on the value $p(X)$: if $X \in p^{-1}(E)$, for $E \in [E_1 - \varepsilon, E_2 + \varepsilon]$, we have

$$T(X) = T_p(E), \quad J(X) = J_p(E),$$

for some functions T_p and J_p. Moreover, with $J_p' = dJ_p/dE$, one has

$$J_p'(E) = T_p(E).$$

Recall that $\xi dx = \sum_{j=1}^n \xi_j dx_j$ is the canonical 1-form on $\mathbb{R}^{2n}$, such that $d(\xi dx) = \sigma$.

Proof. Let $[0, 1] \ni s \longmapsto X(s) \in p^{-1}([E_1 - \varepsilon, E_2 + \varepsilon])$ be a C^1 curve and put

$$\varphi(t, s) := \exp(tH_p)(X(s)), \quad 0 \leq s \leq 1, \ 0 \leq t \leq T(X(s)),$$

so that $\varphi(T(X(s)),s) = \varphi(0,s)$ by $(H3)$. Put

$$\gamma_s := \{\varphi(t,s);\ t \in [0, T(X(s))]\}.$$

Hence γ_s is a closed curve, for all $s \in [0,1]$. We consider now the pull-back $\varphi^*\sigma$ of the symplectic form σ by the map φ. Then

$$\varphi^*\sigma = \alpha(t,s)dt \wedge ds,$$

where, writing φ_* for the tangent map associated with φ,

$$\begin{aligned}\alpha(t,s) &= \sigma\Big(\varphi_*(\frac{\partial}{\partial t}), \varphi_*(\frac{\partial}{\partial s})\Big) = \sigma\Big(H_p, \varphi_*(\frac{\partial}{\partial s})\Big) \quad \text{(by (11.1))}\\ &= -dp(\varphi_*(\frac{\partial}{\partial s})) = -\frac{\partial}{\partial s}(p(\varphi(t,s))).\end{aligned}$$

Since $d(\xi dx) = \sigma$, by the Stokes formula we have

$$\int_{\gamma_s} \xi dx - \int_{\gamma_0} \xi dx = \int_0^s \Big(\int_0^{T(X(s'))} \frac{\partial}{\partial s'}(p(\varphi(t,s')))dt\Big)ds'. \tag{11.3}$$

But by the definition of $\varphi(t,s)$ we have that $p(\varphi(t,s)) = p(\varphi(0,s)) =: \bar{p}(s)$, so that (11.3) reduces to

$$\int_{\gamma_s} \xi dx - \int_{\gamma_0} \xi dx = \int_0^s T(X(s'))\frac{\partial}{\partial s'}(p(\varphi(0,s')))ds' = \int_0^s T(X(s'))d\bar{p}(s'). \tag{11.4}$$

If all the γ_s are contained in the same energy surface $p^{-1}(E)$, we get that

$$\int_{\gamma_s} \xi dx - \int_{\gamma_0} \xi dx = 0,$$

that is, $J(X(s))$ is independent of s. Hence $J(X) = J_p(E)$ and (11.4) shows that

$$J_p'(E) = T_p(E),$$

where $T_p(E) = T(X(0))$. □

We next show that assumptions $(H1)$ and $(H2)$ are fulfilled by smooth functions that are positively homogeneous of some positive degree (we will be interested in the case $2k$, for some $k \in \mathbb{N}$). This follows from Proposition 11.1.4 below. First of all, we have to recall *Euler's identity*.

Lemma 11.1.3. *Let $p \in C^\infty(\mathbb{R}^{2n} \setminus \{0\})$ be positively homogeneous of degree $0 \neq \alpha \in \mathbb{R}$, that is*

$$p(\tau X) = \tau^\alpha p(X),\ \forall \tau > 0,\ \forall X \neq 0. \tag{11.5}$$

Then

$$p(X) = \frac{1}{\alpha}\langle X, \nabla_X p(X)\rangle,\ \forall X \neq 0. \tag{11.6}$$

Proof. Given any $X \neq 0$, we have on the one hand

$$\frac{d}{d\tau}\Big(p(\tau X)\Big)\Big|_{\tau=1} = \langle X, \nabla_X p(X)\rangle,$$

and on the other

$$\frac{d}{d\tau}\Big(p(\tau X)\Big)\Big|_{\tau=1} = \frac{d}{d\tau}\Big(\tau^{\alpha} p(X)\Big)\Big|_{\tau=1} = \alpha p(X).$$

This concludes the proof. □

Proposition 11.1.4. *Let $p \in C^{\infty}(\mathbb{R}^{2n} \setminus \{0\}; \mathbb{R}_+)$ be positively homogeneous of degree $2k$, for some $k \in \mathbb{N}$, that is*

$$p(\tau X) = \tau^{2k} p(X), \ \forall \tau > 0, \ \forall X \neq 0.$$

Let I be any given **closed** *interval of $\mathbb{R}$, with $I \subset (0, +\infty)$. Then*

- *$p^{-1}(I)$ is a* **connected** *set, which is* **compact** *when I is bounded;*
- *For all $E > 0$, the set $p^{-1}(E)$ is a* **smooth compact and connected hypersurface** *of $\mathbb{R}^{2n} \setminus \{0\}$.*

Hence hypotheses $(H1)$ and $(H2)$ are always satisfied in this case, for every closed energy interval $I \subset (0, +\infty)$.

Proof. Write any given $X \neq 0$ using polar coordinates as $X = |X|\dfrac{X}{|X|} = \rho\omega$, with $\rho > 0$ and $\omega \in \mathbb{S}^{2n-1}$. Then

$$p(X) = \rho^{2k} p(\omega), \ \text{ where } 0 < c_1 = \min_{\omega \in \mathbb{S}^{2n-1}} p(\omega) \leq c_2 = \max_{\omega \in \mathbb{S}^{2n-1}} p(\omega).$$

Now, given any $X \in p^{-1}(I)$, we have that $p(X) = \rho^{2k} p(\omega) = E$ for some $E \in I$, whence $\rho = (E/p(\omega))^{1/2k}$. This shows that the map

$$\gamma\colon \mathbb{S}^{2n-1} \times I \longrightarrow p^{-1}(I), \ \gamma(\omega, E) = \left(\frac{E}{p(\omega)}\right)^{\frac{1}{2k}} \omega,$$

is *smooth* and *onto*. Since $\mathbb{S}^{2n-1} \times I$ is connected, this proves that $p^{-1}(I)$ is also connected. When I is in addition also bounded, hence compact, then $\mathbb{S}^{2n-1} \times I$ is compact, and therefore the same holds for $p^{-1}(I)$. This proves the first claim in the statement.

To prove the second claim, let $E > 0$ and take any $X_0 \in p^{-1}(E)$ (then, necessarily, $X_0 \neq 0$). Hence X_0 satisfies the equation $p(X_0) = E$. By Euler's relation (11.6) we have

$$0 < E = p(X_0) = \frac{1}{2k}\langle X_0, \nabla_X p(X_0)\rangle,$$

which implies that $\nabla_X p(X_0) \neq 0$, and concludes the proof. □

Let us now define the *averaged action-integral* associated with a trajectory $[0,T_p(E)] \ni t \longmapsto \gamma_X(t) = \exp(tH_p)(X)$ contained in $p^{-1}(E)$.

Definition 11.1.5. In the dynamical assumptions $(H1),(H2)$ and $(H3)$ we define the **averaged action-integral** $A_p(E) = A_{p,\gamma_X}(E)$ of an integral curve $\gamma_X \subset p^{-1}(E)$, $\gamma_X\colon [0,T(X) = T_p(E)] \ni t \longmapsto \exp(tH_p)(X)$, by

$$A_p(E) := \frac{1}{T(X)}\int_{\gamma_X} \xi dx - E = \frac{J_p(E)}{T_p(E)} - E. \tag{11.7}$$

Notice that, equivalently, since $p(\gamma_X(t)) = E$ for all $t \in [0,T(X)]$,

$$A_p(E) = \frac{1}{T(X)}\int_{\gamma_X} \xi dx - \frac{1}{T(X)}\int_0^{T(X)} p \circ \exp(tH_p)(X)dt. \tag{11.8}$$

By Lemma 11.1.2 the definition of A_p **does not** depend on γ_X, but only on E.

When p satisfies $(H1)$, $(H2)$, $(H3)$ and is such that the period $T_p(E)$ is independent of E, for E belonging to some energy-interval I, then $A_p(E)$ is constant, as shown by the next lemma.

Lemma 11.1.6. *Assume* $(H1)$, $(H2)$, $(H3)$. *Suppose that p is such that for some energy-interval* $I = [E_1,E_2]$ *the period* T_p *is* **independent of** E. *Then* $A_p(E)$ *is* **constant** *for all* $E \in I$.

Proof. Let $T_p(E) = T_0$ for all $E \in I$. Since $J_p'(E) = T_p(E) = T_0$ for all $E \in [E_1,E_2]$, we have that $J_p(E) = T_0 E + J_p(E_1)$ on I. Hence

$$A_p(E) = \frac{J_p(E)}{T_p(E)} - E = \frac{T_0 E}{T_0} - E + \frac{J_p(E_1)}{T_0} = \frac{J_p(E_1)}{T_0}, \quad \forall E \in I.$$

□

It will be useful also to have the following resut, about the behavior of the period $T_p(E)$ and the action $A_p(E)$ when one changes the symbol p to αp, where $\alpha \in \mathbb{R}_+$ is a constant.

Lemma 11.1.7. *Suppose p satisfies hypotheses* $(H1)$, $(H2)$, $(H3)$. *Let* $\alpha \in \mathbb{R}_+$ *be a constant. Put* $q{=}\alpha p$. *Then q satisfies* $(H1)$, $(H2)$, $(H3)$ *on the interval* $[\alpha(E_1-\varepsilon),\alpha(E_2+\varepsilon)]$ *with*

$$T_q(\alpha E) = \frac{T_p(E)}{\alpha}, \quad E \in [E_1-\varepsilon,E_2+\varepsilon].$$

Furthermore,

$$A_q(\alpha E) = \alpha A_p(E), \ \forall E \in [E_1-\varepsilon,E_2+\varepsilon]. \tag{11.9}$$

Proof. We have

$$H_q = \alpha H_p, \quad p^{-1}(E) = q^{-1}(\alpha E), \; \forall E \in [E_1 - \varepsilon, E_2 + \varepsilon].$$

Given any $X \in p^{-1}(E) = q^{-1}(\alpha E)$, $E \in [E_1 - \varepsilon, E_2 + \varepsilon]$, consider the curve

$$\tilde{\gamma}_X : [0, T_p(E)/\alpha] \ni t \longmapsto \exp(\alpha t H_p)(X) = \exp(t H_q)(X).$$

Then, for all $X \in q^{-1}(\alpha E) = p^{-1}(E)$, with $E \in [E_1 - \varepsilon, E_2 + \varepsilon]$,

$$\begin{aligned}\tilde{\gamma}_X(t) &= (x_q(t), \xi_q(t)) = \exp(t H_q)(X) \\ &= \exp(t\alpha H_p)(X) = (x_p(\alpha t), \xi_p(\alpha t)) = \gamma_X(\alpha t), \; \forall t \in [0, T_q(\alpha E)],\end{aligned}$$

where $\gamma_X : [0, T_p(E)] \ni s \longmapsto \exp(sH_p)(X)$. It follows that

$$\exp(\frac{T_p(E)}{\alpha} H_q)(X) = \exp(T_p(E) H_p)(X) = X,$$

which shows that the function

$$q^{-1}([\alpha(E_1 - \varepsilon), \alpha(E_2 + \varepsilon)] \ni X \longmapsto T(X) = T_q(q(X)) = \frac{T_p(p(X))}{\alpha} > 0$$

is smooth. Hence hypotheses $(H1)$, $(H2)$ and $(H3)$ hold for q. Next, one has

$$\begin{aligned}J_q(\alpha E) &= \int_{\tilde{\gamma}_X} \xi \, dx = \int_0^{T_q(\alpha E)} \langle \xi_q(t), \dot{x}_q(t) \rangle dt \\ &= \int_0^{T_q(\alpha E)} \langle \xi_q(t), (\partial_\xi q)(\exp(t H_q)(X)) \rangle dt \\ &= \alpha \int_0^{T_q(\alpha E)} \langle \xi_p(\alpha t), (\partial_\xi p)(\exp(t\alpha H_p)(X)) \rangle dt \\ &\qquad \text{(setting } \alpha t = s\text{)} \\ &= \int_0^{\alpha T_q(\alpha E)} \langle \xi_p(s), \dot{x}_p(s) \rangle ds = \int_{\gamma_X} \xi \, dx = J_p(E).\end{aligned}$$

It thus follows that

$$A_q(\alpha E) = \frac{J_q(\alpha E)}{T_q(\alpha E)} - \alpha E = \frac{J_p(E)}{T_p(E)/\alpha} - \alpha E = \alpha A_p(E),$$

for all $E \in [E_1 - \varepsilon, E_2 + \varepsilon]$, which concludes the proof. □

We next show that whenever there is $T = T_p(E)$ for which hypothesis $(H3)$ holds for a p *positively homogeneous of degree* 2, then T is *independent of* E, and $A_p(E) = 0$.

Lemma 11.1.8. *Suppose* $p\colon \mathbb{R}^{2n}\setminus\{0\} \longrightarrow \mathbb{R}_+$ *is smooth, positively homogeneous of degree* 2 *and that* $(H3)$ *is satisfied on* $[E_1-\varepsilon, E_2+\varepsilon] \subset \mathbb{R}_+$. *Then* T_p **does not** *depend on* $E \in [E_1-\varepsilon, E_2+\varepsilon]$, *and* $A_p(E) = 0$ *for all* $E \in [E_1-\varepsilon, E_2+\varepsilon]$.

Proof. Consider $\gamma_X(t) = \exp(tH_p)(X)$, where $X \in p^{-1}(E)$. In the first place, we notice that

$$\gamma_X^*(\xi dx) = \langle \xi(t), \dot{x}(t)\rangle dt.$$

Now, on the one hand by virtue of the periodicity we have

$$\int_0^{T_p(E)} \frac{d}{dt}\Big(\langle \xi(t), x(t)\rangle\Big)dt = \langle \xi(T_p(E)), x(T_p(E))\rangle - \langle \xi(0), x(0)\rangle = 0,$$

and on the other

$$\int_0^{T_p(E)} \frac{d}{dt}\Big(\langle \xi(t), x(t)\rangle\Big)dt = \int_0^{T_p(E)} \Big(\langle \dot{\xi}(t), x(t)\rangle + \langle \xi(t), \dot{x}(t)\rangle\Big)dt,$$

whence

$$J_p(E) = \int_{\gamma_X} \xi dx = \int_0^{T_p(E)} \langle \xi(t), \dot{x}(t)\rangle dt = -\int_0^{T_p(E)} \langle \dot{\xi}(t), x(t)\rangle dt.$$

Since, by Hamilton's equations,

$$\int_0^{T_p(E)} \langle \xi(t), \dot{x}(t)\rangle dt = \int_0^{T_p(E)} \Big\langle \xi(t), \frac{\partial p}{\partial \xi}(x(t), \xi(t))\Big\rangle dt$$

and

$$\int_0^{T_p(E)} \langle \dot{\xi}(t), x(t)\rangle dt = -\int_0^{T_p(E)} \Big\langle \frac{\partial p}{\partial x}(x(t), \xi(t)), x(t)\Big\rangle dt,$$

it follows that

$$J_p(E) = \frac{1}{2}\int_0^{T_p(E)} \langle \gamma_X(t), (\nabla_X p)(\gamma_X(t))\rangle dt.$$

But Euler's relation (11.6) says that

$$2p(X) = \langle X, (\nabla_X p)(X)\rangle,$$

so that, since $p(\gamma_X(t)) = E$ for all t,

$$J_p(E) = \frac{1}{2}\int_0^{T_p(E)} \langle \gamma_X(t), (\nabla_X p)(\gamma_X(t))\rangle dt = \int_0^{T_p(E)} p(\gamma_X(t))dt = ET_p(E). \tag{11.10}$$

Using Lemma 11.1.2 and Proposition 11.1.4, we therefore get

$$J'_p(E) = T_p(E) = ET'_p(E) + T_p(E), \ \forall E \in [E_1 - \varepsilon, E_2 + \varepsilon],$$

whence

$$T'_p(E) = 0, \ \forall E \in [E_1 - \varepsilon, E_2 + \varepsilon].$$

On the other hand, from (11.10) we have

$$A_p(E) = \frac{J_p(E)}{T_p(E)} - E = \frac{ET_p(E)}{T_p(E)} - E = 0,$$

which concludes the proof. □

Remark 11.1.9. If $p \colon \mathbb{R}^{2n} \setminus \{0\} \longrightarrow \mathbb{R}_+$ is positively homogeneous of degree $2k$, $k \geq 2$, then we have

$$J_p(E) = kET_p(E), \ \forall E > 0,$$

whence, with $T'_p = dT_p/dE$, the equation

$$T_p(E) = kET'_p(E) + kT_p(E), \ \text{i.e.} \ ET'_p(E) = \frac{1-k}{k} T_p(E), \ E > 0.$$

Hence, fixing $E_0 \in [E_1 - \varepsilon, E_2 + \varepsilon] \subset \mathbb{R}_+$ (thus $E_0 > 0$) gives

$$T_p(E) = T_p(E_0) \left(\frac{E}{E_0} \right)^{(1-k)/k}, \ E \in [E_1 - \varepsilon, E_2 + \varepsilon].$$

△

When $n = 1$ and $0 < p$ is positively homogeneous of degree 2 we have that $p^{-1}(E)$ is a periodic trajectory of H_p of period $T > 0$, independent of E, for all $E > 0$. In fact, we have the following proposition.

Proposition 11.1.10. *Let $n = 1$, and let $p \colon \mathbb{R}^2 \setminus \{0\} \longrightarrow \mathbb{R}_+$ be smooth, positively homogenoeus of degree 2. Then all the integral trajectories of H_p lying in $p^{-1}(0, +\infty)$ are periodic with* **least** *period T_p* **independent of** *the energy and given by the formula*

$$T_p = \int_0^{2\pi} \frac{d\theta}{2p(\sin\theta, \cos\theta)}. \tag{11.11}$$

Proof. Let $E > 0$, and take $X_0 \in p^{-1}(E)$. We use polar coordinates

$$\varphi \colon (0, +\infty) \times [0, 2\pi] \longrightarrow \mathbb{R}^2 \setminus \{0\}, \ \varphi(\rho, \theta) = (\rho \sin\theta, \rho \cos\theta) = X.$$

Then

$$\varphi^* \sigma = \rho d\rho \wedge d\theta, \ \varphi^* \{f, g\} = \frac{1}{\rho} \{\varphi^* f, \varphi^* g\}_{\rho, \theta},$$

where $\varphi^* f = f \circ \varphi$, $\{f, g\}$ is the Poisson bracket of f and g, and, finally, $\{\tilde{f}, \tilde{g}\}_{\rho,\theta} := \partial_\rho \tilde{f} \partial_\theta \tilde{g} - \partial_\theta \tilde{f} \partial_\rho \tilde{g}$. Hence

$$\exp(tH_p)(X_0) = \varphi(\rho(t), \theta(t)), \quad X_0 = \varphi(\rho_0, \theta_0),$$

where, with $\dot{\theta} = d\theta/dt$ and $\dot{\rho} = d\rho/dt$, ρ and θ satisfy the equations

$$\begin{cases} \dot{\theta} = \dfrac{1}{\rho} \partial_\rho(\varphi^* p) \\ \dot{\rho} = -\dfrac{1}{\rho} \partial_\theta(\varphi^* p). \end{cases} \tag{11.12}$$

By the homogeneity

$$(\varphi^* p)(\rho, \theta) = p(\rho \sin\theta, \rho \cos\theta) = \rho^2 p(\sin\theta, \cos\theta), \tag{11.13}$$

where for $p(\sin\theta, \cos\theta) = p(\omega)$ we have

$$0 < c_1 = \min_{\omega \in \mathbb{S}^1} p(\omega) \leq c_2 = \max_{\omega \in \mathbb{S}^1} p(\omega).$$

It follows from (11.13) and the first equation in (11.12) that the map $t \longmapsto \theta(t)$ exists for all $t \in \mathbb{R}$ (the derivative being uniformly bounded), and that

$$\dot{\theta} = 2p(\sin\theta, \cos\theta) \geq 2c_1,$$

that is,

$$\mathbb{R} \ni t \longmapsto \theta(t) \in \mathbb{R} \textit{ is an increasing diffeomorphism}, \tag{11.14}$$

and $t \longmapsto \exp(tH_p)(X_0)$ is therefore the closed curve $p^{-1}(E)$. By Lemma 11.1.8 we thus have that the **least period**, given by the relations

$$\exp(T_p(E)H_p)(X_0) = X_0, \text{ and } \exp(tH_p)(X_0) \neq X_0, \ \forall t \in (0, T_p(E)),$$

i.e. given by the condition $\theta(T_p) = \theta_0 + 2\pi$, is independent of the energy. The formula for the least period T_p follows again from the homogeneity of p, for

$$\dot{\theta} = \frac{1}{\rho} \frac{\partial}{\partial\rho} (\varphi^* p) = 2p(\sin\theta, \cos\theta),$$

whence (by the periodicity of $p(\sin\theta, \cos\theta)$)

$$T_p = \int_{\theta_0}^{\theta_0 + 2\pi} \frac{d\theta}{2p(\sin\theta, \cos\theta)} = \int_0^{2\pi} \frac{d\theta}{2p(\sin\theta, \cos\theta)}.$$

□

It is important to notice that from Proposition 11.1.10 it follows that the closed curve $p^{-1}(E)$ is **clockwise** oriented.

It will also be important, when considering the Maslov index of a curve in $\mathbb{R}\times\mathbb{R}$, to be able to establish a-priori the orientation of a periodic trajectory $p^{-1}(E)=\gamma_E=\{\exp(tH_p)(X);\ 0\le t\le T(E)\}\subset\mathbb{R}\times\mathbb{R}$ for p satisfying Assuption 11.1.1.

In this case, one observes that

$$\sigma(\nabla_X p(X),H_p(X)) = dp(X)(\nabla_X p(X)) = |\nabla_X p(X)|^2 > 0,\ \forall X\in\gamma_E,$$

so that

$$\sigma(\nabla_X p,H_p) = -\det[\nabla_x p|H_p] > 0,\ \forall X\in\gamma_E,$$

whence, *in* $\mathbb{R}\times\mathbb{R}$,

- *whenever* $\nabla_x p(X)$ *points* **outwards** *from* γ_E *at* $X\in\gamma_E$, *then this is true at all points of* γ_E, *and the couple* $\{\nabla_X p,H_p\}$ *forms on* γ_E *a reference frame with the orientation opposite to the canonical one, that is,* γ_E *is clockwise oriented;*
- *whereas, whenever* $\nabla_x p(X)$ *points* **inwards** *from* γ_E *at* $X\in\gamma_E$, *then this is true at all points of* γ_E, *and the couple* $\{\nabla_X p,H_p\}$ *forms on* γ_E *a reference frame which is canonically oriented, that is,* γ_E *is counter-clockwise oriented.*

Lemma 11.1.11. *Let* $p\colon\mathbb{R}^{2n}\setminus\{0\}\longrightarrow\mathbb{R}_+$ *be smooth, positively homogeneous of degree* $2k$, $k\in\mathbb{N}$. *Let* $E>0$ *be given, and consider the compact (connected) set* $p^{-1}(E)=\gamma_E$. *Then* $\nabla_X p(X)$ *points outwards from* γ_E, *for all* $X\in\gamma_E$. *Hence, in particular, when* $n=1$ *and* γ_E *is a periodic trajectory of* H_p, *it is* **clockwise** *oriented.*

Proof. We know that $\gamma_E\subset\mathbb{R}^{2n}\setminus\{0\}$ is a smooth compact and connected hypersurface of $\mathbb{R}^{2n}$, whence it is orientable. The vector-field $\gamma_E\ni X\longmapsto\nabla_X p(X)$ is a nowhere-zero normal to γ_E, whence it either points outwards from γ_E at all points of γ_E, or else it points inwards from γ_E at all points of γ_E. Consider the compact set $D_E=\{X\neq 0;\ p(X)\le E\}\cup\{0\}$. Then $\partial D_E=\gamma_E$, and if $E'>E$, by virtue of the homogeneity, we have $D_E\subsetneq D_{E'}$. Hence $\gamma_{E'}$ is contained in the exterior of γ_E. Since moving from γ_E to $\gamma_{E'}$ increases the value of p from E to E', and since $\nabla_X p$ gives the direction of maximal growth of p, we get that $\nabla_X p\big|_{\gamma_E}$ points outwards from γ_E. This concludes the proof. □

Of course, the above lemma also holds when the homogeneity degree of p is a real $\alpha>0$.

Remark 11.1.12. Consider the "Schrödinger" non-homogeneous Hamiltonian in $\mathbb{R}^2$ given by $p(x,\xi)=\xi^2+V(x)$, where $V(x)=(x^2-1)^2$. Consider the equation $V(x)=E$, $E\in(0,1)$. Let $x_{\pm,b}(E)=\pm\sqrt{1+\sqrt{E}}$ the outer solutions ("big", in absolute value), and $x_{\pm,s}(E)=\pm\sqrt{1-\sqrt{E}}$ the inner solutions ("small", in absolute value). It is well known that we then have two periodic motions associated with H_p, whose x-projections lie in $[x_{-,b}(E),x_{-,s}(E)]$ and $[x_{+,s}(E),x_{+,b}(E)]$, respectively, which are possible due to the conditions

$$V'(x_{-,b}(E)) < 0 < V'(x_{-,s}(E)), \; V'(x_{+,s}(E)) < 0 < V'(x_{+,b}(E)),$$

which yield that on the periodic trajectories the vector-field $\nabla_X p$ is pointing outwards. Hence the periodic trajectories are clockwise oriented. △

Let us now use Lemma 11.1.2 to compute the periods of the periodic trajectories associated with the eigenvalues of our NCHO $Q_{(\alpha,\beta)}$, for $\alpha,\beta > 0$, $\alpha\beta > 1$ and $\alpha \neq \beta$. Of course, one may equivalently use (11.11). We have the following result.

Lemma 11.1.13. *Recall that $\alpha,\beta > 0$, $\alpha\beta > 1$, and that $p_0(x,\xi) = (x^2+\xi^2)/2$, $(x,\xi) \in \mathbb{R}^2$. Put $\mu_\pm = (\alpha \pm \beta)/2$. For $E > 0$, the integral curves $\gamma_\pm(E) = \lambda_\pm^{-1}(E)$ of the Hamilton vector-fields associated with the eigenvalues*

$$\lambda_\pm(x,\xi) = \mu_+ p_0(x,\xi) \pm \sqrt{\mu_-^2 p_0(x,\xi)^2 + x^2\xi^2}$$

of $Q_{(\alpha,\beta)}(x,\xi)$, $(x,\xi) \in \mathbb{R}^2 \setminus \{0\}$, are all periodic, with **least** *periods*

$$T_\pm = T_\pm(\alpha,\beta) = \frac{\alpha+\beta}{\sqrt{\alpha\beta(\alpha\beta-1)}}\pi \mp \int_0^{2\pi} \frac{\sqrt{\mu_-^2+\sin^2(2\theta)}}{\alpha\beta-\sin^2(2\theta)}d\theta, \tag{11.15}$$

all **independent of** *the energy E. (When $E=0$, $\gamma_\pm(0) = \{(0,0)\}$.)*

Proof. We have to compute $J'_{\lambda_\pm}(E)$. We use polar coordinates to parametrize the curves $\gamma_\pm(E)$. For $\theta \in [0,2\pi]$, consider

$$(x(\theta),\xi(\theta)) = (\sqrt{2}\rho_\pm(\theta)\sin\theta, \sqrt{2}\rho_\pm(\theta)\cos\theta),$$

where

$$\rho_\pm(\theta) = \left(\frac{E}{\mu_+ \pm \sqrt{\mu_-^2+\sin^2(2\theta)}}\right)^{1/2}.$$

Now, we have that on $\lambda_\pm^{-1}(E)$

$$\begin{aligned}\xi dx\big|_{\lambda_\pm^{-1}(E)} &= 2\rho_\pm(\theta)\cos\theta\, d\Big(\rho_\pm(\theta)\sin\theta\Big)\\ &= \Big(\frac{d\rho_\pm}{d\theta}(\theta)\rho_\pm(\theta)\sin(2\theta) + 2\rho_\pm(\theta)^2\cos^2\theta\Big)d\theta.\end{aligned}$$

Hence

$$\int_{\gamma_\pm(E)} \xi dx = \int_0^{2\pi} \frac{d}{d\theta}\Big(\frac{\rho_\pm(\theta)^2}{2}\Big)\sin(2\theta)d\theta + 2E\int_0^{2\pi} \frac{\cos^2\theta}{\mu_+ \pm \sqrt{\mu_-^2+\sin^2(2\theta)}}d\theta$$

$$= -2\int_0^{2\pi} \frac{\rho_\pm(\theta)^2}{2}\cos(2\theta)d\theta + 2E\int_0^{2\pi} \frac{\cos^2\theta}{\mu_+ \pm \sqrt{\mu_-^2+\sin^2(2\theta)}}d\theta$$

$$= -E\int_0^{2\pi} \frac{\cos^2\theta - \sin^2\theta}{\mu_+ \pm \sqrt{\mu_-^2+\sin^2(2\theta)}}d\theta$$

$$+2E\int_0^{2\pi} \frac{\cos^2\theta}{\mu_+ \pm \sqrt{\mu_-^2+\sin^2(2\theta)}}d\theta$$

$$= E\int_0^{2\pi} \frac{d\theta}{\mu_+ \pm \sqrt{\mu_-^2+\sin^2(2\theta)}}$$

$$= E\left(\mu_+\int_0^{2\pi} \frac{d\theta}{\alpha\beta - \sin^2(2\theta)} \mp \int_0^{2\pi} \frac{\sqrt{\mu_-^2+\sin^2(2\theta)}}{\alpha\beta-\sin^2(2\theta)}d\theta\right).$$

Using the well-known formula

$$\int_0^{2\pi} \frac{d\theta}{\alpha\beta - \sin^2(\theta)} = \frac{2\pi}{\sqrt{\alpha\beta(\alpha\beta-1)}},$$

we thus obtain

$$J_{\lambda_\pm}(E) = E\left(\frac{\alpha+\beta}{\sqrt{\alpha\beta(\alpha\beta-1)}}\pi \mp \int_0^{2\pi} \frac{\sqrt{\mu_-^2+\sin^2(2\theta)}}{\alpha\beta-\sin^2(2\theta)}d\theta\right),$$

whence, finally,

$$T_\pm = T_\pm(\alpha,\beta) = \frac{\alpha+\beta}{\sqrt{\alpha\beta(\alpha\beta-1)}}\pi \mp \int_0^{2\pi} \frac{\sqrt{\mu_-^2+\sin^2(2\theta)}}{\alpha\beta-\sin^2(2\theta)}d\theta,$$

by Lemma 11.1.2. That the $T_\pm$ are the smallest periods follows at once from Proposition 11.1.10 and (11.14). □

We close this section by giving an example of symbol p whose (non-trivial) trajectories are all periodic, with period dependent on the energy, and whose trajectories are counter-clockwise oriented. Consider in $\mathbb{R}^2$ the symbol $p(x,\xi) = e^{-(x^2+\xi^2)}$. Let $E \in (0,1)$ and consider

$$p^{-1}(E) = \{(x,\xi);\, e^{-(x^2+\xi^2)} = E\} = \{(x,\xi);\, x^2+\xi^2 = \ln(1/E)\} =: \gamma_E.$$

Observe that $\nabla_X p = 0$ iff $X = 0$. One computes, at $X_0 = (\sqrt{\ln(1/E)}, 0) \in \gamma_E$,

$$\nabla_X p(X_0) = -2EX_0,$$

which is opposite to the direction $(1,0)$, whence it points inwards from γ_E. By the above discussion, we have that γ_E is counter-clockwise oriented. We now compute the period of the trajectory. We use polar coordinates $(x,\xi)=(\rho\cos\theta,\rho\sin\theta)$. Then Hamilton's equations become, using $\rho^2\equiv\ln(1/E)$,

$$\dot{\rho}=0,\ \ \dot{\theta}=2E>0.$$

The latter equation proves, once more, that γ_E is counter-clockwise oriented. Hence

$$x=\sqrt{\ln(1/E)}\cos(2Et),\ \ \xi=\sqrt{\ln(1/E)}\sin(2Et),$$

with least period $T(E)=\pi/E$. Hence $\gamma_E=\{\exp(tH_p)(X_0);\ 0\le t\le T(E)\}$. Notice, finally, that in this case

$$J(E)=\pi\ln E<0,\ \ E\in(0,1).$$

11.2 The Maslov Index

The classical approach to the Maslov index is due to V. Arnold (see [2] and also Hörmander [26]). We shall follow here (with some differences, though) Duistermaat's book [12], Hörmander's book [29], McDuff-Salamon's book [43], Treves' book [70], and the paper by Robbin and Salamon [62].

We address the reader to those books for a deep study of Symplectic Geometry. We will content ourselves here with giving the statements in a form suited to our purposes, and prove only those statements whose proofs will not carry us "too far away".

The main difference with the approaches in the above-mentioned books is that we shall give the Maslov cycle a different orientation, so as to have that the closed trajectories of the harmonic oscillator, oriented in the clockwise direction (which is natural to the symplectic form $d\xi\wedge dx$ in $\mathbb{R}\times\mathbb{R}$), have Maslov index 2 (our choice is consistent with Arnold's one).

Recall that we are working in phase-space $T^*\mathbb{R}^n=\mathbb{R}^n\times\mathbb{R}^n$, endowed with the canonical symplectic form (see (2.1))

$$\sigma=\sum_{j=1}^{n}d\xi_j\wedge dx_j,$$

where (x,ξ) are symplectic coordinates in $T^*\mathbb{R}^n$. Recall that we shall write a tangent vector $v\in T_{(x,\xi)}T^*\mathbb{R}^n=\mathbb{R}^n\times\mathbb{R}^n$ at a point (x,ξ) as $\begin{bmatrix}\delta x\\ \delta\xi\end{bmatrix}$. Hence

$$\sigma(v,v')=\langle\delta\xi,\delta x'\rangle-\langle\delta\xi',\delta x\rangle,\ \ v=\begin{bmatrix}\delta x\\ \delta\xi\end{bmatrix},\ v'=\begin{bmatrix}\delta x'\\ \delta\xi'\end{bmatrix}.$$

The map $\mathbb{R}^n \times \mathbb{R}^n \ni (x,\xi) \longmapsto x+i\xi \in \mathbb{C}^n$ allows us to identify $T^*\mathbb{R}^n$ with $\mathbb{C}^n$. The Hermitian scalar product in $\mathbb{C}^n$ is defined by

$$\langle v,w\rangle_{\mathbb{C}^n} = \sum_{j=1}^{n}(\delta x_j + i\delta\xi_j)(\delta y_j - i\delta\eta_j),$$

where

$$v = \begin{bmatrix} \delta x \\ \delta\xi \end{bmatrix} \simeq \delta x + i\delta\xi, \quad w = \begin{bmatrix} \delta y \\ \delta\eta \end{bmatrix} \simeq \delta y + i\delta\eta,$$

so that

$$\mathrm{Re}\langle v,w\rangle_{\mathbb{C}^n} = \langle v,w\rangle_{\mathbb{R}^{2n}} = \sum_{j=1}^{n}(\delta x_j\delta y_j + \delta\xi_j\delta\eta_j)$$

is the Euclidean scalar product in $\mathbb{R}^n \times \mathbb{R}^n$, and

$$\mathrm{Im}\langle v,w\rangle_{\mathbb{C}^n} = \sigma(v,w).$$

Throughout this section we shall put $S_n := T^*\mathbb{R}^n = \mathbb{R}^n \times \mathbb{R}^n$.

Definition 11.2.1. Define $\Lambda(n)$ to be the set of n-dimensional real subspaces λ of $\mathbb{R}^n \times \mathbb{R}^n$ such that

$$\sigma(v,v') = 0, \quad \forall v,v' \in \lambda.$$

One calls λ a **Lagrangian subspace**. In other words, denoting by λ^σ the σ-orthogonal of λ, one has that λ is a Lagrangian subspace if $\lambda = \lambda^\sigma$.

Notice hence that, when $\mathbb{R}^n \times \mathbb{R}^n$ is identified with $\mathbb{C}^n$, having a Lagrangian subspace λ means that λ is a *real* subspace of $\mathbb{C}^n$ (of real dimension n) and that λ and $i\lambda$ are orthogonal with respect to the Eucliden scalar product $\langle\cdot,\cdot\rangle_{\mathbb{R}^{2n}}$.

Remark 11.2.2. In general, given a subspace $\lambda \subset S_n$, one says that λ is

- *isotropic* when $\lambda \subset \lambda^\sigma$,
- *Lagrangian* when $\lambda = \lambda^\sigma$,
- *involutive* when $\lambda^\sigma \subset \lambda$. △

The following result gives that $\Lambda(n)$ is a smooth manifold.

Lemma 11.2.3. *The unitary group $U(n)$ acts on $\Lambda(n)$ transitively, with stabilizer the orthogonal group $O(n)$. Hence $\Lambda(n) = U(n)/O(n)$ is a smooth manifold of dimension $n^2 - n(n-1)/2 = n(n+1)/2$.*

Proof. It is clear that if $U \in U(n)$, then $\lambda \in \Lambda(n)$ implies $U\lambda \in \Lambda(n)$. To see that $U(n)$ acts transitively, take $\lambda \in \Lambda(n)$ and let $w_1,\dots,w_n$ be a $\langle\cdot,\cdot\rangle_{\mathbb{R}^{2n}}$-orthonormal basis of λ. Thus $\langle w_j,w_k\rangle_{\mathbb{R}^{2n}} = \delta_{jk}$, $\sigma(w_j,w_k) = 0$, so that $\langle w_j,w_k\rangle_{\mathbb{C}^n} = \delta_{jk}$, and we have a unitary basis of $\mathbb{C}^n$. We have therefore the unitary matrix $U = [w_1|w_2|\dots|w_n]$ (the j-th column of U is the vector w_j) such that $U\mathbb{R}^n = \lambda$, which shows the transitive action. We now have that $U\mathbb{R}^n = \mathbb{R}^n$ if and only if U has real coefficients, that is $U \in O(n)$, which concludes the proof. □

We next study (following Duistermaat [12]) the tangent spaces $T_\lambda \Lambda(n)$ at $\lambda \in \Lambda(n)$. We have first to show that any given Lagrangian subspace λ possesses a Lagrangian complement μ, that is, an element $\mu \in \Lambda(n)$ such that $\lambda \cap \mu = \{0\}$ and $\lambda + \mu = S_n$.

Lemma 11.2.4. *Given any $\lambda \in \Lambda(n)$, there is a Lagrangian complement μ. Notice that a Lagrangian complement is not unique.*

Proof. We start with a subspace $\mu \subset S_n$, with $\dim \mu < n$, such that $\lambda \cap \mu = \{0\}$ and $\mu \subsetneq \mu^\sigma$. Then μ^σ cannot be contained in $\lambda + \mu$, since $\mu^\sigma \subset \lambda + \mu$ yields $\mu \supset \lambda^\sigma \cap \mu^\sigma = \lambda \cap \mu^\sigma$. Hence $\lambda \cap \mu^\sigma = \{0\}$, and since $\dim \mu^\sigma > n$ and $\dim \lambda = n$, this is impossible. Thus we may choose $e \in \mu^\sigma \setminus (\lambda + \mu)$. It follows that $\mu + \mathrm{Span}\{e\} \subset (\mu + \mathrm{Span}\{e\})^\sigma$ (as the reader may easily check) and $(\mu + \mathrm{Span}\{e\}) \cap \lambda = \{0\}$. We can therefore construct by induction a Lagrangian subspace μ such that $\mu \cap \lambda = \{0\}$, and hence conclude the proof. □

Remark 11.2.5. The proof of Lemma 11.2.4 is valid in any given symplectic vector space. Using Lemma 11.2.3, namely the transitivity of the $U(n)$-action, one may prove Lemma 11.2.4 also as follows. We have that $\lambda = U\mathbb{R}^n$, for some $U \in U(n)$. One may then take $\mu = e^{i\theta} U\mathbb{R}^n$, for any fixed $\theta \neq k\pi$, $k \in \mathbb{Z}$, which concludes the proof, and shows also the non-uniqueness of the Lagrangian complement. △

Theorem 11.2.6. *Given any $\lambda \in \Lambda(n)$, let* $\mathrm{Symm}(\lambda)$ *be the vector space of all symmetric bilinear forms on λ. Then*

$$T_\lambda \Lambda(n) \simeq \mathrm{Symm}(\lambda),$$

in a canonical way.

Proof. Let $\lambda, \mu \in \Lambda(n)$ with $\lambda \cap \mu = \{0\}$. Each n-dimensional subspace λ' which is transversal to μ (that is, $\lambda' + \mu = S_n$) is of the form $\lambda' = \{v + Av;\ v \in \lambda\}$, for a linear map $A\colon \lambda \longrightarrow \mu$. Then

$$q_{\lambda'}(v,w) := \sigma(v, Aw), \quad v, w \in \lambda,$$

defines a bilinear form on λ *which is symmetric iff* $\lambda' \in \Lambda(n)$. The latter follows immediately by observing that for $v, w \in \lambda$

$$\sigma(v + Av, w + Aw) = \sigma(v, Aw) - \sigma(w, Av) = 0 \Longleftrightarrow \lambda' \in \Lambda(n).$$

(Notice that $\lambda' \mapsto q_{\lambda'}$ differs from the map constructed by Duistermaat in [12] by a minus sign). This map defines therefore a bijective map

$$q\colon \{\lambda' \in \Lambda(n);\ \lambda' \cap \mu = \{0\}\} =: \Lambda_0(\mu) \longrightarrow \mathrm{Symm}(\lambda),\ q\colon \lambda' \longmapsto q_{\lambda'}, \qquad (11.16)$$

which depends on λ and μ (notice that $q(\lambda) = q_\lambda = 0$). However, its differential at λ does not depend on μ. In fact, let $\tilde{\mu} \in \Lambda(n)$ be such that $\tilde{\mu} \cap \lambda = \{0\}$. Then

$\tilde{\mu} = \{v + Bv;\ v \in \mu\}$, for some linear map $B\colon \mu \longrightarrow \lambda$. If λ' is close enough to λ, then λ' is still transversal to both μ and $\tilde{\mu}$, and we may write

$$\lambda' = \{w + Aw;\ w \in \lambda\} = \{u + \tilde{A}u;\ u \in \lambda\},$$

for certain linear maps $A\colon \lambda \longrightarrow \mu$ and $\tilde{A}\colon \lambda \longrightarrow \tilde{\mu}$. It follows that

$$w + Aw = u + \tilde{A}u, \quad \text{and} \quad \tilde{A}u = v + Bv,$$

for some $v \in \mu$. Hence

$$w + Aw = u + v + Bv,$$

from which it follows that

$$\underbrace{w - Bv - u}_{\in\lambda} = \underbrace{v - Aw}_{\in\mu},$$

which yields, since $\lambda \cap \mu = \{0\}$, that

$$w - Bv = u, \quad \text{and} \quad v = Aw.$$

Hence

$$w + Aw = (w - BAw) + \tilde{A}(w - BAw),$$

and taking the symplectic product with $z \in \lambda$ gives

$$\sigma(z, Aw) = \sigma(z, \tilde{A}(w - BAw)) = \sigma(z, \tilde{A}w) - \sigma(z, \tilde{A}BAw).$$

Now, $\lambda' \to \lambda$ iff $A \to 0$ and $\tilde{A} \to 0$, which gives that

$$\sigma(z, \tilde{A}BAw) = O(\tilde{A}A),$$

i.e.

$$\sigma(z, Aw) = \sigma(z, \tilde{A}w) + \Big(\text{ term vanishing to 2nd order when } \lambda' \to \lambda\Big), \quad \forall w, z \in \lambda,$$

which concludes the proof. □

Remark 11.2.7. From the proof of the above theorem it follows that the pairs $(\Lambda_0(\mu), q)$, as μ varies in $\Lambda(n)$, form an atlas of $\Lambda(n)$. One sees that each $\Lambda_0(\mu)$ is dense and open in $\Lambda(n)$, and that, since the latter is compact, 2^n of the former sets cover $\Lambda(n)$.

Moreover, in the notation of Treves' book [70] one has

$$q_{\lambda'}(v, w) = -\beta^{\lambda}_{\mu,\lambda'}(v, w) = \sigma(P^{\lambda}_{\mu,\lambda'}v, w), \quad v, w \in \lambda,$$

where $P^{\lambda}_{\mu,\lambda'}$ is the projection $\mu \oplus \lambda' \to \mu$ restricted to λ.

In fact, if $\lambda' = \{v + Av;\ v \in \lambda\}$ where $A\colon \lambda \longrightarrow \mu$, with λ, λ', μ as in the theorem, then $Av = -P^{\lambda}_{\mu,\lambda'} v$, for all $v \in \lambda$, and the claim follows from the symmetry of $q_{\lambda'}$.

△

It will be convenient to have the following lemma, which is essentially contained in the previous theorem.

Lemma 11.2.8. *Let $\lambda, \mu \in \Lambda(n)$, with $\lambda \cap \mu = \{0\}$. Then any other given Lagrangian complement α of λ is the graph of a symmetric quadratic form on μ, with radical given by $\mu \cap \alpha$.*

Proof. We shall write μ^* for the (real) dual space of μ. Consider thus a Lagrangian complement $\alpha \subset S_n$ of λ. Then $\alpha = \{v + Av;\ v \in \mu\}$ for a linear map $A\colon \mu \longrightarrow \lambda$. Note that $\operatorname{Ker} A = \mu \cap \alpha$. Since $\lambda \simeq \mu^*$, where the isomorphism is realized through the symplectic form σ by $\lambda \ni v \overset{\sim\sigma}{\longmapsto} (\mu \ni w \longmapsto \sigma(w,v)) \in \mu^*$ (the proof is left to the reader), we may consider the composite map $\tilde{A}\colon \mu \xrightarrow{A} \lambda \xrightarrow{\sim\sigma} \mu^* \in \mu^* \otimes \mu^*$, $\tilde{A}v = \sigma(\cdot, Av)$. Consider now the bilinear form

$$q_\alpha\colon \mu \times \mu \ni (v, v') \longmapsto \sigma(v, Av') = (\tilde{A}v')v \in \mathbb{R}.$$

As $\alpha \in \Lambda(n)$, the form q_α is symmetric, which precisely means that $\tilde{A} \in S^2\mu^*$. Moreover, it is clear that $\operatorname{Rad} q_\alpha = \mu \cap \alpha$. This concludes the proof. □

From Lemma 11.2.8 it follows in particular that when, say, $\lambda = \mathbb{R}^n \times \{0\}$, choosing $\mu = \{0\} \times \mathbb{R}^n$ gives that every Lagrangian complement $\alpha \subset S_n$ of λ is therefore of the form

$$\alpha = \{(Av, v);\ v \in \mathbb{R}^n\}, \quad \text{for a symmetric } n \times n \text{ real matrix } A. \tag{11.17}$$

We next describe (following [43,62]) Lagrangian subspaces of S_n in more explicit terms. One has the following lemma.

Lemma 11.2.9. *Let X, Y be real $n \times n$ matrices, let Z be the $2n \times n$ real matrix $Z = \begin{bmatrix} X \\ Y \end{bmatrix}$, and define $\lambda \subset S_n$ by*

$$\lambda = \operatorname{Im} Z = \{Zu;\ u \in \mathbb{R}^n\}.$$

Then $\lambda \in \Lambda(n)$ iff rank $Z = n$ *and*

$$^t XY = {}^t YX. \tag{11.18}$$

Every element of $\Lambda(n)$ can be written in this form.

Proof. Let $v = Zu, v' = Zu' \in \lambda$, for $u, u' \in \mathbb{R}^n$. Then

$$\sigma(v,v') = \langle Yu, Xu'\rangle - \langle Yu', Xu\rangle = \langle ({}^tXY - {}^tYX)u, u'\rangle = 0, \ \forall u, u' \in \mathbb{R}^n,$$

iff (11.18) holds.

We now prove that given any $\lambda \in \Lambda(n)$ there exists $Z = \begin{bmatrix} X \\ Y \end{bmatrix}$, with rank n and (11.18) fulfilled, such that $\lambda = \operatorname{Im} Z$. In case $\lambda = \lambda_o := \mathbb{R}^n \times \{0\}$ we may take $X = I$ and $Y = 0$, and in case $\lambda = \lambda_v := \{0\} \times \mathbb{R}^n$ we may take $X = 0$ and $Y = I$. So, we may suppose $\lambda \neq \lambda_o, \lambda_v$. Since $S_n = \lambda_o \oplus \lambda_v$, we have that each $w \in \lambda$ is uniquely decomposed as $w = w' + w''$, where $w' \in \lambda_o$ and $w'' \in \lambda_v$. Choose a basis $\{w_1, \ldots, w_n\}$ of λ and define the matrix $Z = \begin{bmatrix} X \\ Y \end{bmatrix}$ by taking $X = [v_1'|\ldots|v_n']$ and $Y = [v_1''|\ldots|v_n'']$. This concludes the proof. □

Notice that Lemma 11.2.9 gives back (11.17), for it suffices to take $X = A$, $Y = I$.

The matrix Z of Lemma 11.2.9 is called a **Lagrangian frame**.

It is useful to recall now that a linear map $\chi\colon S_n \longrightarrow S_n$ is **symplectic** if

$$\sigma(\chi(v), \chi(v')) = \sigma(v, v'), \ \forall v, v' \in S_n,$$

which is equivalent to saying that

$${}^t\chi \begin{bmatrix} 0 & -I \\ I & 0 \end{bmatrix} \chi = \begin{bmatrix} 0 & -I \\ I & 0 \end{bmatrix},$$

which in turn is equivalent to saying that the inverse χ^{-1} of $\chi = \begin{bmatrix} A & B \\ C & D \end{bmatrix}$ has the block-form

$$\chi^{-1} = \begin{bmatrix} {}^tD & -{}^tB \\ -{}^tC & {}^tA \end{bmatrix}. \tag{11.19}$$

Notice that if $\chi\colon S_n \longrightarrow S_n$ is a linear symplectic map, then $\chi(\lambda) \in \Lambda(n)$ if $\lambda \in \Lambda(n)$.

It is now possible to make the quadratic forms in $T_\lambda\Lambda(n)$, $\lambda \in \Lambda(n)$, given by Theorem 11.2.6, more explicit. We have the following result, due to Robbin and Salamon [62] (adapted to our choice of symplectic form and quadratic forms $q_{\lambda'}$).

Theorem 11.2.10. *Let $[a,b] \ni t \longmapsto \lambda(t) \in \Lambda(n)$ be a smooth curve of Lagrangian subspaces, with $\lambda(0) = \lambda$ and $\dot\lambda(0) = \lambda_0 \in T_\lambda\Lambda(n)$.*

(i) Let $\mu \in \Lambda(n)$ be a fixed Lagrangian complement of λ, and for $v \in \lambda$ and t small, define $w(t) \in \mu$ by $v + w(t) \in \lambda(t)$. Then the quadratic form on λ,

$$q_{\lambda,\lambda_0}(v) = \frac{d}{dt}\Big|_{t=0} \sigma(v, w(t)),$$

is independent of the choice of μ.

(ii) Let $[a,b] \ni t \longmapsto Z(t) = \begin{bmatrix} X(t) \\ Y(t) \end{bmatrix}$ *be a Lagrangian frame of* $\lambda(t)$ *as in (i). Then*

$$q_{\lambda,\lambda_0}(v) = \langle Y(0)u, \dot{X}(0)u\rangle - \langle X(0)u, \dot{Y}(0)u\rangle, \ \ v = Z(0)u, \ u \in \mathbb{R}^n. \tag{11.20}$$

(iii) The form q_{λ,λ_0} *is* **natural**, *in the sense that*

$$q_{\chi(\lambda),\chi(\lambda_0)} \circ \chi = q_{\lambda,\lambda_0},$$

for any given symplectic matrix χ.

Proof. We may choose coordinates in S_n in such a way that $\lambda(0) = \lambda = \mathbb{R}^n \times \{0\}$.

We start by proving (i). By Lemma 11.2.8 we then have that any given Lagrangian complement of $\lambda(0)$ is the graph of a real $n \times n$ symmetric matrix A. Hence $\mu = \{(Au,u);\ u \in \mathbb{R}^n\}$, and for small t the Lagrangian subspace $\lambda(t)$ is the graph of a real $n \times n$ symmetric matrix $B(t)$, $\lambda(t) = \{(z, B(t)z);\ z \in \mathbb{R}^n\}$. We therefore have that $v = \begin{bmatrix} z \\ 0 \end{bmatrix}$, $w(t) = \begin{bmatrix} Au(t) \\ u(t) \end{bmatrix}$, and by construction $u(t) = B(t)(z + Au(t))$. Thus $\sigma(v, w(t)) = -\langle u(t), z\rangle$, and

$$q_{\lambda,\lambda_0}(v) = -\langle \dot{u}(0), z\rangle = -\langle \dot{B}(0)z, z\rangle,$$

which is independent of A and proves (i).

To prove (ii), assume that $\mu = \{0\} \times \mathbb{R}^n$. Let $Z(t)$ be a Lagrangian frame for $\lambda(t)$. Then $v = \begin{bmatrix} X(0)u \\ Y(0)u \end{bmatrix}$, and $w(t) = \begin{bmatrix} 0 \\ z(t) \end{bmatrix}$, where, by construction,

$$Y(0)u + z(t) = Y(t)X(t)^{-1}X(0)u$$

(since $\lambda(0) = \lambda = \mathbb{R}^n \times \{0\}$, $X(0) = I$ and therefore $X(t)$ is invertible for small t). Thus $\sigma(v, w(t)) = -\langle z(t), X(0)u\rangle$ and, since

$$X(0)^{-1}\dot{X}(0) + \left.\frac{dX(t)^{-1}}{dt}\right|_{t=0} X(0) = 0,$$

this yields

$$q_{\lambda,\lambda_0}(v) = -\langle \dot{z}(0), X(0)u\rangle = \langle (Y(0)X(0)^{-1}\dot{X}(0) - \dot{Y}(0))u, X(0)u\rangle =$$

(using ${}^tXY = {}^tYX$)

$$= \langle Y(0)u, \dot{X}(0)u\rangle - \langle \dot{Y}(0)u, X(0)u\rangle, \ \ \forall u \in \mathbb{R}^n.$$

This proves (ii).

The proof of (iii) is an immediate consquence of the definition. □

Remark 11.2.11. It is useful to "spell out" the construction of $w(t)$ in terms of Theorem 11.2.6. We take a Lagrangian subspace μ transversal to λ, and $\lambda(t)$ transversal to μ (for small t). We then write $\lambda(t) = \{v + \mathsf{A}(t)v;\ v \in \lambda\}$ for $\mathsf{A}(t)\colon \lambda \longrightarrow \mu$, so that, since $\mu \oplus \lambda(t) = S_n$,

$$\lambda \ni v = \underbrace{v_1(t)}_{\in\mu} + \underbrace{v_2(t)}_{\in\lambda(t)} \Longrightarrow v_2(t) = v - v_1(t),$$

that is

$$w(t) = \mathsf{A}(t)v = -v_1(t) = -P^{\lambda}_{\mu,\lambda(t)}v.$$

Hence

$$q_{\lambda,\lambda_0}(v) = \frac{d}{dt}\Big|_{t=0}\sigma(v,w(t)) = -\frac{d}{dt}\Big|_{t=0}\beta^{\lambda}_{\mu,\lambda(t)}(v,v) = \frac{d}{dt}\Big|_{t=0}\sigma(v,\mathsf{A}(t)v),\ v \in \lambda.$$

$\triangle$

Now, since det: $U(n) \longrightarrow \mathbb{S}^1 = \{z \in \mathbb{C};\ |z| = 1\}$ takes values ± 1 on $O(n)$, we have that $\det^2\colon \Lambda(n) \longrightarrow \mathbb{S}^1$ is well-defined (where, recall, $\det^2\colon \mathsf{M}_n \ni A \longmapsto \det(A^2)$). One has the following theorem (see [2]; see also [43]).

Theorem 11.2.12. *The map* $\det^2\colon \Lambda(n) \longrightarrow \mathbb{S}^1$ *induces an isomorphism of fundamental groups* $\pi_1(\Lambda(n)) \simeq \pi_1(\mathbb{S}^1) = \mathbb{Z}$, *called* **the Maslov index of** $\mathbb{R}^n \times \mathbb{R}^n$.

As a consequence, $H^1(\Lambda(n);\mathbb{Z}) \simeq \mathrm{Hom}(\pi_1(\Lambda(n));\mathbb{Z}) \simeq \mathbb{Z}$, with generator the **Maslov class**, that is, the pull-back by $\det^2$ of the generator of $H^1(\mathbb{S}^1;\mathbb{Z})$. The Maslov class depends only on the chosen symplectic structure (see Hörmander [26, p. 156]).

Following Robbin-Salamon [62], *but with the* **opposite** *choice of symplectic form*, we now explain a way to compute the Maslov index of a curve in $\Lambda(n)$ (their discussion is much more general, but we shall content ourselves with considering only the "vertical" case). Let $\lambda_{\mathrm{v}} = \{0\} \times \mathbb{R}^n$ be the "vertical" Lagrangian subspace of S_n. One has that

$$\Lambda(n) = \bigcup_{k=0}^{n} \Lambda_k(\lambda_{\mathrm{v}}),$$

where $\Lambda_k(\lambda_{\mathrm{v}}) := \{\lambda \in \Lambda(n);\ \dim(\lambda \cap \lambda_{\mathrm{v}}) = k\}$. It turns out (see, e.g., [12] or [62]) that $\Lambda_k(\lambda_{\mathrm{v}})$ is a submanifold of $\Lambda(n)$ of codimension $k(k+1)/2$, which is **connected**. The **Maslov cycle** (determined by λ_{v}) is the algebraic variety (a singular hypersurface of $\Lambda(n)$)

$$\Sigma(\lambda_{\mathrm{v}}) = \overline{\Lambda_1(\lambda_{\mathrm{v}})} = \bigcup_{k=1}^{n} \Lambda_k(\lambda_{\mathrm{v}}),$$

whose regular part is $\Lambda_1(\lambda_{\mathrm{v}})$, and whose tangent space at a point $\lambda \in \Lambda_k(\lambda_{\mathrm{v}})$ ($k \geq 1$, of course) is given by (see Theorem 11.2.6, Lemma 11.2.8 and Theorem 11.2.10)

$$T_{\lambda}\Lambda_k(\lambda_v) = \{\tilde{\lambda} \in T_{\lambda}\Lambda(n);\ q_{\lambda,\tilde{\lambda}}|_{\lambda\cap\lambda_v} = 0\}.$$

The cycle $\Sigma(\lambda_v)$ carries a natural orientation. If $\lambda \in \Lambda_1(\lambda_v)$, we have that $T_{\lambda}\Lambda_1(\lambda_v)$ is identified with the space of all symmetric bilinear forms on λ which vanish on the (one-dimensional) line $\lambda \cap \lambda_v$, which is a hyperplane in $\mathrm{Symm}(\lambda) \simeq T_{\lambda}\Lambda(n)$. The *positive side* of $\Sigma(\lambda_v)$ at λ is then defined to be the half tangent space made out of those symmetric bilinear forms on λ whose restriction to $\lambda \cap \lambda_v$ is *positive definite*. Notice that the orientation introduced here is *opposite* to the usually chosen one in [12, 29, 70].

Example 11.2.13. As a matter of example, let us consider $\Lambda(1) = U(1)/O(1)$. Since $U(1) = \mathbb{S}^1 \subset \mathbb{C}$ and $O(1) = \{-1, +1\}$, the map $z \longmapsto z^2$ realizes the quotient as $\mathbb{S}^1 = \mathbb{P}^1(\mathbb{R})$. The point corresponding to the subspace $\lambda_v = i\mathbb{R}$ gives the Maslov cycle. Hence, given a path $\lambda\colon [a,b] \longrightarrow \Lambda(1)$, the Maslov index $\mu_M(\lambda)$ of the path λ (see Definition 11.2.14 below) counts the number of intersections of the path λ with this point (with the appropriate correction at the endpoints). △

Let $\lambda\colon [a,b] \longrightarrow \Lambda(n)$ be a smooth path of Lagrangian subspaces. Robbin and Salamon define a **crossing** for λ some $t \in [a,b]$ such that $\dim(\lambda(t) \cap \lambda_v) \geq 1$, that is, a time $t \in [a,b]$ such that $\lambda(t) \in \Sigma(\lambda_v)$. Hence, if $Z(t) = \begin{bmatrix} X(t) \\ Y(t) \end{bmatrix}$ is a frame for $\lambda(t)$,

$$t \text{ is a crossing iff } \det X(t) = 0.$$

The set of crossings is obviously **compact**. At each crossing $t \in [a,b]$, one defines the **crossing form**

$$\Gamma(\lambda,t)\colon \mathrm{Ker}\, X(t) \ni u \longmapsto \langle \dot{X}(t)u, Y(t)u\rangle \in \mathbb{R}. \tag{11.21}$$

By Theorem 11.2.10 one has that the crossing form is **natural**, in the sense that

$$\Gamma(\chi(\lambda),t)\circ\chi = \Gamma(\lambda,t), \quad \text{for all symplectic matrices } \chi \text{ such that } \chi(\lambda_v) = \lambda_v$$

(notice that, in the block-form (11.19) of χ, the condition $\chi(\lambda_v) = \lambda_v$ corresponds to having $B = 0$).

A crossing t is **regular** when $\Gamma(\lambda,t)$ is **nonsingular**. It can be shown that regular crossings are **isolated**.

Definition 11.2.14 (Robbin and Salamon). Let $\lambda\colon [a,b] \longrightarrow \Lambda(n)$ be a smooth curve with only regular crossings. One defines the Maslov index of the curve λ by

$$\mu_M(\lambda) = \frac{1}{2}\mathrm{sign}\,\Gamma(\lambda,a) + \sum_{\substack{t\in(a,b)\\ t \text{ is a crossing}}} \mathrm{sign}\,\Gamma(\lambda,t) + \frac{1}{2}\mathrm{sign}\,\Gamma(\lambda,b), \tag{11.22}$$

where sign is the signature of the form (the number of positive eigenvalues minus the number of negative eigenvalues).

The following lemmas (due to Robbin and Salamon; see [62]) are fundamental.

Lemma 11.2.15. *Suppose that $\lambda_j\colon [a,b] \longrightarrow \Lambda(n)$, $j=1,2$, with $\lambda_1(a)=\lambda_2(a)$ and $\lambda_1(b)=\lambda_2(b)$, have only regular crossings. If λ_1 and λ_2 are homotopic with fixed endpoints then they have the same Maslov index.*

Lemma 11.2.16. *Every Lagrangian path $\lambda\colon [a,b] \longrightarrow \Lambda(n)$ is homotopic with fixed endpoints to one having only regular crossings.*

The lemmas yield the following theorem, which characterizes μ_M.

Theorem 11.2.17 (Robbin and Salamon). *The Maslov index μ_M is defined for every continuous path in $\Lambda(n)$ and has the following properties.*

- *(Naturality) For all symplectic matrices χ such that $\chi(\lambda_v)=\lambda_v$,*
$$\mu_M(\chi(\lambda)) = \mu_M(\lambda).$$
- *(Catenation) For $c \in (a,b)$,*
$$\mu_M(\lambda) = \mu_M(\lambda|_{[a,c]}) + \mu_M(\lambda|_{[c,b]}).$$
- *(Direct sum) If $n=n'+n''$, we identify $\Lambda(n')\times\Lambda(n'')$ with a submanifold of $\Lambda(n)$ in the following natural way: if $\lambda' = \operatorname{Im} Z' = \operatorname{Im}\begin{bmatrix} X' \\ Y' \end{bmatrix} \in \Lambda(n')$ and $\lambda'' = \operatorname{Im} Z'' = \operatorname{Im}\begin{bmatrix} X'' \\ Y'' \end{bmatrix} \in \Lambda(n'')$, then $\lambda' \oplus \lambda''$ is identified with $\lambda = \operatorname{Im} Z$, where $Z = \left[\begin{array}{cc} X' & 0 \\ 0 & X'' \\ \hline Y' & 0 \\ 0 & Y'' \end{array}\right]$. Then, identifying $\lambda_v' \oplus \lambda_v''$ with λ_v, where λ_v', resp. λ_v'', is the "vertical" Lagrangian subspace of $\mathbb{R}^{n'}\times\mathbb{R}^{n'}$, resp. $\mathbb{R}^{n''}\times\mathbb{R}^{n''}$, one has*
$$\mu_M(\lambda' \oplus \lambda'') = \mu_M(\lambda') + \mu_M(\lambda'').$$
- *(Homotopy) Two paths $\lambda_j\colon [a,b] \longrightarrow \Lambda(n)$, $j=1,2$, with $\lambda_1(a)=\lambda_2(a)$ and $\lambda_1(b)=\lambda_2(b)$ are homotopic with fixed endpoints iff they have the same Maslov index.*
- *(Zero) For every path $\lambda\colon [a,b] \longrightarrow \Lambda_k(\lambda_v)$ one has $\mu_M(\lambda)=0$.*
- *(Normalization) For the loop $\lambda\colon [0,\pi] \longrightarrow \Lambda(1)$ defined by*
$$\lambda(t) = \begin{bmatrix} \cos t & \sin t \\ -\sin t & \cos t \end{bmatrix} \mathbb{R}\times\{0\} \subset S_1 = \mathbb{R}\times\mathbb{R},$$
one has $\mu_M(\lambda)=1$.

Remark 11.2.18. Theorem 11.2.17 is actually a particular case of a more generale theorem proved in [62, Theorem 2.4, p. 831]. Robbin and Salamon define a Maslov index of a curve $\lambda\colon [a,b] \longrightarrow \Lambda(n)$ with respect to any fixed $\lambda_0 \in \Lambda(n)$, denoted by $\mu_M(\lambda;\lambda_0)$ (hence $\mu_M(\lambda) = \mu_M(\lambda;\lambda_v)$). In our framework, we may define $\mu_M(\lambda;\lambda_0)$ as follows. Let $\chi\colon S_n \longrightarrow S_n$ be linear symplectic such that $\chi(\lambda_v) = \lambda_0$. We define

$$\mu_M(\lambda;\lambda_0) := \mu_M(\chi^{-1}(\lambda)).$$

To show that this definition is consistent, take two linear symplectic maps $\chi_j\colon S_n \longrightarrow S_n$, such that $\chi_j(\lambda_v) = \lambda_0$, $j = 1,2$. Then

$$(\chi_2^{-1} \circ \chi_1)(\lambda_v) = \lambda_v,$$

whence, by the *naturality* property of Theorem 11.2.17,

$$\mu_M(\chi_1^{-1}(\lambda)) = \mu_M((\chi_2^{-1} \circ \chi_1)(\chi_1^{-1}(\lambda))) = \mu_M(\chi_2^{-1}(\lambda)),$$

which proves the consistency of the definition. △

Remark 11.2.19. It is important to notice that with this choice of Maslov index, the Hörmander index $s(\mu_1,\mu_2,\lambda_1,\lambda_2)$ (see [12], or [29], or [70]) of Lagrangian subspaces λ_1,λ_2 and μ_1,μ_2, such that λ_j is transverse to μ_k, $j,k = 1,2$, we have

$$s(\mu_1,\mu_2,\lambda_1,\lambda_2) = -\mu_M(\gamma),$$

where γ is a closed curve in $\Lambda(n)$ consisting of an arc of Lagrangian subspaces from λ_1 to λ_2 transversal to μ_1, followed by an arc of Lagrangian subspaces from λ_2 to λ_1 transversal to μ_2. △

As a matter of example, let us now compute $\mu_M(\lambda)$ where λ is the Lagrangian loop given in the last item of the previous theorem. In this case $\Lambda(1) \simeq \mathbb{S}^1$ and the Maslov cycle is a point. Hence

$$\lambda(t) = \{\begin{bmatrix} u\cos t \\ -u\sin t \end{bmatrix};\ u \in \mathbb{R}\},\ \ t \in [0,\pi],$$

$\lambda(0) = \lambda(\pi)$, and the Maslov cycle is therefore given by $\{\lambda(\pi/2)\}$. We have to compute the crossing forms at the crossing times in $[0,\pi]$. One has that $X(t) = \cos t$ and $Y(t) = -\sin t$, so that

$$\det X(t) = 0, t \in [0,\pi] \Longleftrightarrow t = \pi/2,$$

and, according to (11.21),

$$\Gamma(\lambda,\pi/2)(u) = (-\sin(\pi/2))^2 u^2 = 0 \Longleftrightarrow u = 0,$$

that is, $t = \pi/2$ is a regular crossing, with signature 1, whence $\mu_M(\lambda) = 1$.

Remark 11.2.20. The choice of the normalization condition in Theorem 11.2.17 above is the natural one. In fact, we are considering $S_1 = \mathbb{R}_x \times \mathbb{R}_\xi$, with symplectic form $\sigma = d\xi \wedge dx$, which is the opposite of the area form. Hence, the induced orientation on the circle $x^2 + \xi^2 = 1$ is the clockwise one. Notice, moreover, that

$$\begin{bmatrix} Q(t,X) \\ P(t,X) \end{bmatrix} = \begin{bmatrix} \cos t & \sin t \\ -\sin t & \cos t \end{bmatrix} \begin{bmatrix} x \\ \xi \end{bmatrix}, \ X = \begin{bmatrix} x \\ \xi \end{bmatrix},$$

is the integral trajectory issuing from X of the Hamiltonian vector field H_{p_0} associated with the harmonic oscillator $p_0(x,\xi) = (\xi^2 + x^2)/2$ in $\mathbb{R}^2$. The tangent lines to these curves are automatically curves of Lagrangian subspaces, with Maslov index 2. In fact, let $X_0 = (1,0)$. With $\gamma_{X_0}(t) = (Q(t,X_0), P(t,X_0))$, consider

$$[0,2\pi] \ni t \longmapsto \lambda(t) = \text{Span}\{\dot{\gamma}_{X_0}(t)\} = \text{Span}\{\begin{bmatrix} \sin t \\ \cos t \end{bmatrix}\} \in \Lambda(1).$$

Then the crossing points, that are all regular, occur for $t_0 = 0, \pi, 2\pi$, and there we have $\Gamma(\lambda, t_0)(u) = (\cos t_0)^2 u^2 = u^2$, whence the claim. The case $X_0 \neq (1,0)$ follows using the transitivity of the action of linear symplectomorphisms. △

We next compute the Maslov index of the (periodic) bicharacteristics, *suitably lifted to a loop in* $\Lambda(2n)$, of the harmonic oscillator $p_0(X) = |X|^2/2$, $X = (x,\xi) \in S_n = \mathbb{R}^n \times \mathbb{R}^n$. Recall that the integral trajectories $t \longmapsto \Phi^t(X) := \exp(tH_{p_0})(X)$ of H_{p_0} exist for all times, and they are given by

$$\Phi^t(X) = (x\cos t + \xi \sin t, \xi \cos t - x \sin t), \ \ t \in \mathbb{R},$$

i.e. by

$$\begin{bmatrix} (\cos t)I & (\sin t)I \\ (-\sin t)I & (\cos t)I \end{bmatrix} \begin{bmatrix} x \\ \xi \end{bmatrix} =: \begin{bmatrix} Q(t,X) \\ P(t,X) \end{bmatrix},$$

where I is the $n \times n$ identity matrix. Hence they are all periodic (with period 2π). In the first place we have now to lift them so as to give rise to a loop of Lagrangian manifolds, where, recall, a *Lagrangian manifold* is a smooth manifold whose tangent spaces are all Lagrangian. To this purpose we note that the map $X \longmapsto \Phi^t(X)$, for every fixed t, is a *symplectomorphism* of S_n, that is, it is a diffeomorphism and its tangent map is a linear symplectic transformation at each point (the proof is left to the reader). Hence the set $L(t)$ constructed out of its graph,

$$L(t) := \{(Q(t,X), x, P(t,X), -\xi);\ X \in S_n\} \subset S_{2n},$$

is for all t a Lagrangian manifold of S_{2n} endowed with the symplectic form $\sigma \oplus \sigma$. We lift $t \longmapsto \Phi^t(X)$ to the curve $\hat{\gamma}_X \colon [0,2\pi] \ni t \mapsto \hat{\gamma}_X(t) = (Q(t,X), x, P(t,X), -\xi)$ $\in L(t)$. Since $Q(t,X)$ and $P(t,X)$ are linear in X, the tangent space to $L(t)$ at $\hat{\gamma}_X(t) \in L(t)$ is given by

$$\lambda(t) = \{(Q(t,\delta X), \delta x, P(t,\delta X), -\delta\xi);\ \delta X = \begin{bmatrix} \delta x \\ \delta\xi \end{bmatrix} \in S_n\} \in \Lambda(2n),$$

so that $[0,2\pi] \ni t \longmapsto \lambda(t) \in \Lambda(2n)$ is a loop in $\Lambda(2n)$. A corresponding Lagrangian frame for $\lambda(t)$ is given by

$$\mathsf{Z}(t) = \begin{bmatrix} \mathsf{X}(t) \\ \mathsf{Y}(t) \end{bmatrix}, \ \mathsf{X}(t) = \begin{bmatrix} (\cos t)I & (\sin t)I \\ I & 0 \end{bmatrix}, \ \mathsf{Y}(t) = \begin{bmatrix} (-\sin t)I & (\cos t)I \\ 0 & -I \end{bmatrix}.$$

(Since

$${}^t\mathsf{X}(t)\mathsf{Y}(t) = \begin{bmatrix} (-\sin t \cos t)I & (\cos^2 t - 1)I \\ (-\sin^2 t)I & (\sin t \cos t)I \end{bmatrix} = {}^t\mathsf{Y}(t)\mathsf{X}(t),$$

$\lambda(t)$ is indeed Lagrangian, for all t.) We nex take the canonical symplectic basis $(e_1, \ldots, e_n, \varepsilon_1, \ldots, \varepsilon_n)$ of S_n, where e_j has all coordinates zero except for the j-th one which is 1, and ε_j has all the coordinates zero except for the $n+j$-th one which is 1, $j = 1, \ldots, n$. Now, for $t \in [0,2\pi]$,

$$\det \mathsf{X}(t) = 0 \iff \sin t = 0 \iff t = 0, \pi, 2\pi,$$

with

$$\operatorname{Ker} \mathsf{X}(t_0) = \operatorname{Span}\{\varepsilon_1, \ldots, \varepsilon_n\}, \ t_0 = 0, \pi, 2\pi.$$

Since $\dot{\mathsf{X}}(t) = \begin{bmatrix} (-\sin t)I & (\cos t)I \\ 0 & 0 \end{bmatrix}$, we have that the crossing form at $t_0 = 0, \pi, 2\pi$ is

$$\Gamma(\lambda, t_0)(u) = \langle \dot{\mathsf{X}}(t_0)u, \mathsf{Y}(t_0)u \rangle = \langle \begin{bmatrix} 0 & 0 \\ 0 & I \end{bmatrix} u, u \rangle = |u''|^2, \ u = \begin{bmatrix} 0 \\ u'' \end{bmatrix} \in \operatorname{Ker} \mathsf{X}(t_0),$$

where $u'' \in \mathbb{R}^n$. It follows that $0, \pi, 2\pi$ are regular crossings with signature 1, and therefore that

$$\mu_M(\lambda) = \frac{1}{2} + 1 + \frac{1}{2} = 2. \tag{11.23}$$

In some cases one takes a lift of the periodic curve as a periodic integral curve of $\partial/\partial t + H_p$ (where p is a symbol of the kind we are interested in; in the preceding case we considered $p = p_0$). Thus, the problem is: *do we obtain the same value for the Maslov index*? The next proposition gives a positive answer to the problem.

Proposition 11.2.21. *Let p be a smooth positive symbol such that its bicharacteristic flow $(t,X) \longmapsto \exp(tH_p)(X)$ exists for all $(t,X) \in \mathbb{R} \times S_n$. Let $X_0 = (x_0, \xi_0) \in p^{-1}(E)$, and suppose that $\gamma : t \longmapsto \Phi^t(X_0) = \exp(tH_p)(X_0) = (Q(t,X_0), P(t,X_0))$ be periodic with least period $T(E) > 0$. Consider for every $t \in [0, T(E)]$ the lift $\hat{\gamma}(t) = (Q(t,X_0), x_0, P(t,X_0), -\xi_0)$ of γ to*

$$L(t) := \{(Q(t,X), x, P(t,X), -\xi);\ X \in S_n\} \subset S_{2n},$$

and the lift $\tilde{\gamma}(t) = (t, Q(t,X_0), x_0, \tau = -E, P(t,X_0), -\xi_0)$ of γ as an integral trajectory of $\partial/\partial t + H_p$ (the Hamiltonian vector field of $\tau + p(X)$) issuing from $(t_0 = 0, x_0, \tau_0 = -p(X_0) = -E, \xi_0)$, contained in the Lagrangian manifold

$$\tilde{L} := \{(t, Q(t,X), x, \tau, P(t,X), -\xi);\ \tau = -p(X),\ (t,X) \in \mathbb{R} \times S_n\} \subset S_{2n+1}.$$

Let $[0,T(E)] \ni t \mapsto \lambda(t) = T_{\hat{\gamma}(t)}L(t)$ *and* $[0,T(E)] \ni t \mapsto \tilde{\lambda}(t) = T_{\tilde{\gamma}(t)}\tilde{L}$ *be the corresponding loops in* $\Lambda(2n)$ *and* $\Lambda(2n+1)$ *(for the latter we identify* $T(E)$ *with* 0*). Then*

$$\mu_M(\lambda) = \mu_M(\tilde{\lambda}).$$

Proof. It is clear that $L(t)$ and $\tilde{L}$ are Lagrangian submanifolds of S_{2n} and S_{2n+1}, respectively (in the case of $\tilde{L}$ one uses the fact that $p(\Phi^t(X)) = p(X)$, for all $t \in \mathbb{R}$). We rewrite

$$\tilde{L} = \{(t, Q(t,X), x, -p(X), P(t,X), -\xi);\ (t,X) \in \mathbb{R} \times S_n\}.$$

At first we consider $\lambda(t)$. In this case (denoting by Q'_x the Jacobian matrix of Q with respect to x, etc.)

$$\mathsf{X}(t) = \begin{bmatrix} Q'_x(t,X_0) & Q'_\xi(t,X_0) \\ I & 0 \end{bmatrix},\quad \mathsf{Y}(t) = \begin{bmatrix} P'_x(t,X_0) & P'_\xi(t,X_0) \\ 0 & -I \end{bmatrix}.$$

Hence

$$\operatorname{Ker}\mathsf{X}(t) = \{u = \begin{bmatrix} \delta x = 0 \\ \delta\xi \end{bmatrix};\ \delta\xi \in \operatorname{Ker} Q'_\xi(t,X_0)\},$$

and for the crossing points, i.e. those times $t \in [0,T(E)]$ for which $\operatorname{Ker}\mathsf{X}(t) \neq \{0\}$, i.e. $\operatorname{Ker} Q'_\xi(t,X_0) \neq \{0\}$, the crossing form is

$$\Gamma(\lambda,t)(u) = \langle \partial_t Q'_\xi(t,X_0)\delta\xi, P'_\xi(t,X_0)\delta\xi \rangle,\ u = \begin{bmatrix} 0 \\ \delta\xi \end{bmatrix} \in \operatorname{Ker}\mathsf{X}(t).$$

We now consider $\tilde{\lambda}(t)$. In this case, by writing p'_ξ for the row-vector of ξ-derivatives of p etc.,

$$\tilde{\mathsf{X}}(t) = \begin{bmatrix} 1 & 0 & 0 \\ {}^t p'_\xi(\Phi^t(X_0)) & Q'_x(t,X_0) & Q'_\xi(t,X_0) \\ 0 & I & 0 \end{bmatrix},$$

$$\tilde{\mathsf{Y}}(t) = \begin{bmatrix} 0 & -p'_x(X_0) & -p'_\xi(X_0) \\ {}^t p'_x(\Phi^t(X_0)) & P'_x(t,X_0) & P'_\xi(t,X_0) \\ 0 & 0 & -I \end{bmatrix}.$$

Hence

$$\operatorname{Ker}\tilde{\mathsf{X}}(t) = \{\tilde{u} = \begin{bmatrix} \delta t = 0 \\ \delta x = 0 \\ \delta\xi \end{bmatrix};\ \delta\xi \in \operatorname{Ker} Q'_\xi(t,X_0)\},$$

and for the crossing points, i.e. those $t \in [0, T(E)]$ for which $\operatorname{Ker}\tilde{X}(t) \neq \{0\}$, the crossing form is

$$\tilde{\Gamma}(\tilde{\lambda}, t)(\tilde{u}) = \langle \partial_t Q'_\xi(t, X_0)\delta\xi, P'_\xi(t, X_0)\delta\xi \rangle, \quad \tilde{u} = \begin{bmatrix} 0 \\ 0 \\ \delta\xi \end{bmatrix} \in \operatorname{Ker}\tilde{X}(t).$$

We therefore have that

- *$t \in [0, T(E)]$ is a crossing point for λ iff it is a crossing point for $\tilde{\lambda}$;*
- *the respective crossing forms there are equal;*
- *the crossing point is regular for λ iff it is regular for $\tilde{\lambda}$.*

Hence, if λ does not have any crossings with λ_v, then neither has $\tilde{\lambda}$ with $\tilde{\lambda}_v = \{0\} \times \mathbb{R}^{2n+1}$, and $\mu_M(\lambda) = \mu_M(\tilde{\lambda}) = 0$.

So we may suppose there is at least one crossing time t_0, necessarily the same for both λ and $\tilde{\lambda}$. Define, for each $s \in [0, 1]$, the loop in $\Lambda(2n+1)$ defined by $[0, T(E)] \ni t \longmapsto \tilde{\lambda}_s(t) = \operatorname{Im}\tilde{Z}_s(t)$, where $\tilde{Z}_s(t) = \begin{bmatrix} \tilde{X}_s(t) \\ \tilde{Y}_s(t) \end{bmatrix}$, with

$$\tilde{X}_s(t) = \begin{bmatrix} 1 & 0 & 0 \\ s^t p'_\xi(\Phi^t(X_0)) & Q'_x(t, X_0) & Q'_\xi(t, X_0) \\ 0 & I & 0 \end{bmatrix},$$

$$\tilde{Y}_s(t) = \begin{bmatrix} 0 & -s p'_x(X_0) & -s p'_\xi(X_0) \\ s^t p'_x(\Phi^t(X_0)) & P'_x(t, X_0) & P'_\xi(t, X_0) \\ 0 & 0 & -I \end{bmatrix}.$$

It is then clear from the above discussion that

- *for all $s \in [0, 1]$, $t \in [0, T(E)]$ is a crossing for $\tilde{\lambda}_s$ iff it is a crossing for $\tilde{\lambda} = \tilde{\lambda}_1$, and the respective crossing forms there are all equal (hence independent of $s \in [0, 1]$).*

We now show that

$$\mu_M(\tilde{\lambda}) = \mu_M(\tilde{\lambda}_0).$$

In the first place we observe that the curve $[0, 1] \ni s \longmapsto \tilde{\lambda}_s(t_0) \subset \Lambda_k(\tilde{\lambda}_v)$ where $k = \dim \operatorname{Ker} Q'_\xi(t_0, X_0)$ (since $\tilde{\lambda}_s(t_0) \cap \tilde{\lambda}_v$ is the injective image of $\operatorname{Ker} Q'_\xi(t_0, X_0)$), and therefore, by the *zero* property, has Maslov index 0. Next, one constructs a loop $\hat{\lambda}$ that starts at $\tilde{\lambda}_0(t_0)$, follows $\tilde{\lambda}_s(t_0)$, $s \in [0, 1]$, to reach $\tilde{\lambda}_1(t_0) = \tilde{\lambda}(t_0)$, then follows $\tilde{\lambda}(t)$ (with its orientation) till it comes back to $\tilde{\lambda}(t_0)$, and finally follows $\tilde{\lambda}_s(t_0)$, $s \in [0, 1]$, backward from $\tilde{\lambda}(t_0)$ to $\tilde{\lambda}_0(t_0)$. Hence, by the *catenation* and *zero* properties, we have that $\mu_M(\tilde{\lambda}) = \mu_M(\hat{\lambda})$. Since $\tilde{\lambda}_0$ is homotopic to $\hat{\lambda}$ with fixed endpoints (by using a riparametrization of the family $(t, s) \mapsto \tilde{\lambda}_s(t)$), we have, by the *homotopy property*, that $\mu_M(\hat{\lambda}) = \mu_M(\tilde{\lambda}_0)$, which proves the claim.

Let now $[0,T(E)]\times[0,1]\ni(t,s)\mapsto\lambda_s(t)=\operatorname{Im}\mathsf{Z}_s(t)$, $\mathsf{Z}_s(t)=\begin{bmatrix}\mathsf{X}_s(t)\\\mathsf{Y}_s(t)\end{bmatrix}$, be a homotopy in $\Lambda(2n)$ of the loop $\lambda=\lambda_0$ to a loop λ_1 (with fixed endpoints) with only regular crossings. Then $\mu_M(\lambda)=\mu_M(\lambda_1)$. Now,

$$[0,T(E)]\times[0,1]\ni(t,s)\longmapsto\mathsf{Z}(t,s):=\begin{bmatrix}1&0\\0&\mathsf{X}_s(t)\\\hline 0&0\\0&\mathsf{Y}_s(t)\end{bmatrix}$$

defines a homotopy (with fixed endpoints) between $\tilde{\lambda}_0$ and the lift $\operatorname{Im}\mathsf{Z}(t,1)$ of λ_1 to $\Lambda(2n+1)$. Hence $\mu_M(\tilde{\lambda}_0)=\mu_M(\operatorname{Im}\mathsf{Z}(\cdot,1))$. Since $\operatorname{Im}\mathsf{Z}(\cdot,1)$ has only regular crossings, which, by the above considerations, are the same as those of λ_1, with equal respective crossing forms there, we have that $\mu_M(\operatorname{Im}\mathsf{Z}(\cdot,1))=\mu_M(\lambda_1)$, and this concludes the proof. □

Proposition 11.2.21 and the previous discussion thus justify the following definition of Maslov index of a periodic bicharacteristic of a smooth, real-valued function p.

Definition 11.2.22. Let the bicharacteristic flow of p be defined for all $(t,X)\in\mathbb{R}\times S_n$. Let $X_0\in p^{-1}(E)$, and let $\gamma_{X_0}=\{\exp(tH_p)(X_0)=(Q(t,X_0),P(t,X_0));\ 0\le t\le T(X_0)\}$ be a periodic trajectory, with least period $T(X_0)>0$. Consider the lifts $[0,T(X_0)]\ni t\longmapsto\lambda(t)\in\Lambda(2n)$ and $[0,T(X_0)]\ni t\longmapsto\tilde{\lambda}(t)\in\Lambda(2n+1)$ of γ_{X_0} to $\Lambda(2n)$ and $\Lambda(2n+1)$, respectively, given in Proposition 11.2.21. One defines the Maslov index of γ_{X_0} by

$$\mu_M(\gamma_{X_0})=\mu_M(\lambda)=\mu_M(\tilde{\lambda}). \tag{11.24}$$

Remark 11.2.23. Suppose p is a smooth, real-valued function of which we only know that it satisfies Assumption 11.1.1 for an energy interval $[E_1-\varepsilon,E_2+\varepsilon]$, with no other information on its bicharacteristic flow for other energy values. Take $E\in[E_1,E_2]$. Let $X_0\in p^{-1}(E)$ and let $\gamma_{X_0}\subset p^{-1}(E)$ be a periodic trajectory through X_0, with least period $T(E)>0$. Since the bicharacteristics of p, when $X\in p^{-1}([E_1-\varepsilon,E_2+\varepsilon])$, exist for all times, as done in Proposition 11.2.21 one may consider the Lagrangian submanifold of S_{2n}, resp. S_{2n+1},

$$L(t):=\{(Q(t,X),x,P(t,X),-\xi);\ X\in p^{-1}(E_1-\varepsilon,E_2+\varepsilon)\},$$

resp.

$$\tilde{L}:=\{(t,Q(t,X)x,\tau,P(t,X),-\xi);\ \tau=-p(X),\ (t,X)\in\mathbb{R}\times p^{-1}(E_1-\varepsilon,E_2+\varepsilon)\},$$

and hence may construct, on recalling the lifts $\hat{\gamma}_{X_0}$ and $\tilde{\gamma}_{X_0}$ of γ_{X_0} considered in Proposition 11.2.21, the loops $[0,T(E)]\ni t\longmapsto\lambda(t)=T_{\hat{\gamma}_{X_0}(t)}L(t)\in\Lambda(2n)$ and

$[0,T(E)] \ni t \longmapsto \tilde{\lambda}(t) = T_{\tilde{\gamma}_{X_0}(t)}\tilde{L} \in \Lambda(2n+1)$ (where, again, for the latter we identify $T(E)$ with 0). Hence the Maslov index of a periodic trajectory $\gamma_{X_0} \subset p^{-1}(E)$ is defined according to Definition 11.2.22. △

From Lemma 11.1.11 and the foregoing discussion we immediately have the following fundamental result.

Proposition 11.2.24. *Let $n = 1$. Let p be a smooth, real-valued function satisfying Assumption 11.1.1 for an energy interval $[E_1 - \varepsilon, E_2 + \varepsilon]$. Let $E \in [E_1, E_2]$. Then $\gamma_{X_0} := p^{-1}(E)$, $X_0 \in p^{-1}(E)$, is homotopic to a circle, with orientation depending on whether $\nabla_X p\big|_{\gamma_{X_0}}$ points outwards from γ_{X_0} (clockwise orientation) or not (counter-clockwise orientation), and $\mu_M(\gamma_{X_0}) = 2$ in the former case, $\mu_M(\gamma_{X_0}) = -2$ in the latter.*

When $n \geq 1$, $p\colon \mathbb{R}^{2n} \setminus 0 \longrightarrow \mathbb{R}_+$ is smooth, positively homogeneous of positive degree, and $\gamma_{X_0} \subset p^{-1}(E)$ is a periodic trajectory, then $\mu_M(\gamma_{X_0}) = 2$.

Remark 11.2.25. As shown by Duistermaat in [11], using the fact that for fixed t the map $X \longmapsto \Phi^t(X) = \exp(tH_p)(X) = (Q(t;X), P(t;X))$ is a symplectomorphism (hence its Jacobian $d_X\Phi^t(X_0)$ at any given X_0 is a linear symplectic map), one may compute the Maslov index of a periodic curve γ_{X_0} as in Definition 11.2.22 (or Remark 11.2.23) by computing the Maslov index of the curve $[0, T(X_0)] \ni t \longmapsto (d_X\Phi^t(X_0))^{-1}(\lambda_v) \in \Lambda(n)$. Since $d_X\Phi^t(X_0) = \begin{bmatrix} Q'_x(t;X_0) & Q'_\xi(t;X_0) \\ P'_x(t;X_0) & P'_\xi(t;X_0) \end{bmatrix}$, we have

$$(d_X\Phi^t(X_0))^{-1} = \begin{bmatrix} {}^tP'_\xi(t;X_0) & -{}^tQ'_\xi(t;X_0) \\ -{}^tP'_x(t;X_0) & {}^tQ'_x(t;X_0) \end{bmatrix},$$

so that

$$\lambda(t) = \{\begin{bmatrix} -{}^tQ'_\xi(t;X_0)\delta\xi \\ {}^tQ'_x(t;X_0)\delta\xi \end{bmatrix};\ \delta\xi \in \mathbb{R}^n\}.$$

Notice that

$$\lambda(t) \cap \lambda_v \neq \{0\} \Longleftrightarrow \operatorname{Ker}{}^tQ'_\xi(t;X_0) \neq \{0\} \Longleftrightarrow \det Q'_\xi(t;X_0) \neq 0,$$

i.e. iff $\operatorname{Ker} Q'_\xi(t;X_0) \neq \{0\}$. △

11.3 Notes

Maslov introduced his index (see Maslov [42]) for studying problems related to quantum mechanics (high-frequency approximation and semiclassical approximation). It was Lax who noted that the asymptotic expansions of geometrical optics could be used to construct solutions (at least locally) to some PDEs. The point is to approximate solutions to a particular PDE by formal solutions of the kind

$$e^{i\omega\phi}\sum_{j\geq 0}\omega^{-j}a_j,\ \ \omega\to+\infty,$$

or more generally of the kind

$$\sum_1^m (2\pi)^{-n}\iint e^{i(\phi_k(x,y,\xi)-\langle y,\xi\rangle)}a_k(x,y,\xi)f(y)dyd\xi,$$

with $a_k\sim\sum_{j\geq 0}a_{k,j}$, where ϕ, ϕ_j are (in general) real *phase-functions* and where the terms a_j and $a_{k,j}$ are the *transport terms*. This gave rise to the modern theory of Fourier Integral Operators, which is due to Hörmander. The theory necessarily requires for the definition of the principal symbol the use of the *Keller-Maslov* bundle: a complex line bundle, whose trivializations are constructed in terms of the Maslov index. We address the reader to the fundamental "FIO" paper by Hörmander [26] (and the second fundamental "FIO" paper by Duistermaat and Hörmander [9]) and the books about FIO theory by Hörmander [29, 30], Duistermaat [12], Treves [70], Helffer [17] (for the theory in the framework of global symbols), and Robert [65] (for the theory in the semiclassical setting), and to the nice papers of Eckmann-Sénéor [14] and Marsden-Weinstein [41] for elementary applications of the Maslov index to Schrödinger equations. The geometric discussion of caustics can by found in Duistermaat [10]. Constructions of semiclassical approximations can also be found in the book by Mishchenko, Shatalov and Sternin [44].

Chapter 12
Localization and Multiplicity of a Self-Adjoint Elliptic 2×2 Positive NCHO in $\mathbb{R}^n$

Using the machinery developed in the past chapters, we can now describe the beautiful connection between the spectrum of a *self-adjoint elliptic* 2×2 *positive* NCHO in n variables, and the periods of the *bicharacteristics* of the principal symbol (i.e., in the case of systems, the periods of the bicharacteristic curves associated with the eigenvalues of the principal symbol). This connection is a very deep result and its history, in the *scalar* case, is based on fundamental papers by Chazarain, Coline de Verdière, Duistermaat and Guillemin, Ivrii, Helffer and Robert (see also Cardoso and Mendoza [4]), and many others (for such results in the semiclassical setting, see Chazarain [5], Dimassi-Sjöstrand [7], Dozias [8], Helffer-Robert [18, 20], Ivrii [34], Robert [65]). However, very few are the results for systems (see Ivrii [34], Parmeggiani [52, 55]). Our results here (see Section 12.3) for elliptic NCHOs in $\mathbb{R}^n$, whose symbols have distinct eigenvalues, are based on the approach followed in Parmeggiani [55]. In Section 12.4 we shall specialize these results to NCHOs of the kind $Q^{\mathrm{w}}_{(\alpha,\beta)}(x,D)$ (with $\alpha,\beta > 0$ and $\alpha\beta > 1$), and in the final Section 12.5 we shall give a slightly more general result (namely, when the subprincipal part of the diagonalization has constant average on periodic bicharacteristics).

We now briefly describe the strategy we will be using. Since the eigenvalues of a diagonalizable second order GPD system are not smooth at $(0,0) \in \mathbb{R}^{2n}$, we shall consider a special set of symbols (which is still meaningful, for its Weyl-quantization contains our preferred NCHOs $Q^{\mathrm{w}}_{(\alpha,\beta)}(x,D)$ for $\alpha,\beta > 0$, $\alpha \neq \beta$ and $\alpha\beta > 1$), whose eigenvalues are distinct and can be regularized keeping them distinct (so as not to destroy the diagonalizability property). This will allow us to construct a *reference operator* that can be h-Weyl quantized, and whose spectral properties, for large energies, will be analyzed by *semiclassical* techniques through a full decoupling modulo $O(h^\infty)$ and by using the well-known and very precise results for scalar operators (recalled in Section 12.2) due (mainly) to Helffer and Robert. The spectral properties of the reference operator, such as *clustering* and multiplicity of the large eigenvalues, approximate in a precise way (by the results of Chapter 10, Section 10.2) those of the h-quantization of the system under study. Finally, we will use Lemma 9.4.3 to get rid of the semiclassical parameter h and obtain the result for the NCHO under study in the usual Weyl-quantization.

A. Parmeggiani, *Spectral Theory of Non-Commutative Harmonic Oscillators: An Introduction*, Lecture Notes in Mathematics 1992,
DOI 10.1007/978-3-642-11922-4_12, © Springer-Verlag Berlin Heidelberg 2010

12.1 The Set $\mathscr{Q}_2$ and Its Semiclassical Deformation

In this section, we consider a particular set of symbols of 2 × 2 NCHOs that contains our $Q^{\mathrm{w}}_{(\alpha,\beta)}(x,D)$ (for $\alpha,\beta > 0$, $\alpha \neq \beta$ and $\alpha\beta > 1$), and show how to construct a *reference operator*, whose spectral properties (for large energies) can be studied by means of Semiclassical Analysis.

Definition 12.1.1. We say that $A_2 = \begin{bmatrix} a_{11} & a_{12} \\ a_{21} & a_{22} \end{bmatrix} \in S_{\mathrm{cl}}(m^2, g; \mathsf{M}_2)$ belongs to $\mathscr{Q}_2$ if

$$\begin{cases} X \longmapsto A_2(X) \text{ is an } \mathsf{M}_2\text{-valued homogeneous polynomial in } X \text{ of degree } 2, \\ A_2(X) = A_2(X)^* > 0 \text{ and } \det A_2(X) \approx |X|^4,\ \forall X \in \mathbb{R}^{2n} \setminus \{0\}, \\ a_{11}(X) \neq a_{22}(X),\ \ \forall X \in \mathbb{R}^{2n} \setminus \{0\}. \end{cases} \tag{12.1}$$

Notice that since $\mathbb{R}^{2n} \setminus \{0\}$ is connected for all $n \geq 1$, the last condition may be rephrased by saying that

$$\text{either } a_{11}(X) > a_{22}(X), \ \text{ or } \ a_{22}(X) > a_{11}(X), \ \ \forall X \in \mathbb{R}^{2n} \setminus \{0\}.$$

We shall be interested mainly in the following subclass of $\mathscr{Q}_2$.

Definition 12.1.2. We denote by $\mathscr{Q}_2^s$ the subset of all $A_2 \in \mathscr{Q}_2$ such that

$$\text{either } a_{12}(X) \in \mathbb{R} \text{ for all } X \in \mathbb{R}^{2n}, \text{ or } a_{12}(X) \in i\mathbb{R} \text{ for all } X \in \mathbb{R}^{2n}. \tag{12.2}$$

Remark 12.1.3. For all $\alpha \neq \beta$ with $\alpha,\beta > 0$ and $\alpha\beta > 1$ we have that

$$Q_{(\alpha,\beta)} \in \mathscr{Q}_2^s \subset \mathscr{Q}_2.$$

Notice that when $\alpha = \beta > 1$ the operator $Q^{\mathrm{w}}_{(\alpha,\alpha)}(x,D)$ is completely understood: it is unitarily equivalent to a scalar harmonic oscillator. Also, when $A_2 = A_2^* \in S_{\mathrm{cl}}(m^2, g; \mathsf{M}_2)$ fulfills condition (12.2) with $a_{12}(X) = \gamma \tilde{a}_{12}(X)$, where either $\gamma \in \mathbb{R}$ or $\gamma \in i\mathbb{R}$ and where $\tilde{a}_{12}$ is real-valued, and with, in addition, $a_{11}(X) = a_{22}(X)$ for all X, we may find a **constant** unitary transformation $U \colon \mathbb{C}^2 \longrightarrow \mathbb{C}^2$ such that

$$U^* A_2(X) U = \begin{bmatrix} a_{11}(X) - \tilde{a}_{12}(X) & 0 \\ 0 & a_{11}(X) + \tilde{a}_{12}(X) \end{bmatrix}, \ \ \forall X \in \mathbb{R}^{2n},$$

whence the spectral properties of its Weyl-quantization may be studied by scalar techniques. This is the reason why we are interested in generalizations of the case $\alpha \neq \beta$, which gives rise to the "strange" requirement in the definition of $\mathscr{Q}_2$ that $a_{11}(X) \neq a_{22}(X)$ for all $X \neq 0$.

Of course, it is a very interesting problem to understand the case in which $a_{11}(X) = a_{22}(X)$ for some $0 \neq X \in \mathbb{R}^{2n}$.

We remark that the case in which condition (12.2) is not fulfilled will be (partially) studied in Section 12.5 below. △

Let

$$\mathbb{G}_2 := \text{subgroup of } GL_2(\mathbb{C}) \text{ generated by } I, J, K \text{ and } \mathbb{S}^1,$$

where, recall,

$$I = \begin{bmatrix} 1 & 0 \\ 0 & 1 \end{bmatrix}, \; J = \begin{bmatrix} 0 & -1 \\ 1 & 0 \end{bmatrix}, \; K = \begin{bmatrix} 0 & 1 \\ 1 & 0 \end{bmatrix},$$

and where

$$\mathbb{S}^1 = \{\omega \in \mathbb{C}; \; |\omega| = 1\}$$

is embedded in $GL_2(\mathbb{C})$ by the map $\omega \longmapsto \omega I$. Note that $U \in \mathbb{G}_2$ implies that $UU^* = U^*U = I$.

It is then immediate to prove the following lemma.

Lemma 12.1.4. *We have the following "invariance" property of the set $\mathscr{Q}_2$: given any $A_2 \in \mathscr{Q}_2$*

$$U^*A_2U \in \mathscr{Q}_2, \; A_2 \circ \kappa \in \mathscr{Q}_2, \tag{12.3}$$

for all $U \in \mathbb{G}_2$ and all $\kappa \colon \mathbb{R}^{2n} \longrightarrow \mathbb{R}^{2n}$ **linear** *symplectic transformations.*

The same holds true for $\mathscr{Q}_2^s$.

Remark 12.1.5. Notice that in Lemma 12.1.4 we consider only **linear** symplectic maps, instead of general affine symplectic transformations. This is due to the fact that we wish to preserve homogeneous polynomials of degree 2. △

Let hence $A_2 \in \mathscr{Q}_2$ with $A_2(X) = \begin{bmatrix} a_{11}(X) & a_{12}(X) \\ \overline{a_{12}(X)} & a_{22}(X) \end{bmatrix}$. Put

$$a_{\pm}(X) := \frac{a_{11}(X) \pm a_{22}(X))}{2},$$

and let

$$\lambda_{\pm}(X) := a_{+}(X) \pm \sqrt{a_{-}(X)^2 + |a_{12}(X)|^2}, \; X \in \mathbb{R}^{2n}, \tag{12.4}$$

be the eigenvalues of $A_2(X)$.

Lemma 12.1.6. *Let $A_2 \in \mathscr{Q}_2$. Then for the discriminant we have*

$$\delta(X) := a_{-}(X)^2 + |a_{12}(X)|^2 \approx p_0(X)^2, \; \forall X \in \mathbb{R}^{2n}, \tag{12.5}$$

where, recall, $p_0(X) = |X|^2/2$. Hence, since A_2 belongs to $\mathscr{Q}_2$, we have that $\lambda_{\pm}$ are smooth in $\mathbb{R}^{2n} \setminus \{0\}$, positively homogeneous of degree 2 and

$$0 < \lambda_{-}(X) < \lambda_{+}(X), \; \forall X \in \mathbb{R}^{2n} \setminus \{0\}.$$

By first restricting to $\mathbb{S}^{2n-1}$*, we therefore get by the homogeneity that for* $0 < c_1 < c_2$

$$c_1 p_0(X) \leq \lambda_-(X) < \lambda_+(X) \leq c_2 p_0(X), \ \forall X \in \mathbb{R}^{2n} \setminus \{0\}. \tag{12.6}$$

Proof. We need only prove (12.5). By the third condition in (12.1) we have

$$\frac{c_0}{4} \leq a_-(\omega)^2 = (a_{11}(\omega) - a_{22}(\omega))^2 \leq \frac{C_0}{4}, \ \forall \omega \in \mathbb{S}^{2n-1},$$

and

$$|a_{12}(\omega)|^2 \leq C_0', \ \forall \omega \in \mathbb{S}^{2n-1},$$

for suitable $c_0, C_0, C_0' > 0$. The result therefore follows by homogeneity. □

Define now, for an energy $E \geq 0$, the sets

$$\Sigma_\pm(E) := \{X \in \mathbb{R}^{2n};\ \lambda_\pm(X) = E\}.$$

Since $\lambda_\pm$ are continuous and $\lambda_\pm(X) \longrightarrow +\infty$ as $|X| \to +\infty$, the sets $\Sigma_\pm(E)$ are *compact* for all $E \geq 0$ and they both reduce to $\{0\}$ when $E = 0$ (by 12.6)).

Consider next, for $A_2 \in \mathscr{Q}_2$, the operator $A_2^{\mathrm{w}}(x, hD)$ thought of as an unbounded self-adjoint operator

$$A_2^{\mathrm{w}}(x, hD) \colon B^2(\mathbb{R}^n; \mathbb{C}^2) \subset L^2(\mathbb{R}^n; \mathbb{C}^2) \longrightarrow L^2(\mathbb{R}^n; \mathbb{C}^2), \tag{12.7}$$

which thus has a discrete spectrum, bounded from below, made of a diverging (to $+\infty$) sequence of eigenvalues with finite multiplicities.

We want to understand the large eigenvalues of $A_2^{\mathrm{w}}(x, D)$ (i.e. with $h = 1$). Our strategy will be as follows:

- **Step 1:** We study the large eigenvalues of the h-Weyl quantization $A_2^{\mathrm{w}}(x, hD)$ given in (12.7), which has the same domain $B^2(\mathbb{R}^n; \mathbb{C}^2)$ and properties as $A_2^{\mathrm{w}}(x, D)$, and whose spectrum is linked to that of $A_2^{\mathrm{w}}(x, D)$ through Corollary 9.4.5 by the relation

$$\lambda_j \in \mathrm{Spec}(A_2^{\mathrm{w}}(x, D)) \Longleftrightarrow h\lambda_j = \lambda_j(h) \in \mathrm{Spec}(A_2^{\mathrm{w}}(x, hD)), \ j \geq 1.$$

- **Step 2:** We approximate $A_2^{\mathrm{w}}(x, hD)$ by a *reference* operator $A_r^{\mathrm{w}}(x, hD)$, whose symbol $A_r(X)$ has the same eigenvalues of $A_2(X)$ for large X, that can be h^∞-diagonalized into a couple of scalar operators whose principal symbols are the eigenvalues of $A_2(X)$ for large X.
- **Step 3:** We use the precise, well-known, theorems, which we will recall in Section 12.2, that describe the location (and in some cases the multiplicity) of the spectrum of scalar h-pseudodifferential operators in terms of the dynamical quantities, related to the principal and subprincipal symbols, that were discussed in Chapter 11.

- **Step 4:** We use the results of Chapter 10, Section 10.2 to transfer the information about the spectrum of the diagonalization of $A_r^{\mathrm{w}}(x,hD)$ to information about the spectrum of $A_2^{\mathrm{w}}(x,hD)$. By **Step 1** we hence obtain information about the spectrum of $A_2^{\mathrm{w}}(x,D)$.

Following Evans-Zworski [15] and Parmeggiani [55], we now prepare the ground for the localization of large eigenvalues.

The first result concerns uniform h^∞-estimates.

Proposition 12.1.7. *Let $E>0$. Let $a\in S_\delta^0(1;\mathsf{M}_2)$ with $\operatorname{supp} a\subset K$, where $K\subset\mathbb{R}^{2n}$ is a* **compact** *set* **independent of** *h such that*

$$K\cap\Big(\Sigma_+(E)\cup\Sigma_-(E)\Big)=\emptyset,$$

and let $\gamma:=\operatorname{dist}(K,\Sigma_+(E)\cup\Sigma_-(E))$. Let $u(h)\in L^2(\mathbb{R}^n;\mathbb{C}^2)$ solve the eigenvalue equation $A_2^{\mathrm{w}}(x,hD)u(h)=\lambda(h)u(h)$. If $|\lambda(h)-E|<\varepsilon$, for some $\varepsilon\in(0,1/2]$ sufficiently small (compared to E and to γ), then

$$\|a^{\mathrm{w}}(x,hD)u(h)\|_0=O(h^\infty)\|u(h)\|_0$$

(where the constants in $O(h^\infty)$ are allowed to depend on E and γ).

Proof. The set $\Sigma_+(E)\cup\Sigma_-(E)=:K'$ is compact. We may therefore find $\chi\in C_0^\infty(\mathbb{R}^{2n})$, with $0\le\chi\le 1$, such that $\chi\equiv 1$ near K' and $\chi\equiv 0$ near K. Define

$$b(X)=b_h(X)=A_2(X)-\lambda(h)I+i\chi(X)I\in S_0^0(m^2;\mathsf{M}_2).$$

If ε is sufficiently small (with respect to E and γ) we then have that

$$|\det b(X)|^2=\Big(|\lambda_+(X)-\lambda(h)|^2+\chi(X)^2\Big)\Big(|\lambda_-(X)-\lambda(h)|^2+\chi(X)^2\Big)>0,$$

for all $X\in\mathbb{R}^{2n}$, and

$$C^{-1}\le\frac{|\det b(X)|^2}{m(X)^4}\le C,$$

for some $C=C_{E,\gamma}>0$. Hence we may find $c\in S_0^0(m^{-2};\mathsf{M}_2)$ and $r\in S_0^0(1;\mathsf{M}_2)$ such that

$$c^{\mathrm{w}}(x,hD)b^{\mathrm{w}}(x,hD)=I+r^{\mathrm{w}}(x,hD),\ \text{ and }\ r^{\mathrm{w}}=O(h^\infty),$$

where by $r^{\mathrm{w}}=O(h^\infty)$ we mean (as before in Section 10.2) $\|r^{\mathrm{w}}\|_{L^2\to L^2}=O(h^\infty)$. Then

$$a^{\mathrm{w}}(x,hD)c^{\mathrm{w}}(x,hD)b^{\mathrm{w}}(x,hD)=a^{\mathrm{w}}(x,hD)+O(h^\infty),$$

where

$$b^{\mathrm{w}}(x,hD)=A_2^{\mathrm{w}}(x,hD)-\lambda(h)+i\chi^{\mathrm{w}}(x,hD).$$

We now claim that

$$a^{\mathrm{w}}(x,hD)c^{\mathrm{w}}(x,hD)\chi^{\mathrm{w}}(x,hD) = O(h^{\infty}). \tag{12.8}$$

When $\delta \in [0,1/2)$ the claim immediately follows, for we have

$$\operatorname{supp} a \cap \operatorname{supp} \chi = \emptyset \text{ uniformly in } h.$$

When $\delta = 1/2$, we have that for any given $N \in \mathbb{Z}_+$,

$$c^{\mathrm{w}}(x,hD)\chi^{\mathrm{w}}(x,hD) = (c\sharp\chi)_N^{\mathrm{w}}(x,hD) + O(h^{N+1}),$$

where

$$(c\sharp\chi)_N(X) := \sum_{k=0}^{N} \frac{1}{k!}\left(\frac{ih}{2}\sigma(D_X;D_Y)\right)^k c(X)\chi(Y)|_{X=Y}$$

has compact support (in fact $\operatorname{supp}(c\sharp\chi)_N \subset \operatorname{supp}\chi$), uniformly disjoint from $\operatorname{supp} a$. We are hence in a position to use (9.10) and obtain (by the continuity) the claim also when $\delta = 1/2$.

In all cases, on the other hand, from $A_2^{\mathrm{w}}(x,D)u(h) = \lambda(h)u(h)$ it follows

$$\begin{aligned} a^{\mathrm{w}}(x,hD)u(h) &= a^{\mathrm{w}}(x,hD)c^{\mathrm{w}}(x,hD)b^{\mathrm{w}}(x,hD)u(h) + O(h^{\infty})u(h) \\ &= ia^{\mathrm{w}}(x,hD)c^{\mathrm{w}}(x,hD)\chi^{\mathrm{w}}(x,hD)u(h) + O(h^{\infty})u(h) = O(h^{\infty})u(h), \end{aligned}$$

which concludes the proof of the proposition. □

We next prove the following estimate.

Proposition 12.1.8. *Let $u(h) \in L^2(\mathbb{R}^n;\mathbb{C}^2)$ be an eigenfunction of the operator $A_2^{\mathrm{w}}(x,hD)$ belonging to $\lambda(h)$. Let $a \in S_0^0(1;\mathsf{M}_2)$ be independent of h. Suppose that, for $R>0$, $\operatorname{supp} a \subset \{X \in \mathbb{R}^{2n}; |X|^2 \le R/c_2\}$ and $\lambda(h) > R+1$. Then*

$$\|a^{\mathrm{w}}(x,hD)u(h)\|_0 = O\left(\left(\frac{h}{\lambda(h)}\right)^{\infty}\right)\|u(h)\|_0.$$

Proof. Let $E > R+1$ be such that $|\lambda(h) - E| \le \varepsilon E$, where $\varepsilon \in (0,1/2]$ is sufficiently small, so as to have

$$(1-\varepsilon)E > R+1.$$

Since $A_2(\sqrt{E}X) = EA_2(X)$, we get from (9.5) that

$$U_E^{-1}A_2^{\mathrm{w}}(x,hD)U_E = EA_2^{\mathrm{w}}(x,\tilde{h}D), \quad \text{where } \tilde{h} = \frac{h}{E},$$

and where U_E is the L^2-isometry given in (9.4). Now, writing $\tilde{a}(X) := a(\sqrt{E}X)$, we have

$$|\partial_X^{\alpha}\tilde{a}(X)| \le C_{\alpha}E^{|\alpha|/2} \le C_{\alpha}\tilde{h}^{-|\alpha|/2},$$

that is, $\tilde{a} \in S^0_{1/2}(1; \mathsf{M}_2)$ **with respect to the parameter** $\tilde{h}$, and

$$\operatorname{supp}\tilde{a} \subset \{X \in \mathbb{R}^{2n};\ |X|^2 \le R/(c_2E)\} \subset \{X \in \mathbb{R}^{2n};\ |X|^2 \le (1-\varepsilon)/c_2\} =: K,$$

while $|\lambda(\tilde{h}) - 1| < \varepsilon$, where $\lambda(\tilde{h}) = \lambda(h)/E \in \operatorname{Spec}(A_2^{\mathrm{w}}(x, \tilde{h}D))$ (see Corollary 9.4.5). Hence

$$K \cap \Big(\Sigma_+(1) \cup \Sigma_-(1)\Big) = \emptyset,$$

for, by (12.6), we have that

$$\Sigma_\pm(1) \subset \{X \in \mathbb{R}^{2n};\ |X|^2 \ge 2/c_2\}.$$

Therefore $\tilde{u}(\tilde{h}) := U_E^{-1}u(h)$ is an eigenfunction of $A_2^{\mathrm{w}}(x,\tilde{h}D)$ belonging to the eigenvalue $\lambda(\tilde{h})$, and by Proposition 12.1.7, applied to the $\tilde{h}$-Weyl quantization $\tilde{a}^{\mathrm{w}}(x,\tilde{h}D)$ of $\tilde{a}$ (by shrinking ε if necessary), we get that

$$\|\tilde{a}^{\mathrm{w}}(x,\tilde{h}D)\tilde{u}(\tilde{h})\|_0 = O(\tilde{h}^\infty)\|\tilde{u}(\tilde{h})\|_0$$

(now the constants in $O(\tilde{h}^\infty)$ are *universal* constants), which is, through the L^2-isometry U_E (see (9.4)), equivalent to

$$\|a^{\mathrm{w}}(x,hD)u(h)\|_0 = O\left(\left(\frac{h}{E}\right)^\infty\right)\|u(h)\|_0.$$

Since $E/2 \le (1-\varepsilon)E \le \lambda(h) \le (1+\varepsilon)E \le 3E/2$ (and $1/2 \le 1-\varepsilon < 1+\varepsilon \le 3/2$), we finally obtain

$$\|a^{\mathrm{w}}(x,hD)u(h)\|_0 = O\left(\left(\frac{h}{\lambda(h)}\right)^\infty\right)\|u(h)\|_0,$$

which concludes the proof. □

We hence obtain the following crucial corollary, that will make it possible to study the large eigenvalues of $A_2^{\mathrm{w}}(x,hD)$, and hence those of $A_2^{\mathrm{w}}(x,D)$.

Corollary 12.1.9. *Let $R > 0$. Consider the orthogonal projection*

$$\Pi\colon L^2(\mathbb{R}^n;\mathbb{C}^2) \longrightarrow \operatorname{Span}\{u(h);\ A_2^{\mathrm{w}}(x,hD)u(h) = \lambda(h)u(h),\ \lambda(h) \le R+1\}.$$

Let $a = a^* \in S^0_0(1; \mathsf{M}_2)$ be independent of h, with $\operatorname{supp} a \subset \{X \in \mathbb{R}^{2n};\ |X|^2 < R/c_2\}$. Then

$$\|a^{\mathrm{w}}(x,hD)(I-\Pi)\|_{L^2\to L^2} = O(h^\infty), \quad \|(I-\Pi)a^{\mathrm{w}}(x,hD)\|_{L^2\to L^2} = O(h^\infty).$$

Proof. Let $\mathrm{Spec}(A_2^{\mathrm{w}}(x,hD)) = \{-\infty < \lambda_1(h) \le \lambda_2(h) \le \ldots \to +\infty\}$, with repetitions according to multiplicity, and let $\{u_j(h)\}_{j\ge 1}$ be an **othonormal** basis of $L^2(\mathbb{R}^n;\mathbb{C}^2)$ made of eigenfunctions of $A_2^{\mathrm{w}}(x,hD)$, where $u_j(h)$ belongs to $\lambda_j(h)$, $j \ge 1$. We may write

$$I - \Pi = \sum_{\lambda_j(h) > R+1} u_j(h)^* \otimes u_j(h),$$

where, recall, $\big(u_j(h)^* \otimes u_j(h)\big)v = (v, u_j(h))u_j(h)$. Then

$$a^{\mathrm{w}}(x,hD)(I-\Pi) = \sum_{\lambda_j(h)>R+1} u_j(h)^* \otimes \Big(a^{\mathrm{w}}(x,hD)u_j(h)\Big),$$

and

$$\|a^{\mathrm{w}}(x,hD)(I-\Pi)\|_{L^2\to L^2} \le \left(\sum_{\lambda_j(h)>R+1} \|a^{\mathrm{w}}(x,hD)u_j(h)\|_0^2 \right)^{1/2}. \tag{12.9}$$

By Theorem 4.4.1, and Corollary 9.4.5, we have for the eigenvalues of $A_2^{\mathrm{w}}(x,hD)$ that for some $c > 0$,

$$\lambda_j(h) \ge c j^{1/n} h,$$

whence by Proposition 12.1.8 (and recalling that we consider the eigenvalues $\lambda_j(h)$ for which $\lambda_j(h) > R+1$) we obtain, for each $M < N$,

$$\begin{aligned}\|a^{\mathrm{w}}(x,hD)u_j(h)\|_0 &\le C_N \left(\frac{h}{\lambda_j(h)}\right)^N \\ &\le C_N h^M \left(\frac{h}{\lambda_j(h)}\right)^{N-M} \le \frac{C_N}{c^{N-M}} h^M j^{-(N-M)/n}.\end{aligned}$$

Hence we choose $N = M+n+1$, so that the sum on the right-hand side of (12.9) converges. We therefore get that for any given $M \in \mathbb{Z}_+$ there exists $C'_M > 0$ such that

$$\|a^{\mathrm{w}}(x,hD)(I-\Pi)\|_{L^2\to L^2} \le C'_M h^M.$$

Since $a^{\mathrm{w}}(x,hD) = a^{\mathrm{w}}(x,hD)^*$, $(I-\Pi)^* = I-\Pi$, and $\|A\|_{L^2\to L^2} = \|A^*\|_{L^2\to L^2}$ (for any given L^2-bounded operator A), we also obtain the second desired inequality. □

Let us now construct the *reference operator* $A_r^{\mathrm{w}}(x,hD)$, with symbol $A_r(X)$ whose eigenvalues are *distinct and smooth* for all $X \in \mathbb{R}^{2n}$ and coincide with those of $A_2(X)$ for large X. Notice that by Theorem 9.2.1, $A_r^{\mathrm{w}}(x,hD)$ thus possesses an h^∞-diagonalization. Recall from (12.6) that ($\lambda_\pm$ being continuous)

$$\lambda_\pm(X) \longrightarrow +\infty, \text{ as } |X| \to +\infty,$$

and

$$\lambda_+(X) \leq \frac{c_2}{c_1}\lambda_-(X), \ \forall X \in \mathbb{R}^{2n}. \tag{12.10}$$

It thus follows that, for $\varepsilon_0 > 0$ fixed,

$$\{X \in \mathbb{R}^{2n};\ \lambda_+(X) \leq \varepsilon_0\} \subset \{X \in \mathbb{R}^{2n};\ c_1 p_0(X) \leq \varepsilon_0\}, \tag{12.11}$$

$$\{X \in \mathbb{R}^{2n};\ c_1 p_0(X) \leq \varepsilon_0\} \cap \{X \in \mathbb{R}^{2n};\ c_1 p_0(X) \geq 2\varepsilon_0\} = \emptyset, \tag{12.12}$$

and

$$\{X \in \mathbb{R}^{2n};\ \lambda_-(X) \geq 2\frac{c_2}{c_1}\varepsilon_0 =: \varepsilon_1\} \subset \{X \in \mathbb{R}^{2n};\ c_1 p_0(X) \geq 2\varepsilon_0\}. \tag{12.13}$$

Notice that $\varepsilon_1 \geq 2\varepsilon_0$. Hence, by (12.11), (12.12) and (12.13), we may take $\chi \in C^\infty(\mathbb{R}^{2n})$, with $0 \leq \chi \leq 1$, such that

$$\chi \equiv 0 \text{ when } \lambda_+ \leq \varepsilon_0, \ \chi \equiv 1 \text{ when } \lambda_- \geq \varepsilon_1.$$

Put $\chi_1 = \chi$ and $\chi_2 = 1 - \chi$, so that $\chi_1 + \chi_2 = 1$ and $\chi_2 \in C_0^\infty(\mathbb{R}^{2n})$, with

$$\operatorname{supp}\chi_2 \subset \{X \in \mathbb{R}^{2n};\ \lambda_-(X) \leq \varepsilon_1\}, \text{ and } 0 \leq \chi_2 \leq 1. \tag{12.14}$$

We are now ready to define the *reference operator.*

Definition 12.1.10. Let $A_2 \in \mathscr{Q}_2$. When $a_{11}(X) > a_{22}(X)$ for all $X \neq 0$, define the **reference operator** $A_r^{\mathrm{w}}(x,hD)$ by the h-Weyl quantization of

$$A_r(X) = \chi_1(X)A_2(X) + \chi_2(X)\begin{bmatrix} 1 & 0 \\ 0 & 1/2 \end{bmatrix}, \ X \in \mathbb{R}^{2n}. \tag{12.15}$$

In case $a_{11}(X) < a_{22}(X)$ for all $X \neq 0$, define $A_r^{\mathrm{w}}(x,hD)$ by the h-Weyl quantization of

$$A_r(X) = \chi_1(X)A_2(X) + \chi_2(X)\begin{bmatrix} 1/2 & 0 \\ 0 & 1 \end{bmatrix}.$$

Hence $A_r \in S(m^2, g; \mathsf{M}_2)$.

Remark 12.1.11. It is worth remarking once more that we need to use A_r in order to regularize the eigenvalues of A_2, that are not smooth at the origin. As shown below, our construction yields that the eigenvalues of A_r still do not cross, and that they are smooth everywhere. Hence, we shall be in a position to use the h^∞-diagonalization granted by Theorem 9.2.1. △

To fix ideas we shall assume from now on, with no loss of generality by Lemma 12.1.4 and Lemma 8.0.1, that $a_{11}(X) > a_{22}(X)$ for all $X \neq 0$.

Recall that

$$a_\pm(X) = \frac{a_{11}(X) \pm a_{22}(X)}{2}.$$

Then the eigenvalues of $A_2(X)$ are given by

$$\lambda_\pm(X) = a_+(X) \pm \sqrt{a_-(X)^2 + |a_{12}(X)|^2},$$

and those of $A_r(X)$ are given by

$$l_\pm(X) = a_+(X)\chi_1(X) + \frac{3}{4}\chi_2(X) \pm \sqrt{\Big(a_-(X)\chi_1(X) + \frac{1}{4}\chi_2(X)\Big)^2 + \chi_1(X)^2|a_{12}(X)|^2},$$

for, in our case, the discriminant is given by

$$\chi_1^2 a_+^2 + \frac{9}{16}\chi_2^2 + \frac{3}{2}\chi_1\chi_2 a_+ - \chi_1^2 a_{11}a_{22} + \chi_1|a_{12}|^2 - \chi_1\chi_2\Big(a_{22} + \frac{1}{2}a_{11}\Big) - \frac{1}{2}\chi_2^2$$

$$= \chi_1^2 a_-^2 + \frac{1}{16}\chi_2^2 + \chi_1\chi_2\frac{a_{11} - a_{22}}{4} + \chi_1^2|a_{12}|^2 = \Big(\chi_1 a_- + \frac{1}{4}\chi_2\Big)^2 + \chi_1^2|a_{12}|^2.$$

We have the following proposition.

Proposition 12.1.12. *We have*

$$A_r = A_r^* \in S(m^2, g; \mathsf{M}_2) \text{ is } \textbf{globally positive elliptic}, \tag{12.16}$$

i.e.

$$\langle A_r(X)v, v\rangle_{\mathbb{C}^2} \approx m(X)^2|v|_{\mathbb{C}^2},\ \forall v \in \mathbb{C}^2,\ \forall X \in \mathbb{R}^{2n}.$$

Moreover, upon denoting by $l_\pm(X)$ the eigenvalues of $A_r(X)$, we have

$$l_\pm \in C^\infty(\mathbb{R}^{2n}),\ 0 < l_-(X) < l_+(X),\ \forall X \in \mathbb{R}^{2n}, \tag{12.17}$$

$$l_\pm\big|_{\lambda_- \geq \varepsilon_1} = a_+(X) \pm \sqrt{a_-(X)^2 + |a_{12}(X)|^2}\,\Big|_{\lambda_- \geq \varepsilon_1} = \lambda_\pm\big|_{\lambda_- \geq \varepsilon_1}, \tag{12.18}$$

$$l_\pm\big|_{\lambda_+ \leq \varepsilon_0} = \frac{3}{4} \pm \frac{1}{4}, \tag{12.19}$$

$$l_\pm(X) \approx p_0(X),\ \text{for}\ |X| \geq c > 0. \tag{12.20}$$

Hence the $l_\pm \in S(m^2, g)$ are **globally positive elliptic** *symbols and, in particular,*

$$l_\pm(X) \longrightarrow +\infty,\ \text{as}\ |X| \to +\infty.$$

Proof. We have, for all $v \in \mathbb{C}^2$,

$$\langle A_r(X)v, v\rangle_{\mathbb{C}^2} = \chi_1(X)\langle A_2(X)v, v\rangle_{\mathbb{C}^2} + \chi_2(X)\langle \begin{bmatrix} 1 & 0 \\ 0 & 1/2 \end{bmatrix} v, v\rangle_{\mathbb{C}^2}.$$

Since $0 \le \chi_1, \chi_2 \le 1$, and $\chi_1 + \chi_2 \equiv 1$, and since $\lambda_\pm(X) \approx p_0(X)$, we immediately obtain the existence of $C > 0$ such that on the one hand

$$\langle A_r(X)v, v\rangle_{\mathbb{C}^2} \le \Big(\chi_1(X)\lambda_+(X) + \chi_2(X)\Big)|v|^2_{\mathbb{C}^2} \le Cm(X)^2|v|^2_{\mathbb{C}^2}, \tag{12.21}$$

and on the other

$$\langle A_r(X)v, v\rangle_{\mathbb{C}^2} \ge \Big(\chi_1(X)\lambda_-(X) + \frac{1}{2}\chi_2(X)\Big)|v|^2_{\mathbb{C}^2} \ge C^{-1}m(X)^2|v|^2_{\mathbb{C}^2}, \tag{12.22}$$

for all $X \in \mathbb{R}^{2n}$, and all $v \in \mathbb{C}^2$. This proves (12.16).

Now, the eigenvalues of $A_r(X)$ are given by

$$l_\pm(X) = a_+(X)\chi_1(X) + \frac{3}{4}\chi_2(X) \pm \sqrt{\delta_0(X)}, \tag{12.23}$$

where the discriminant δ_0 has the form

$$\delta_0(X) = \Big(a_-(X)\chi_1(X) + \frac{1}{4}\chi_2(X)\Big)^2 + \chi_1(X)^2|a_{12}(X)|^2 > 0 \tag{12.24}$$

by construction. In fact, we have that when $X = 0$, $\delta_0(0) = \chi_2(0)^2/16 > 0$, and when $X \ne 0$ (on recalling that we are supposing $a_{11}(X) - a_{22}(X) = 2a_-(X) > 0$ for all $X \ne 0$)

$$\delta_0(X) = 0 \Longrightarrow a_-(X)\chi_1(X) + \frac{1}{4}\chi_2(X) = 0 \Longrightarrow \chi_1(X) = \chi_2(X) = 0,$$

which is impossible, for $\chi_1(X) + \chi_2(X) = 1$. Hence, from (12.23), (12.24), (12.21) and (12.22), we obtain (12.17), (12.18), (12.19) and (12.20). This concludes the proof of the proposition. □

By Proposition 9.4.6 we hence have the following fact concerning the reference operator $A_r^{\mathrm{w}}(x, hD)$.

Lemma 12.1.13. *For all $h \in (0,1]$, the operator $A_r^{\mathrm{w}}(x, hD)$, as an unbounded operator in $L^2(\mathbb{R}^n;\mathbb{C}^2)$ with domain $B^2(\mathbb{R}^n;\mathbb{C}^2)$, is self-adjoint with a discrete spectrum bounded from below, made of a diverging (to $+\infty$) sequence of eigenvalues with finite multiplicities.*

Proof. Since $A_r^{\mathrm{w}}(x, hD) = A_2^{\mathrm{w}}(x, hD) + R^{\mathrm{w}}(x, hD)$, where

$$R(X) = \chi_2(X)\left(\begin{bmatrix} 1 & 0 \\ 0 & 1/2 \end{bmatrix} - A_2(X)\right) \tag{12.25}$$

has compact support (equal to the support of χ_2), it follows that we may apply Proposition 9.4.6. This concludes the proof. □

Remark 12.1.14. Note that Lemma 12.1.13 may also be proved directly by Proposition 12.1.12 using a parametrix construction for $A_r^{\mathrm{w}}(x,hD)$. △

In addition, from (12.18) and the previous considerations, we also obtain the following basic lemma.

Lemma 12.1.15. *Put*

$$E_0 := \max\left\{1, 2\frac{c_2}{c_1}\varepsilon_1, \max_{\lambda_- \leq 2\varepsilon_1} l_\pm\right\}. \tag{12.26}$$

Then the subset Ω of $T^*\mathbb{R}^n$,

$$\Omega := \{X;\ \lambda_-(X) \geq E_0\} \subset \{X;\ \lambda_\pm(X) \geq E_0\} \subset \{X;\ \lambda_-(X) \geq 2\varepsilon_1\},$$

is foliated by energy-surfaces

$$\Sigma_\pm(E) = \{X;\ \lambda_\pm(X) = E\} = \{X;\ l_\pm(X) = E\},\ \ E \geq E_0.$$

At this point, we are in a position to measure the distance of $\mathrm{Spec}(A_2^{\mathrm{w}}(x,hD))$ to $\mathrm{Spec}(A_r^{\mathrm{w}}(x,hD))$. Fix the energy $E_0' = E_0 + 10$, and consider the orthogonal projector Π associated with $A_2^{\mathrm{w}}(x,hD)$, considered in Corollary 12.1.9, with $R+1 = E_0'$. Since

$$E_0' - 1 > E_0 \geq 2\frac{c_2}{c_1}\varepsilon_1 = 4\left(\frac{c_2}{c_1}\right)^2 \varepsilon_0,$$

we get in particular that

$$\mathrm{supp}\,\chi_2 \subset \{X;\ \lambda_-(X) \leq \varepsilon_1\} \subset \{X;\ |X|^2 < (E_0' - 1)/c_2\},$$

whence the same is true for the symbol $R(X)$ in (12.25). Hence, we have the existence of a constant $c_\chi > 0$ *independent of* $h \in (0,1]$, such that

$$\|R^{\mathrm{w}}(x,hD)\|_{L^2 \to L^2} \leq c_\chi,\ \ \forall h \in (0,1]. \tag{12.27}$$

Since

$$\Pi A_r^{\mathrm{w}}(x,hD)\Pi = \Pi A_2^{\mathrm{w}}(x,hD)\Pi + \Pi R^{\mathrm{w}}(x,hD)\Pi, \tag{12.28}$$

and Π *commutes with* $A_2^{\mathrm{w}}(x,hD)$ (for it is associated with the eigenfunctions of $A_2^{\mathrm{w}}(x,hD)$), by the Spectral Theorem we get

$$\|\Pi A_r^{\mathrm{w}}(x,hD)\Pi\|_{L^2 \to L^2} \leq E_0' + c_\chi. \tag{12.29}$$

Furthermore, from Corollary 12.1.9 we have that

$$(I - \Pi)A_2^{\mathrm{w}}(x,hD)(I - \Pi) = (I - \Pi)A_r^{\mathrm{w}}(x,hD)(I - \Pi) + O(h^\infty), \tag{12.30}$$

and

$$\|\Pi A_r^{\mathrm{w}}(x,hD)(I-\Pi)\|_{L^2\to L^2} = \|(I-\Pi)A_r^{\mathrm{w}}(x,hD)\Pi\|_{L^2\to L^2} = O(h^\infty). \quad (12.31)$$

We may therefore use Theorem 10.2.1 from Section 10.2, and obtain the following result.

Theorem 12.1.16. *Let $A_2 \in \mathscr{Q}_2$, and let*

$$\mathrm{Spec}(A_2^{\mathrm{w}}(x,hD)) = \{-\infty < \lambda_1(h) \le \lambda_2(h) \le \ldots \to +\infty\}$$

and

$$\mathrm{Spec}(A_r^{\mathrm{w}}(x,hD)) = \{-\infty < \lambda_1^r(h) \le \lambda_2^r(h) \le \ldots \to +\infty\},$$

with repetitions according to multiplicity. Let $E, E' > 0$ with $E > E' \ge E_0 + 10^2 + c_\chi$. For any given integer $N \ge 1$ there exists $h_0 = h(N,E,E') \in (0,1]$ such that

$$\mathrm{Spec}(A_2^{\mathrm{w}}(x,hD)) \cap (E,+\infty) \subset \bigcup_{E'<\lambda^r(h)\in\mathrm{Spec}(A_r^{\mathrm{w}}(x,hD))} [\lambda^r(h)-h^N, \lambda^r(h)+h^N], \quad (12.32)$$

for all $h \in (0,h_0]$. More precisely, given any $j \ge 1$ so large that $\lambda_j(h) > E$, then, with the **same** *j,*

$$|\lambda_j(h) - \lambda_j^r(h)| \le h^N. \quad (12.33)$$

We shall use the theorem with $N = 10$.

By the h^∞-decoupling Theorem 9.2.1 we next have the following result for $A_r^{\mathrm{w}}(x,hD)$.

Proposition 12.1.17. *For any given $j \in \mathbb{Z}_+$ we may find h-independent symbols $E_{-2j} \in S(m^{-2j}, g; \mathsf{M}_2)$ and $\Lambda_{2-2j}^{\pm} \in S(m^{2-2j}, g)$ such that*

$$\Lambda_2^{\pm}(X) = l_{\pm}(X), \ \forall X \in \mathbb{R}^{2n}, \quad (12.34)$$

and such that, for any given $N \in \mathbb{Z}_+$, upon writing

$$E_N(h) := \sum_{j=0}^{N} h^j E_{-2j}, \ \Lambda_{\pm,N}(h) := \sum_{j=0}^{N} h^j \Lambda_{2-2j}^{\pm},$$

we have

$$A_r^{\mathrm{w}}(x,hD) = E_N^{\mathrm{w}}(h) \begin{bmatrix} \Lambda_{+,N}^{\mathrm{w}}(h) & 0 \\ 0 & \Lambda_{-,N}^{\mathrm{w}}(h) \end{bmatrix} E_N^{\mathrm{w}}(h)^* + h^{N+1}\tilde{R}_N^{\mathrm{w}}(h), \quad (12.35)$$

where

$$E_N^{\mathrm{w}}(h)E_N^{\mathrm{w}}(h)^* = I + h^{N+1}R_N^{\mathrm{w}}(h) = E_N^{\mathrm{w}}(h)^*E_N^{\mathrm{w}}(h), \quad (12.36)$$

with

$$\|R_N^{\mathrm{w}}(h)\|_{L^2\to L^2} \le C_N,\ \|\tilde{R}_N^{\mathrm{w}}(h)\|_{L^2\to L^2} \le C_N,\ \forall h\in(0,1], \tag{12.37}$$

where

$$\tilde{R}_N \in S_0^0(m^{2-2(N+1)},g;\mathsf{M}_2),\ \text{and}\ R_N \in S_0^0(m^{-2(N+1)},g;\mathsf{M}_2).$$

Hence, in particular, by Proposition 9.3.1, when $N\ge 1$,

$$R_N^{\mathrm{w}}(x,hD)\colon L^2(\mathbb{R}^n;\mathbb{C}^2)\longrightarrow B^2(\mathbb{R}^n;\mathbb{C}^2)\ \text{is } \textbf{continuous},\ \forall h\in(0,1]. \tag{12.38}$$

We are now in a position to prove the following result, that allows us to approximate the spectrum of $A_r^{\mathrm{w}}(x,hD)$ by the spectrum of the *scalar* operators $\Lambda_{\pm,N}^{\mathrm{w}}(h)$.

Theorem 12.1.18. *There exists* $h_0\in(0,1]$ *sufficiently small, and for all* $h\in(0,h_0]$ *an* L^2*-bounded operator* $E(h)$ *and scalar positive elliptic symbols* $\Lambda_\pm(h)\in S_{0,\mathrm{cl}}^0(m^2,g)$ *for which (12.34) holds, such that*

$$A_r^{\mathrm{w}}(x,hD) = E(h)\begin{bmatrix}\Lambda_+^{\mathrm{w}}(h) & 0\\ 0 & \Lambda_-^{\mathrm{w}}(h)\end{bmatrix}E(h)^* + h^5\tilde{R}^{\mathrm{w}}(h),\ \forall h\in(0,h_0], \tag{12.39}$$

where $\|\tilde{R}^{\mathrm{w}}(h)\|_{L^2\to L^2}\le C$ *for all* $h\in(0,h_0]$, *and such that*

$$E(h)E(h)^* = I = E(h)^*E(h),\ \forall h\in(0,h_0], \tag{12.40}$$

and

$$E(h)u, E(h)^*u\in B^2(\mathbb{R}^n;\mathbb{C}^2),\ \forall u\in B^2(\mathbb{R}^n;\mathbb{C}^2),\ \forall h\in(0,h_0]. \tag{12.41}$$

In particular

$$\Big(A_r^{\mathrm{w}}(x,hD)u,u\Big) = \Big(\begin{bmatrix}\Lambda_+^{\mathrm{w}}(h) & 0\\ 0 & \Lambda_-^{\mathrm{w}}(h)\end{bmatrix}E(h)^*u, E(h)^*u\Big) + O(h^5)\|u\|_0^2, \tag{12.42}$$

for all $u\in B^2$, *for all* $h\in(0,h_0]$.

Proof. Take $N=4$ in Proposition 12.1.17. Hence (12.36) gives the existence of $h_0\in(0,1]$ such that

$$h\|R_4^{\mathrm{w}}(x,hD)\|_{L^2\to L^2}\le\frac{1}{2},\ \forall h\in(0,h_0],$$

whence the operator

$$I+h^5R_4^{\mathrm{w}}(x,hD)>0,\ \forall h\in(0,h_0],$$

and we are allowed to take the inverse of its positive square root, i.e. the h-dependent operator in $L^2(\mathbb{R}^n;\mathbb{C}^2)$

$$
\begin{aligned}
S(h) &:= \Big(I + h^5 R_4^{\mathrm{w}}(x,hD)\Big)^{-1/2} \\
&= \sum_{k\geq 0} \binom{-1/2}{k} h^{5k} R_4^{\mathrm{w}}(x,hD)^k =: I + h^5 R(h), \;\; h \in (0,h_0].
\end{aligned}
$$

It is important to notice that

$$
R(h) = R_4^{\mathrm{w}}(x,hD) R_0(h) = R_0(h) R_4^{\mathrm{w}}(x,hD),
$$

where

$$
R_0(h) := \sum_{k\geq 1} \binom{-1/2}{k} \Big(h^5 R_4^{\mathrm{w}}(x,hD)\Big)^{k-1} : L^2(\mathbb{R}^n;\mathbb{C}^2) \longrightarrow L^2(\mathbb{R}^n;\mathbb{C}^2)
$$

is continuous with $\|R_0(h)\|_{L^2\to L^2} \leq C$ for all $h \in (0,h_0]$. By (12.38) we therefore have that $R(h)u \in B^2$, for all $u \in L^2$. Setting

$$
E(h) := S(h) E_4^{\mathrm{w}}(h) \;\text{ and }\; \Lambda_\pm^{\mathrm{w}}(h) = \Lambda_{\pm,4}^{\mathrm{w}}(h)
$$

completes the proof. □

Set now

$$
\Lambda^{\mathrm{w}}(h) = \mathrm{diag}(\Lambda_+^{\mathrm{w}}(h), \Lambda_-^{\mathrm{w}}(h)),
$$

and let

$$
\mathrm{Spec}(\Lambda^{\mathrm{w}}(h)) = \{-\infty < \tilde{\lambda}_1(h) \leq \tilde{\lambda}_2(h) \leq \ldots \to +\infty\},
$$

with repetitions according to multiplicity (that the spectrum of $\Lambda^{\mathrm{w}}(h)$ is discrete is granted by Remark 9.4.8). By the Minimax Principle (Theorem 4.1.1) we thus obtain the following result, which is interesting in its own right, since it represents the crucial, intermediate step in achieving the result for $A_2^{\mathrm{w}}(x,D)$ (which is Theorem 12.4.1 below).

Theorem 12.1.19. *There exists $h_0 \in (0,1]$ and a constant $C > 0$ such that*

$$
\mathrm{dist}\Big(\mathrm{Spec}(A_r^{\mathrm{w}}(x,hD)), \bigcup_{\pm} \mathrm{Spec}(\Lambda_\pm^{\mathrm{w}}(h))\Big) \leq Ch^5, \tag{12.43}
$$

for all $h \in (0,h_0]$. More precisely, for each $j \geq 1$,

$$
|\lambda_j^r(h) - \tilde{\lambda}_j(h)| \leq Ch^5, \;\; \forall h \in (0,h_0]. \tag{12.44}
$$

Remark 12.1.20. The error $O(h^5)$ in the approximation (12.44) is sufficient because, as it will be seen below, the knowledge of the large eigenvalues of $\Lambda_\pm^{\mathrm{w}}(h)$ is within an error $O(h^2)$. Of course, any fixed $N \geq 4$ can be taken, yielding in (12.43) and (12.44) an error $O(h^N)$. △

Remark 12.1.21. Using (9.35) of Corollary 9.2.6 we have, for the subprincipal terms $\Lambda_0^{\pm}$ of the operators $\Lambda_{\pm}^{\mathrm{w}}(h)$,

$$\Lambda_0^{\pm}(X) = \frac{1}{2}\mathrm{Im}\Big(\langle\{e_0^*, l_{\pm}\}(X)e_0(X)w_{\pm}, w_{\pm}\rangle\Big) + \frac{1}{2}\mathrm{Im}\Big(\langle e_0^*(X)\{A_r, e_0\}(X)w_{\pm}, w_{\pm}\rangle\Big), \tag{12.45}$$

$X \in \mathbb{R}^{2n}$, where $w_{\pm}$ is the canonical basis of $\mathbb{C}^2$ ($w_+ = \begin{bmatrix}1\\0\end{bmatrix}$, $w_- = \begin{bmatrix}0\\1\end{bmatrix}$), and $e_0^*A_re_0 = \mathrm{diag}(l_+, l_-)$, e_0 being a matrix whose columns are given by the normalized eigenvectors of A_r.

In particular, when $X \in \{X \in \mathbb{R}^{2n};\ \lambda_-(X) \geq \varepsilon_1\}$ we have $A_2(X) = A_r(X)$, whence

$$\Lambda_0^{\pm}(X) = \frac{1}{2}\mathrm{Im}\Big(\langle\{e_0^*, \lambda_{\pm}\}(X)e_0(X)w_{\pm}, w_{\pm}\rangle\Big) + \frac{1}{2}\mathrm{Im}\Big(\langle e_0^*(X)\{A_2, e_0\}(X)w_{\pm}, w_{\pm}\rangle\Big), \tag{12.46}$$

where now $e_0^*A_2e_0 = \mathrm{diag}(\lambda_+, \lambda_-)$. (Recall that **no** subprincipal terms are present in the symbols of $A_2^{\mathrm{w}}(x, hD)$ and of $A_r^{\mathrm{w}}(x, hD)$.) △

We have thus seen how the spectrum of $A_r^{\mathrm{w}}(x, hD)$ is approximated by the spectrum of the diagonal system $\Lambda^{\mathrm{w}}(h)$, and thus by the spectra of the **scalar** operators $\Lambda_{\pm}^{\mathrm{w}}(h)$. In the next section we recall some results about the spectrum of scalar h-pseudodifferential operators, which will then be used in Section 12.3 to obtain a description of the spectrum of the 2×2 systems we are interested in.

12.2 Localization and Multiplicity of the Spectrum in the Scalar Case

We now recall, in a form tailored to our purposes, some results about the localization, and multiplicity, of the spectrum for *scalar* h-pseudodifferential operators.

In the first place we recall the Dyn'kin-Helffer-Sjöstrand formula (see Dimassi-Sjöstrand [7] and also the Appendix).

Given $f \in C_0^{\infty}(\mathbb{R})$ one can always find an **almost-analytic extension** $\tilde{f} \in C_0^{\infty}(\mathbb{C})$, that is a function $\tilde{f}$ such that

$$\tilde{f}\big|_{\mathbb{R}} = f, \ \text{and}\ \forall N \in \mathbb{Z}_+,\ \exists C_N > 0,\ \left|\frac{\partial \tilde{f}}{\partial \bar{z}}(z)\right| \leq C_N|\mathrm{Im}\, z|^N, \tag{12.47}$$

where $\partial/\partial\bar{z} = 2^{-1}(\partial/\partial x + i\partial/\partial y)$.

One has the following theorem (the Dyn'kin-Helffer-Sjöstrand formula, a proof of which will be given in the Appendix).

Theorem 12.2.1. *Let P be a self-adjoint operator (possibly unbounded) on a Hilbert space H. Let $f \in C_0^2(\mathbb{R})$ and let $\tilde{f} \in C_0^1(\mathbb{C})$ be an extension of f for which $\frac{\partial \tilde{f}}{\partial \bar{z}}(z) = O(|\mathrm{Im}\, z|)$. Then*

$$f(P) = -\frac{1}{\pi} \iint_{\mathbb{C}} \frac{\partial \tilde{f}}{\partial \bar{z}}(z)(z-P)^{-1} L(dz). \tag{12.48}$$

Here $L(dz) = dxdy$ is the Lebesgue measure on $\mathbb{C}$, and the integral in (12.48) converges as a Riemann integral for functions with values in $\mathscr{L}(H,H)$.

Let now $p \in S_0^0(m^\mu)$ be *real-valued*. We are interested in the case $\mu > 0$, and therefore shall assume that $p+i$ be elliptic (that is, recall, $|p(X)+i| \gtrsim m(X)^\mu$). Let hence $P = p^{\mathrm{w}}(x,hD;h)$ be the corresponding operator acting as an unbounded operator (since $\mu > 0$) on L^2. By using Beals' characterization of h-pseudodifferential operators (see Theorem 9.1.11), from Theorem 12.2.1 one has the following result (see [7]).

Theorem 12.2.2. *Let $f \in C_0^\infty(\mathbb{R})$. Then there exists $h_0 \in (0,1]$ such that $f(P) = F^{\mathrm{w}}(x,hD;h)$, where $F \in S_0^0(m^{-k})$, $h \in (0,h_0]$, for every $k \in \mathbb{N}$.*

Moreover, when $p \sim p_\mu + hp_{\mu-1} + h^2 p_{\mu-2} + \ldots$ in $S_0^0(m^\mu)$, with the $p_{\mu-j}$ independent of h, we get $F \sim F_0 + hF_1 + \ldots$ in $S_0^0(m^{-1})$, with the F_j independent of h, where

$$F_0(X) = f(p_\mu(X)),\ \ F_1(X) = p_{\mu-1}(X)F'(p_\mu(X)),\ \ \forall X \in \mathbb{R}^{2n}.$$

It is also useful to have the following result for positive elliptic classical semiclassical symbols, due to Robert [64].

Theorem 12.2.3. *Let $p \sim p_\mu + hp_{\mu-2} + \ldots \in S_{0,\mathrm{cl}}^0(m^\mu, g)$ be positive elliptic, $\mu > 0$. Let $f \in C^\infty(\mathbb{R})$ fulfill, for some $r \in \mathbb{R}$, the inequalities*

$$\forall k \in \mathbb{Z}_+,\ \exists C_k > 0,\ \text{such that}\ |f^{(k)}(\lambda)| \le C_k(1+|\lambda|)^{r-k}.$$

Let $P(h) = p^{\mathrm{w}}(x,hD;h)$. Then $f(P(h)) = F^{\mathrm{w}}(x,hD;h)$, where the symbol $F \in S_{0,\mathrm{cl}}^0(m^{r\mu}, g)$, and

$$F_{r\mu} = f(p_\mu),\ \ F_{r\mu-2} = p_{\mu-2} f'(p_\mu).$$

One has also a formula for the successive terms (see Robert [64, p. 77]).

We now have the following fundamental theorem, that we prove for classical semiclassical symbols in $S_{0,\mathrm{cl}}^0(m^\mu, g)$, due (in a much more general form) to Helffer and Robert [20] (see also Dimassi-Sjöstrand [7] and Dozias [8]).

Theorem 12.2.4. *Let $p \in S_{0,\mathrm{cl}}^0(m^\mu, g)$, $\mu > 0$, $p \sim p_\mu + hp_{\mu-2} + \ldots$, be* **real valued**, *with principal symbol $p_\mu \approx m^\mu$ satisfying hypotheses $(H1)$, $(H2)$ and $(H3)$*

(i.e. Assumption 11.1.1) for an energy interval $[E_1 - \varepsilon, E_2 + \varepsilon]$, *some* $\varepsilon \in (0,1]$. *Suppose in addition that:*

- $(H4)$ *The period* $T_{p_\mu}(E)$ *is a constant* T *for all* $E \in [E_1 - \varepsilon, E_2 + \varepsilon]$;
- $(H5)$ *The subprincipal symbol* $p_{\mu-2}$ *is identically zero on* $p_\mu^{-1}([E_1 - \varepsilon, E_2 + \varepsilon])$.

From hypothesis $(H4)$ *it follows by Lemma 11.1.6 that the action* A_{p_μ} *is constant on* $[E_1, E_2]$. *Let hence* $A_{p_\mu}(E) = \delta$, *for all* $E \in [E_1, E_2]$. *Let* $\alpha \in \mathbb{Z}$ *be the Maslov index of the trajectories*

$$\gamma_X = \left\{\exp(tH_{p_\mu})(X);\, t \in [0,T]\right\},\ X \in p_\mu^{-1}(E_1 - \varepsilon, E_2 + \varepsilon).$$

(The Maslov index is a constant α *for it is locally constant and* $p_\mu^{-1}(E_1 - \varepsilon, E_2 + \varepsilon)$ *is connected.) Let* $P(h) = p^{\mathrm{w}}(x, hD; h)$. *Then there exist* $C_0 > 0$ *and* $h_0 \in (0,1]$ *such that*

$$\mathrm{Spec}(P(h)) \cap [E_1, E_2] \subset \bigcup_{k \in \mathbb{Z}} I_k(h), \tag{12.49}$$

for all $h \in (0, h_0]$, *where*

$$I_k(h) = \left[\frac{2\pi}{T}\left(k + \frac{\alpha}{4}\right)h + \delta - C_0 h^2, \frac{2\pi}{T}\left(k + \frac{\alpha}{4}\right)h + \delta + C_0 h^2\right].$$

Proof. We start off by remarking that, since

$$p_\mu(X) \longrightarrow +\infty, \text{ as } |X| \to +\infty,$$

$p_\mu^{-1}([E_1 - \varepsilon, E_2 + \varepsilon])$ is a compact set.

Furthermore, by a result of Helffer and Robert (see [65]), there exists $h_0 \in (0,1]$ such that $P(h)$, as an unbounded operator in L^2 with domain $\mathscr{S}$, is essentially self-adjoint, with compact resolvent for all $h \in (0, h_0]$ (alternatively, when $\mu \geq 1$ one may use Remark 9.4.8 to conclude that $P(h)$, as an unbounded operator in L^2 with domain B^μ, is self-adjoint with a discrete spectrum bounded from below, for all $h \in (0, h_0]$).

Define the set

$$\begin{aligned} \Psi\varepsilon(E_1, E_2) := \Big\{ & \psi \in C_0^\infty(\mathbb{R});\ \mathrm{supp}\,\psi \subset (E_1 - \varepsilon, E_2 + \varepsilon), 0 \leq \psi \leq 1, \\ & \psi \equiv 1 \text{ in a neighborhood of } [E_1, E_2]\Big\}. \end{aligned} \tag{12.50}$$

Take hence $\psi \in \Psi_\varepsilon(E_1, E_2)$. Then by Theorem 12.2.2, $\Psi(h) := \psi(P(h))$ is an h-pseudodifferential operator with principal symbol $\psi(p_\mu)$ and subprincipal symbol $p_{\mu-2}\psi'(p_\mu) \equiv 0$, in view of $(H5)$. Take now $\psi, \tilde{\psi} \in \Psi_\varepsilon(E_1, E_2)$, with $\psi\tilde{\psi} = \psi$, and consider (following Chazarain [5] and Helffer and Robert [20])

$$U_\psi(t) := e^{-ith^{-1}P(h)}\Psi(h) = e^{-ith^{-1}P(h)\tilde{\Psi}(h)}\Psi(h). \tag{12.51}$$

Notice that $P(h)$, $\Psi(h)$ and $\tilde{\Psi}(h)$ all commute. Equation (12.51) follows from the Spectral Theorem, for

$$\begin{aligned} e^{-ith^{-1}P(h)}\Psi(h) &= \int_{\mathrm{Spec}(P(h))} e^{-ith^{-1}\lambda}\psi(\lambda)dE(\lambda) \\ &= \int_{\mathrm{Spec}(P(h))} e^{-ith^{-1}\lambda\tilde{\psi}(\lambda)}\psi(\lambda)dE(\lambda) = e^{-ith^{-1}P(h)\tilde{\Psi}(h)}\Psi(h), \end{aligned}$$

since $\tilde{\psi} \equiv 1$ on the support of ψ. The importance of formula (12.51) lies in the fact that as long as we are interested in the spectral properties of $P(h)$ in the interval $[E_1, E_2]$, we may replace $P(h)$ by the L^2**-bounded** operator $P(h)\tilde{\Psi}(h)$. It is well-known after Helffer and Robert (see, e.g., Dimassi-Sjöstrand [7], Ivrii [34], and Robert [65]) that for a time $T_0 > 0$ sufficiently small, one can construct an h-Fourier integral operator (h-Fio) $F(t)$ which approximates $e^{-ih^{-1}P(h)\tilde{\Psi}(h)}$ for $|t| \leq T_0$. The Schwartz-kernel $F(t;x,y)$ of $F(t)$ is written as

$$F(t;x,y) = (2\pi h)^{-n} \int e^{ih^{-1}(S(t,x,\eta)-\langle y,\eta\rangle)} a(t,x,y,\eta;h)d\eta, \tag{12.52}$$

where the phase-function S is the solution of the Hamilton-Jacobi equation

$$\partial_t S(t,x,\eta) + \left(p_\mu \tilde{\psi}(p_\mu)\right)(x, \nabla_x S(t,x,\eta)) = 0, \ S\big|_{t=0} = \langle x,\eta\rangle,$$

and amplitude

$$a(t,x,y,\eta;h) \sim \sum_{j\geq 0} h^j a_j(t,x,y,\eta),$$

where the a_j satisfy the transport equations with initial conditions

$$a_0\big|_{t=0} = 1, \quad \text{and} \quad a_j\big|_{t=0} = 0, \ j \geq 1.$$

Hence, an h-Fio approximation of $U_\psi(t)$ for $|t| \leq T_0$ has a kernel of the kind (12.52) (with the same phase-function S), and

$$\|F(t)\Psi(h) - U_\psi(t)\|_{L^2\to L^2} = O(h^\infty), \text{ uniformly in } |t| \leq T_0.$$

Next, for $t \in [kT_0, (k+1)T_0]$, one writes

$$U_\psi(t) = \left(e^{-ih^{-1}T_0 P(h)\tilde{\Psi}(h)}\tilde{\Psi}(h)\right)^k U_\psi(t-kT_0) = U_{\tilde{\psi}}(T_0)^k U_\psi(t-kT_0),$$

One then h-Fio-approximates $U_\psi(t)$ on the intervals $[kT_0, (k+1)T_0]$ by using

$$F_k(t)\Psi(h) := \left(F(T_0)\tilde{\Psi}(h)\right)^k F(t-kT_0)\Psi(h),$$

and has that for any given $N \in \mathbb{Z}_+$

$$\|F_k(t)\Psi(h) - U_\psi(t)\|_{L^2 \to L^2} = O(h^{N+1}), \text{ uniformly in } t \in [kT_0, (k+1)T_0].$$

Considering $U_\psi(T)$, and using the h-Fio $F_k(t)\Psi(h)$ one now has the following lemma, due to Chazarain, Helffer and Robert.

Lemma 12.2.5. *The operator $U_\psi(T)$ is an h-pseudodifferential operator with symbol $u \sim u_0 + hu_1 + h^2 u_2 + \dots$ belonging to $S^0_{\mathrm{cl}}(1)$, where the u_j are supported in the bounded set $p_\mu^{-1}(E_1 - \varepsilon, E_2 + \varepsilon)$. Its principal symbol is given by*

$$u_0(X) = \psi(p_\mu(X))e^{-2\pi i \sigma_0(h)}, \quad X \in \mathbb{R}^{2n},$$

with

$$\sigma_0(h) = \frac{\alpha}{4} + \frac{T}{2\pi}\frac{\beta}{h}, \tag{12.53}$$

where $\alpha \in \mathbb{Z}$ is the Maslov index of the closed trajectories

$$\gamma_X := \{\exp(tH_{p_\mu})(X);\, t \in [0,T]\},\, X \in p_\mu^{-1}([E_1 - \varepsilon, E_2 + \varepsilon]),$$

and

$$\beta = A_{p_\mu}(E) + h\frac{1}{T}\int_0^T p_{\mu-2} \circ \exp(tH_{p_\mu})(X)dt = A_{p_\mu}(E) =: \delta. \tag{12.54}$$

Recall that the action A_{p_μ} is constant on $[E_1 - \varepsilon, E_2 + \varepsilon]$, and that we assumed $p_{\mu-2}\big|_{p_\mu^{-1}([E_1-\varepsilon,E_2+\varepsilon])} = 0$. (Recall that the Maslov index is a constant α because it is locally constant and the set $p_\mu^{-1}([E_1 - \varepsilon, E_2 + \varepsilon])$ is connected.)

Lemma 12.2.5 is a consequence of the following lemma for operators for which $T = 2\pi$, due to Helffer and Robert [20] (see also Dimassi-Sjöstrand [7, Theorem 15.4, p. 203]).

Lemma 12.2.6. *Suppose $Q(h)$ is an operator satisfying the same hypotheses as $P(h)$, with respect to the energy interval $[\tilde{E}_1 - \tilde{\varepsilon}, \tilde{E}_2 + \tilde{\varepsilon}]$. Suppose that $T_q(\tilde{E}) = 2\pi$, for all $\tilde{E} \in [\tilde{E}_1 - \tilde{\varepsilon}, \tilde{E}_2 + \tilde{\varepsilon}]$. Let $\psi_0 \in \Psi_{\tilde{\varepsilon}}(\tilde{E}_1, \tilde{E}_2)$, and consider $\Psi_0(h) = \psi_0(Q(h))$ and $U_{\psi_0}(t) = e^{-ith^{-1}Q(h)}\Psi_0(h)$. Then the operator $U_{\psi_0}(2\pi)$ is an h-pseudodifferential operator with symbol $v \sim v_0 + hv_1 + \dots$ belonging to $S^0_{\mathrm{cl}}(1)$, where the v_j are supported in the bounded set $q_\mu^{-1}(\tilde{E}_1 - \tilde{\varepsilon}, \tilde{E}_2 + \tilde{\varepsilon})$. Its principal symbol is given by*

$$v_0(X) = \psi_0(q_\mu(X))e^{-2\pi i \sigma_0(h)}, \quad X \in \mathbb{R}^{2n},$$

with

$$\sigma_0(h) = \frac{\alpha}{4} + \frac{\tilde{\beta}}{h}, \tag{12.55}$$

where $\alpha \in \mathbb{Z}$ is the Maslov index of the closed trajectories

$$\tilde{\gamma}_X := \{\exp(tH_{q_\mu})(X);\, t \in [0, 2\pi]\},\ X \in q_\mu^{-1}([\tilde{E}_1 - \tilde{\varepsilon}, \tilde{E}_2 + \tilde{\varepsilon}]),$$

and

$$\tilde{\beta} = A_{q_\mu}(\tilde{E}) + h\frac{1}{2\pi}\int_0^{2\pi} q_{\mu-2} \circ \exp(tH_{q_\mu})(X)dt = A_{q_\mu}(\tilde{E}) =: \tilde{\delta}.$$

Recall that the action A_{q_μ} is constant on $[\tilde{E}_1 - \tilde{\varepsilon}, \tilde{E}_2 + \tilde{\varepsilon}]$, that by assumption $q_{\mu-2}\big|_{q_\mu^{-1}([\tilde{E}_1 - \tilde{\varepsilon}, \tilde{E}_2 + \tilde{\varepsilon}])} = 0$, and that the Maslov index is a constant α (it is locally constant and $q_\mu^{-1}([\tilde{E}_1 - \tilde{\varepsilon}, \tilde{E}_2 + \tilde{\varepsilon}])$ is connected).

Proof (that Lemma 12.2.5 is a consequence of Lemma 12.2.6). One considers the operator $Q(h) := \frac{T}{2\pi}P(h)$. Then by Lemma 11.1.7 the principal and subprincipal symbols of $Q(h)$, which are $q_\mu = \frac{T}{2\pi}p_\mu$ and $q_{\mu-2} = \frac{T}{2\pi}p_{\mu-2}$, respectively, satisfy the same assumptions with respect to the energy interval $[\tilde{E}_1 - \tilde{\varepsilon}, \tilde{E}_2 + \tilde{\varepsilon}] = \frac{T}{2\pi}[E_1 - \varepsilon, E_2 + \varepsilon]$, with *constant* period $T_q(\tilde{E}) = 2\pi$ for all $\tilde{E} \in [\tilde{E}_1 - \tilde{\varepsilon}, \tilde{E}_2 + \tilde{\varepsilon}]$. If we consider $\psi_0(\tilde{\lambda}) := \psi(\frac{2\pi}{T}\tilde{\lambda})$, for $\tilde{\lambda} \in [\tilde{E}_1 - \tilde{\varepsilon}, \tilde{E}_2 + \tilde{\varepsilon}]$ and $\psi \in \Psi_\varepsilon(E_1, E_2)$, we then have that $\psi_0 \in \Psi_{\tilde{\varepsilon}}(\tilde{E}_1, \tilde{E}_2)$. Therefore

$$\Psi_0(h) = \psi_0(Q(h)) = \psi(\frac{2\pi}{T}Q(h)) = \psi(P(h)) = \Psi(h),$$

and

$$U_{\psi_0}(t) = e^{-ith^{-1}Q(h)}\Psi_0(h) = e^{-ith^{-1}\frac{T}{2\pi}P(h)}\Psi(h) = U_\psi(\frac{T}{2\pi}t).$$

Thus

$$U_{\psi_0}(2\pi) = U_\psi(T),$$

and by Lemma 12.2.6 we have

$$v_0(X) = \psi_0(q_\mu(X))e^{-2\pi\sigma_0(h)} = \psi(p_\mu(X))e^{-2\pi\sigma_0(h)} = u_0(X),\ X \in \mathbb{R}^{2n},$$

where $\sigma_0(h)$ is defined as in (12.55), and the u_j are supported in $p_\mu^{-1}(E_1 - \varepsilon, E_2 + \varepsilon)$. That α is also the Maslov index of the closed trajectories $\gamma_X \subset p_\mu^{-1}([E_1 - \varepsilon, E_2 + \varepsilon])$ follows immediately from the fact that the Maslov index is locally constant, that the set

$$p_\mu^{-1}([E_1 - \varepsilon, E_2 + \varepsilon]) = q_\mu^{-1}([\tilde{E}_1 - \tilde{\varepsilon}, \tilde{E}_2 + \tilde{\varepsilon}])$$

is *connected* by assumption, and that for any given $X \in p_\mu^{-1}([E_1 - \varepsilon, E_2 + \varepsilon])$,

$$\begin{aligned}\tilde{\gamma}_X &= \{\exp(tH_{q_\mu})(X);\, t \in [0, 2\pi]\} = \{\exp(\frac{T}{2\pi}tH_{p_\mu})(X);\, t \in [0, 2\pi]\} \\ &= \{\exp(sH_{p_\mu})(X);\, s \in [0, T]\} = \gamma_X,\end{aligned}$$

that is, any given closed integral curve $\tilde{\gamma}_X$ (as above) of H_{q_μ} contained in q_μ^{-1} $([\tilde{E}_1 - \tilde{\varepsilon}, \tilde{E}_2 + \tilde{\varepsilon}])$ is a riparametrization by $[0, 2\pi] \ni t \longmapsto s = (T/2\pi)t \in [0, T]$ of an integral curve γ_X (as above) of H_{p_μ} contained in $p_\mu^{-1}([E_1-\varepsilon, E_2+\varepsilon])=q_\mu^{-1}([\tilde{E}_1-\tilde{\varepsilon}, \tilde{E}_2+\tilde{\varepsilon}])$. Now, on the one hand

$$\begin{aligned}\frac{1}{2\pi}\int_0^{2\pi} q_{\mu-2}\circ\exp(tH_{q_\mu})(X)dt &= \frac{1}{2\pi}\int_0^{2\pi}\frac{T}{2\pi}p_{\mu-2}\circ\exp(t\frac{T}{2\pi}H_{p_\mu})(X)dt\\ &= \frac{1}{2\pi}\int_0^{T} p_{\mu-2}\circ\exp(sH_{p_\mu})(X)ds\\ &= \frac{T}{2\pi}\frac{1}{T}\int_0^{T} p_{\mu-2}\circ\exp(sH_{p_\mu})(X)ds,\end{aligned}$$

and on the other, again by Lemma 11.1.7,

$$A_{q_\mu}(\frac{T}{2\pi}E) = \frac{T}{2\pi}A_{p_\mu}(E).$$

We therefore get

$$\begin{aligned}\tilde{\beta} &= A_{q_\mu}(\frac{T}{2\pi}E) + h\frac{1}{2\pi}\int_0^{2\pi} q_{\mu-2}\circ\exp(tH_{q_\mu})(X)dt\\ &= \frac{T}{2\pi}A_{p_\mu}(E) + h\frac{T}{2\pi}\frac{1}{T}\int_0^{T} p_{\mu-2}\circ\exp(sH_{p_\mu})(X)ds = \frac{T}{2\pi}\beta,\end{aligned}$$

from which we also conclude for the constants $\tilde{\delta}$ and δ that $\tilde{\delta} = \frac{T}{2\pi}\delta$. This completes the proof of the lemma. □

It therefore follows that

$$e^{-2\pi i h^{-1}\left(\frac{T}{2\pi}P(h)-h\sigma_0(h)\right)}\Psi(h) = \Psi(h) + hW(h),$$

where $W(h)$ is a 0th-order h-pseudodifferential operator whose symbol belongs to $S^0_{\mathrm{cl}}(1)$ (and has compact support), with $\|W(h)\|_{L^2\to L^2} = O(1)$ uniformly in $h \in (0, h_0]$. Take now another $\phi \in \Psi_\varepsilon(E_1, E_2)$ such that $\psi\phi = \phi$. Then, by composing with $\Phi(h) := \phi(P(h))$, and by noting that $[\Phi(h), W(h)] = 0$, we obtain

$$e^{-2\pi i h^{-1}\left(\frac{T}{2\pi}P(h)-h\sigma_0(h)\right)}\Phi(h) = \Phi(h)\Big(I + hW(h)\Big).$$

Hence, for h_0 sufficiently small, we have $h\|W(h)\|_{L^2\to L^2} < 1/2$ for all $h \in (0, h_0]$, and we can consider

$$\Big(I + hW(h)\Big)^{-1} = I + hW'(h),$$

where $W'(h)$ is an h-pseudodifferential operator of order 0. It follows that we may take (by possibly shrinking h_0) the logarithm

$$R(h) = \frac{1}{2\pi i}\log\Big(I + hW'(h)\Big) = \frac{1}{2\pi i}\sum_{k\geq 0}\frac{(-1)^k}{k+1}h^{k+1}W'(h)^{k+1},$$

which turns out to be a 0th-order h-pseudodifferential operator (with symbol belonging to $S^0_{\mathrm{cl}}(1)$). We have

$$[R(h), P(h)\tilde{\Psi}(h)] = 0, \text{ and } \|R(h)\|_{L^2 \to L^2} = O(1), \ \forall h \in (0, h_0].$$

It thus follows that

$$e^{-2\pi i h^{-1}\left(\frac{T}{2\pi}P(h) - h\sigma_0(h) - h^2 R(h)\right)}\Phi(h) = \Phi(h), \ \forall h \in (0, h_0]. \tag{12.56}$$

Put now

$$\tilde{P}(h) := P(h) - h\frac{2\pi}{T}\sigma_0(h) - h^2\frac{2\pi}{T}R(h),$$

and consider the *compact* (self-adjoint) operators $P(h)\Psi(h)$ and $\tilde{P}(h)\Psi(h)$. Since they *commute*, they possess a common basis of eigenfunctions, that we denote by $\{u_j(h)\}_{j\in\mathbb{N}}$, where

$$P(h)\Psi(h)u_j(h) = \mu_j(h)u_j(h), \ \tilde{P}(h)\Psi(h)u_j(h) = \nu_j(h)u_j(h), \ j \in \mathbb{N}.$$

From the definition of $\tilde{P}(h)\Psi(h)$ we have that there exists $C_0 > 0$ such that, using (12.53) and (12.54),

$$|\mu_j(h) - \nu_j(h) - \frac{2\pi}{T}\frac{\alpha}{4}h - \delta| \leq C_0 h^2, \ \forall j \in \mathbb{N}. \tag{12.57}$$

On the other hand, from (12.56) it follows that for h_0 sufficiently small and for all $h \in (0, h_0]$,

$$\nu_j(h) \in \frac{2\pi}{T}h\mathbb{Z}. \tag{12.58}$$

Using (12.58) in (12.57) concludes the proof. □

Remark 12.2.7. It is important to notice that if h_0 is sufficiently small, then

$$I_k(h) \cap I_{k'}(h) = \emptyset, \ \forall h \in (0, h_0], \ \forall k \neq k'.$$

△

Remark 12.2.8. Notice in particular that the spectrum of $P(h)$ inside $[E_1, E_2]$ is discrete. One may find in Robert [65, Theorem III-4, p. 129, and Proposition III-13, p. 145] more general conditions that ensure the discreteness of the spectrum inside an energy band $[E_1, E_2]$. △

One has the following corollary, that deals with the case when the subprincipal symbol $p_{\mu-2}$ is a *non-zero constant* in a region $p_\mu^{-1}([E_1 - \varepsilon, E_2 + \varepsilon])$.

Corollary 12.2.9. *Under the assumptions of Theorem 12.2.4, with $\mu \geq 2$ and with hypothesis $(H5)$ replaced by the condition:*

- *the subprincipal symbol* $p_{\mu-2}$ *is a* **non-zero constant** c_0 *in* $p_\mu^{-1}([E_1-\varepsilon,E_2+\varepsilon])$, *one has the same conclusion of Theorem 12.2.4, that is*

$$\mathrm{Spec}(P(h))\cap[E_1,E_2]\subset\bigcup_{k\in\mathbb{Z}}I_k(h),\ \forall h\in(0,h_0],$$

where this time

$$I_k(h)=\Big[\frac{2\pi}{T}\Big(k+\frac{\alpha}{4}\Big)h+c_0h+\delta-C_0h^2,\frac{2\pi}{T}\Big(k+\frac{\alpha}{4}\Big)h+c_0h+\delta+C_0h^2\Big].$$

Notice that in this case $\beta=A_{p_\mu}(E)+hc_0=\delta+hc_0$, *and that, again, if* h_0 *is chosen sufficiently small, then* $I_k(h)\cap I_{k'}(h)=\emptyset$, *for all* $h\in(0,h_0]$ *and all* $k\neq k'$.

Proof. Let $P(h)=p^{\mathrm{w}}(x,hD;h)$, and let $P_0(h):=P(h)-hp^{\mathrm{w}}_{\mu-2}(x,hD)$. Notice that $P(h)$ and $P_0(h)$ have the same principal symbol p_μ. Assume, in the first place, that $p_{\mu-2}(X)\equiv c_0$ for all X. Then $P(h)=P_0(h)+hc_0$, and they have the **same** spectral family. Using the proof of the theorem (recall that $A_{p_\mu}(E)=\delta$), one therefore obtains, with $\phi\in\Psi_\varepsilon(E_1,E_2)$, that

$$e^{-2\pi ih^{-1}(\frac{T}{2\pi}P_0(h)-h(\frac{\alpha}{4}+\frac{T}{2\pi}\delta h^{-1})-h^2R(h))}\phi(P_0(h))=\phi(P_0(h)),\ \forall h\in(0,h_0],$$

that is,

$$e^{-2\pi ih^{-1}(\frac{T}{2\pi}P(h)-\frac{T}{2\pi}hc_0-h(\frac{\alpha}{4}+\frac{T}{2\pi}\delta h^{-1})-h^2R(h))}\phi(P_0(h))=\phi(P_0(h)),\ \forall h\in(0,h_0],$$

which (by the proof of Theorem 12.2.4, since every operator which commutes with $P_0(h)$ also commutes with $P(h)$) proves the corollary in case the subprincipal part is a constant everywhere.

In case we only know that $p_{\mu-2}$ is a constant c_0 in $p_\mu^{-1}([E_1-\varepsilon,E_2+\varepsilon])$ we proceed as follows. We write $P(h)=P_0(h)+hc_0+h(p^{\mathrm{w}}_{\mu-2}(x,hD)-c_0)$ and put $b(X):=p_{\mu-2}(X)-c_0$. Then $\mathrm{supp}(b)\cap p_\mu^{-1}([E_1-\varepsilon,E_2+\varepsilon])=\emptyset$. Hence, from Theorem 12.2.2 it follows that

$$b^{\mathrm{w}}(x,hD)\psi(P(h))=O(h^\infty),\ \text{and}\ \psi(P(h))=\psi(P_0(h)+hc_0)+h^2r^{\mathrm{w}}(x,hD;h),$$

where, by Robert's Theorem 12.2.3, $r\in S^0_{0,\mathrm{cl}}(m^{-k},g)$, for all $k\geq 1$. We thus obtain that

$$\begin{aligned}P(h)\psi(P(h))&=\Big(P_0(h)+hc_0+hb^{\mathrm{w}}(x,hD)\Big)\psi(P(h))\\&=\Big(P_0(h)+hc_0\Big)\psi(P(h))+O(h^\infty)\\&=\Big(P_0(h)+hc_0\Big)\psi(P_0(h)+hc_0)+O(h^2),\end{aligned}$$

which, by virtue of the minimax principle for compact self-adjoint operators and the first part of the proof, gives the claim. □

One has also the following corollary, due to Helffer and Robert [20], that deals with the case when the period T is not a constant in $[E_1-\varepsilon,E_2+\varepsilon]$.

Corollary 12.2.10. *Let $p \in S^0_{0,\mathrm{cl}}(m^\mu, g)$, $\mu > 0$, $p \sim p_\mu + hp_{\mu-2} + \dots$, be* **real valued**, *with principal symbol $p_\mu \approx m^\mu$ satisfying hypotheses (H1), (H2), (H3) (see Assumption 11.1.1) and (H5) (see Theorem 12.2.4) for an energy interval $[E_1 - \varepsilon, E_2 + \varepsilon]$, for some $\varepsilon \in (0,1]$. Let $\alpha \in \mathbb{Z}$ be the Maslov index of the trajectories $\gamma_X \subset p_\mu^{-1}(E_1 - \varepsilon, E_2 + \varepsilon)$. Then there exist constants $\delta \in \mathbb{R}$, $C_0 > 0$ and $h_0 \in (0,1]$ such that, with $P(h) = p^{\mathrm{w}}(x, hD; h)$,*

$$J\Big(\mathrm{Spec}(P(h)) \cap [E_1, E_2]\Big) \subset \bigcup_{k \in \mathbb{Z}} \Big[\Big(k + \frac{\alpha}{4}\Big)h + \delta - C_0 h^2, \Big(k + \frac{\alpha}{4}\Big)h + \delta + C_0 h^2\Big], \tag{12.59}$$

for all $h \in (0, h_0]$, where $J = J_{p_\mu}$ is the map defined in Lemma 11.1.2.

Proof. In the first place we reduce matters to the case $T(E) = 2\pi$ for all $E \in [E_1 - \varepsilon, E_2 + \varepsilon]$. Consider the function $J = J_{p_\mu}$ and extend it to a map $\tilde{J}\colon \mathbb{R} \longrightarrow \mathbb{R}$ such that

$$\tilde{J} \text{ is strictly increasing}, \ \tilde{J}(E) = \begin{cases} a_1 E + b_1, & \text{for } E \leq E_1 - 2\varepsilon, \\ a_2 E + b_2, & \text{for } E \geq E_2 + 2\varepsilon. \end{cases}$$

Define then the operator

$$Q(h) = \frac{1}{2\pi} \tilde{J}(P(h)).$$

Using Robert's Theorem 12.2.3 we have that $Q(h) = q^{\mathrm{w}}(x, hD; h)$, where $q \in S^0_{0,\mathrm{cl}}(m^\mu, g)$, $q \sim q_\mu + hq_{\mu-2} + \dots$, with the $q_{\mu-2j}$ independent of h, and

$$q_\mu = \tilde{J}(p_\mu), \quad q_{\mu-2} = p_{\mu-2}\tilde{J}'(p_\mu).$$

Hence

$$q_{\mu-2}(X) = 0, \ \forall X \in p_\mu^{-1}([E_1 - \varepsilon, E_2 + \varepsilon]).$$

Since

$$H_{q_\mu} = \frac{\tilde{J}'(p_\mu)}{2\pi} H_{p_\mu},$$

it follows that

$$\exp(tH_{q_\mu}) \text{ is periodic with period } 2\pi \text{ in a neighborhood of } q_\mu^{-1}(\tilde{J}([E_1, E_2])).$$

It thus follows from Lemma 11.1.6 that the (averaged) action A_{q_μ} of the integral trajectories of H_{q_μ} is a constant δ (the one that appears in the statement of the theorem) for all energies belonging to such a neighborhood.

Remark that, $\tilde{J}$ being a smooth diffeomorphism from a neighborhood of $[E_1, E_2]$ to a neighborhood of $\tilde{J}([E_1, E_2]) =: [\tilde{E}_1, \tilde{E}_2]$, the properties of the point spectrum of $P(h)$ in $[E_1, E_2]$ are immediately deduced from those of $Q(h)$ in $[\tilde{E}_1, \tilde{E}_2]$: in fact,

$$\lambda \in \mathrm{Spec}(P(h)) \cap [E_1, E_2] \Longleftrightarrow \frac{J(\lambda)}{2\pi} \in \mathrm{Spec}(Q(h)) \cap [\tilde{E}_1, \tilde{E}_2].$$

Hence the corollary follows from Theorem 12.2.4. □

If one assumes that the trajectories in $p_\mu^{-1}([E_1-\varepsilon,E_2+\varepsilon])$ are all periodic with *least* period T, *independent of* $E\in[E_1-\varepsilon,E_2+\varepsilon]$, that is

- $t\longmapsto \exp(tH_{p_\mu})(X)$ is periodic with period $T>0$ for all $X\in p_\mu^{-1}([E_1-\varepsilon,E_2+\varepsilon])$
- and
$$\exp(tH_{p_\mu})(X)\neq X,\ \forall t\in(0,T),\ \forall X\in p_\mu^{-1}([E_1-\varepsilon,E_2+\varepsilon]),$$

then one can obtain information on the number

$$N_{P(h)}(I_k(h))=\mathrm{card}\{\lambda(h);\ \lambda(h)\in\mathrm{Spec}(P(h))\cap I_k(h)\ \textit{with multiplicity}\},\quad (12.60)$$

that is, the number of the eigenvalues of $P(h)$, repeated according to multiplicity, which belong to $I_k(h)$.

In order to obtain an asymptotic formula for the behavior of $N_{P(h)}(I_k(h))$ as $h\to 0+$, we have to review what the *Leray-Liouville* measure is.

Definition 12.2.11. Let $f\colon \Omega\subset\mathbb{R}^{2n}\longrightarrow\mathbb{R}$ be a smooth function (Ω an open set) such that, for some $\lambda\in\mathbb{R}$, the set $f^{-1}(\lambda)\neq\emptyset$ is **compact**, and $df(\rho)\neq 0$ for all $\rho\in f^{-1}(\lambda)$. One defines the **Leray-Liouville** $(2n-1)$**-form** $L_{f,\lambda}$ by the equation

$$L_{f,\lambda}\wedge df=dX_1\wedge\ldots\wedge dX_{2n},\ \text{ on } f^{-1}(\lambda),\qquad (12.61)$$

where $dX_1\wedge\ldots\wedge dX_{2n}$ is the positive volume-form in $\mathbb{R}^{2n}_X$ associated with the coordinates $X_1,\ldots,X_{2n}$. By $L_{f,\lambda}(dX)$ we shall denote the associated **(positive) Leray-Liouville measure**. Notice that (12.61) determines $L_{f,\lambda}$ up to multiples of df, for if $L_{f,\lambda}\wedge df=dX_1\wedge\ldots\wedge dX_{2n}$ then also $\left(L_{f,\lambda}+\omega\wedge df\right)\wedge df= dX_1\wedge\ldots\wedge dX_{2n}$, for any given $(2n-2)$-form ω. We shall hence always **normalize** $L_{f,\lambda}$ so as to have that the $(2n-2)$-form

$$L_{f,\lambda}(\nabla_X f,\cdot)=\mathrm{i}_{\nabla_X f}L_{f,\lambda}=0.\qquad (12.62)$$

Here $\mathrm{i}_{\nabla_X f}$ is the contraction operator by $\nabla_X f$, where $\nabla_X f$ denotes the gradient vector-field associated with df through the Euclidean scalar product: $df(\rho)(v)=\langle\nabla_X f(\rho),v\rangle$, for all $\rho\in\mathbb{R}^{2n}$ and all $v\in T_\rho\mathbb{R}^{2n}$.

Lemma 12.2.12. *The normalization condition (12.62) fixes $L_{f,\lambda}$ uniquely.*

Proof. We work at any given point of $f^{-1}(\lambda)$. Let L_1, L_2 be such that

$$L_1\wedge df=L_2\wedge df=dX_1\wedge\ldots\wedge dX_{2n},\quad\text{and}\quad \mathrm{i}_{\nabla_X f}L_j=0,\ j=1,2.$$

Then $(L_1-L_2)\wedge df=0$. Hence, L_1 and L_2 being normalized, we get

$$\begin{aligned}0&=\mathrm{i}_{\nabla_X f}\Big((L_1-L_2)\wedge df\Big)\\&=\Big(\mathrm{i}_{\nabla_X f}(L_1-L_2)\Big)\wedge df+(-1)^{2n-1}(L_1-L_2)\wedge\mathrm{i}_{\nabla_X f}df\\&=(-1)^{2n-1}|\nabla_X f|^2(L_1-L_2),\end{aligned}$$

which proves the claim, for by hypothesis $\nabla_X f\neq 0$. □

The following lemma will be useful in the proof of Theorem 12.2.16 below.

Lemma 12.2.13. *Let $\alpha \in \mathbb{R}_+$. Given f and λ as in Definition 12.2.11 above, let $g := \alpha f$. Then for the respective Leray-Liouville $(2n-1)$-forms associated with g and f we have on $g^{-1}(\alpha\lambda) = f^{-1}(\lambda)$ the scaling property*

$$L_{\alpha f,\alpha\lambda} = \frac{1}{\alpha} L_{f,\lambda}.$$

Proof. Consider the smooth hypersurface $S = f^{-1}(\lambda) = g^{-1}(\alpha\lambda)$ of $\mathbb{R}^{2n}$. At any given point $\rho \in S$ we have $dg = \alpha df$ (and $\nabla_X g = \alpha \nabla_X f$), whence

$$L_{f,\lambda} \wedge df = dX_1 \wedge \ldots \wedge dX_{2n} = L_{g,\alpha\lambda} \wedge dg = \alpha L_{g,\alpha\lambda} \wedge df.$$

Since $L_{f,\lambda}$ and $\alpha L_{g,\alpha\lambda}$ are both *normalized*, that is

$$\mathrm{i}_{\nabla_X f} L_{f,\lambda} = 0, \;\; \mathrm{i}_{\nabla_X f}(\alpha L_{g,\alpha\lambda}) = \mathrm{i}_{\nabla_X g} L_{g,\alpha\lambda} = 0,$$

using Lemma 12.2.12 completes the proof. □

We may also write a more explicit expression of $L_{f,\lambda}$. To this aim, recall first of all that for a finite dimensional vectore space V, if $\omega_1, \ldots, \omega_k \in V^*$ and $v_1, \ldots, v_k \in V$, then

$$(\omega_1 \wedge \ldots \wedge \omega_k)(v_1, v_2, \ldots, v_k) = \det\big[\omega_j(v_{j'})\big]_{1 \le j,j' \le k}. \tag{12.63}$$

We have the following formula for $L_{f,\lambda}$ (which yields in particular another proof of Lemma 12.2.13).

Proposition 12.2.14. *Let f and λ be as in Definition 12.2.11. Put*

$$\omega^{(j)} := dX_1 \wedge \ldots \wedge \widehat{dX_j} \wedge \ldots \wedge dX_{2n}, \;\; 1 \le j \le 2n,$$

where the "hat" means that that term has been omitted. For all $\rho \in S = f^{-1}(\lambda)$ and all $v_1, v_2, \ldots, v_{2n-1} \in T_\rho \mathbb{R}^{2n}$ one has

$$L_{f,\lambda}(v_1, \ldots, v_{2n-1}) = \frac{1}{|\nabla_X f|^2} \sum_{j=1}^{2n} (-1)^j \frac{\partial f}{\partial X_j} \omega^{(j)}(v_1, \ldots, v_{2n-1}), \tag{12.64}$$

or, equivalently,

$$L_{f,\lambda} = -\frac{1}{|\nabla_X f|} \mathrm{i}_{\nabla_X f/|\nabla_X f|}(dX_1 \wedge \ldots \wedge dX_{2n}). \tag{12.65}$$

Recall that we take the normalization $\mathrm{i}_{\nabla_X f} L_{f,\lambda} = 0$. In particular, when $n = 1$ we have

$$L_{f,\lambda} = \frac{1}{|\nabla_X f|^2} \Big(\frac{\partial f}{\partial X_2} dX_1 - \frac{\partial f}{\partial X_1} dX_2 \Big), \tag{12.66}$$

where now the normalization condition can be written as $L_{f,\lambda}(\nabla_X f) = 0$.

Proof. We put $v^{(\ell)} = (v_1, \ldots, \widehat{v_\ell}, \ldots, v_{2n-1}) \in \underbrace{\mathbb{R}^{2n} \times \ldots \times \mathbb{R}^{2n}}_{2n-2 \text{ times}}$, $1 \leq \ell \leq 2n-1$, where the "hat" means that that term has been omitted. Since $\{\omega^{(j)}\}_{1 \leq j \leq 2n}$ is a basis of the space of $(2n-1)$-forms, we may write

$$L_{f,\lambda} = \sum_{j=1}^{2n} \alpha_j \omega^{(j)}. \tag{12.67}$$

We now use (12.63). We have on the one hand

$$\begin{aligned}
(L_{f,\lambda} \wedge df)(\nabla_X f, v_1, \ldots, v_{2n-1}) &= \sum_{j=1}^{2n} \alpha_j \big(\omega^{(j)} \wedge df\big)(\nabla_X f, v_1, \ldots, v_{2n-1}) \\
&= \sum_{j=1}^{2n} \alpha_j \left(-|\nabla_X f|^2 \omega^{(j)}(v_1, \ldots, v_{2n-1}) + \sum_{\ell=1}^{2n-1} (-1)^{\ell+1} \omega^{(j)}(\nabla_X f, v^{(\ell)}) df(v_\ell) \right) \\
&= -|\nabla_X f|^2 \sum_{j=1}^{2n} \alpha_j \omega^{(j)}(v_1, \ldots, v_{2n-1}) \\
&\quad + \sum_{\ell=1}^{2n-1} (-1)^{\ell+1} \left(\sum_{j=1}^{2n} \alpha_j \omega^{(j)}(\nabla_X f, v^{(\ell)}) \right) df(v_\ell) \\
&\quad \text{(recalling (12.67))} \\
&= -|\nabla_X f|^2 L_{f,\lambda}(v_1, \ldots, v_{2n-1}) + \sum_{\ell=1}^{2n-1} (-1)^{\ell+1} L_{f,\lambda}(\nabla_X f, v^{(\ell)}) df(v_\ell) \\
&\quad \text{(by the normalization condition)} \\
&= -|\nabla_X f|^2 L_{f,\lambda}(v_1, \ldots, v_{2n-1}).
\end{aligned}$$

On the other hand,

$$(dX_1 \wedge \ldots \wedge dX_{2n})(\nabla_X f, v_1, \ldots, v_{2n-1}) = \sum_{j=1}^{2n} (-1)^{j+1} \frac{\partial f}{\partial X_j} \omega^{(j)}(v_1, \ldots, v_{2n-1}).$$

Therefore

$$L_{f,\lambda}(v_1, \ldots, v_{2n-1}) = -\frac{1}{|\nabla_X f|^2} \sum_{j=1}^{2n} (-1)^{j+1} \frac{\partial f}{\partial X_j} \omega^{(j)}(v_1, \ldots, v_{2n-1}). \tag{12.68}$$

Notice that if $\tilde{L}_{f,\lambda}$ denotes the right-hand side of (12.68), then, for any given $v_1, \ldots, v_{2n-2} \in \mathbb{R}^{2n}$,

$$\begin{aligned}
&\tilde{L}_{f,\lambda}(\nabla_X f, v_1, \ldots, v_{2n-2}) \\
&\qquad = -\frac{1}{|\nabla_X f|^2} (dX_1 \wedge \ldots \wedge dX_{2n})(\nabla_X f, \nabla_X f, v_1, \ldots, v_{2n-2}) = 0,
\end{aligned}$$

that is, the normalization condition is indeed fulfilled. Specializing to the case $n = 1$ gives (12.66) from (12.64). This concludes the proof. □

As regards the Leray-Liouville measure associated with $f^{-1}(\lambda)$ we have the following corollary.

Corollary 12.2.15. *Let f and λ be as in Definition 12.2.11. Recall that the* **canonical volume form** *associated with $\{f = \lambda\}$ is (for every $\rho \in S$) given by*

$$\mathsf{vol}_{f=\lambda} = \mathsf{i}_{\nabla_X f/|\nabla_X f|}(dX_1 \wedge \ldots \wedge dX_{2n})\Big|_{f=\lambda} \tag{12.69}$$

(the restriction is the pull-back by the natural inclusion map). Then, for the Leray-Liouville measure we have

$$L_{f,\lambda}(dX) = \frac{1}{|\nabla_X f|}\mathsf{vol}_{f=\lambda}. \tag{12.70}$$

In particular,

$$\int_{f=\lambda} L_{f,\lambda}(dX) > 0.$$

Proof. The proof follows immediately from (12.65). □

We are ready for the "multiplicity" theorem, due to Helffer and Robert (see [20]).

Theorem 12.2.16. *With the notation and assumptions of Theorem 12.2.4, suppose*

$$(H6) \qquad \exp(tH_{p_\mu})(X) \neq X, \ \forall t \in (0,T), \ \forall X \in p_\mu^{-1}([E_1 - \varepsilon, E_2 + \varepsilon]),$$

that is, $T > 0$ is the least period in $p_\mu^{-1}([E_1 - \varepsilon, E_2 + \varepsilon])$. Let $\psi \in \Psi_\varepsilon(E_1, E_2)$ (see (12.50)). Then there exist real-valued functions $\Gamma_j \in C_0^\infty(E_1 - \varepsilon, E_2 + \varepsilon)$ such that for $\frac{2\pi}{T}(k + \frac{\alpha}{4})h + \delta \in [E_1, E_2]$

$$N_{P(h)}(I_k(h)) \sim \sum_{j \geq 1} \Gamma_j\Big(\frac{2\pi}{T}(k + \frac{\alpha}{4})h + \delta\Big)h^{j-n} \ (h \to 0+), \tag{12.71}$$

where, for $j = 1$,

$$\Gamma_1(\lambda) = (2\pi)^{-n}\psi(\lambda)\frac{2\pi}{T}\int_{p_\mu(X)=\lambda} L_{p_\mu,\lambda}(dX). \tag{12.72}$$

When $n = 1$ one has that $p_\mu^{-1}(E)$ is a smooth closed curve $\gamma(E)$ with Maslov index $\alpha = \pm 2$ (for $\gamma(E)$ is homotopic to a circle then), the sign depending on the orientation of $\gamma(E)$. Since $-2 \equiv 2 \bmod 4$, we may take $\alpha = 2$, so that for h_0 sufficiently small

$$N_{P(h)}(I_k(h)) = 1, \ \forall (k,h) \in \mathbb{Z} \times (0, h_0] \text{ with } \frac{2\pi}{T}(k + \frac{1}{2})h + \delta \in [E_1, E_2]. \tag{12.73}$$

Proof (sketch; see Helffer-Robert [20] and Dimassi-Sjöstrand [7]). Observe in the first place that if $\frac{2\pi}{T}(k+\frac{\alpha}{4})h+\delta \in [E_1,E_2]$, then for h_0 sufficiently small,

$$N_{P(h)}(I_k(h)) = \sum_{\lambda\in \mathrm{Spec}(P(h))\cap I_k(h)} \psi(\lambda) =: N_{k,\psi}(h), \ \forall h \in (0,h_0].$$

Following a by now classical idea of Colin De Verdière, take a real valued $\chi \in C_0^\infty$ $(-3\pi/2, 3\pi/2)$ such that

$$\sum_{j\in\mathbb{Z}} \chi(t-2\pi j) = 1. \tag{12.74}$$

Recall from the proof of Theorem 12.2.4 that, with

$$Q(h) = \frac{T}{2\pi}P(h), \quad \tilde{Q}(h) = \frac{T}{2\pi}P(h) - h\sigma_0(h) - h^2R(h),$$

and $\psi_0(\tilde{\lambda}) = \psi(\frac{2\pi}{T}\tilde{\lambda})$, one has

$$\sum_{\lambda\in \mathrm{Spec}(P(h))\cap I_k(h)} \psi(\lambda) = \sum_{\tilde{\lambda}\in \mathrm{Spec}(Q(h))\cap \frac{T}{2\pi}I_k(h)} \psi_0(\tilde{\lambda}),$$

and

$$\mathrm{Spec}(\tilde{Q}(h)) \cap \frac{T}{2\pi}[E_1-\varepsilon, E_2-\varepsilon] \subset h\mathbb{Z}, \ \forall h \in (0,h_0].$$

Hence

$$\begin{aligned}\mathrm{Tr}\Big(\psi_0(Q(h))e^{-ith^{-1}\tilde{Q}(h)}\Big) &= \sum_{k\in\mathbb{Z}}\Bigg(\sum_{\tilde{\lambda}\in \mathrm{Spec}(Q(h))\cap \frac{T}{2\pi}I_k(h)} \psi_0(\tilde{\lambda})\Bigg) e^{-itk} \\ &= \sum_{k\in\mathbb{Z}}\Bigg(\sum_{\lambda\in \mathrm{Spec}(P(h))\cap I_k(h)} \psi(\lambda)\Bigg) e^{-itk}, \end{aligned} \tag{12.75}$$

whence the $N_{k,\psi}(h)$ are the Fourier-coefficients of (12.75), that is,

$$\begin{aligned} N_{k,\psi}(h) &= \frac{1}{2\pi}\int_0^{2\pi} e^{ikt}\,\mathrm{Tr}\Big(\psi_0(Q(h))e^{-ith^{-1}\tilde{Q}(h)}\Big)dt \quad \text{(using (12.74))} \\ &= \frac{1}{2\pi}\int_{\mathbb{R}} e^{ih^{-1}t\tau}\chi(t)\,\mathrm{Tr}\Big(\psi_0(Q(h))e^{ithR(h)}e^{-ith^{-1}Q(h)}\Big)dt, \end{aligned} \tag{12.76}$$

where

$$\tau := (k+\frac{\alpha}{4})h + \frac{T}{2\pi}\delta \in \frac{T}{2\pi}(E_1-\varepsilon, E_2+\varepsilon).$$

At this point, hypothesis $(H6)$ ensures that $t=0$ is the *unique* period of the integral curves of H_{p_μ} in $p_\mu^{-1}(E_1-\varepsilon, E_2+\varepsilon)$, and stationary-phase arguments (see Dimassi-Sjöstrand [7]) show that

$$\mathrm{Tr}\Big(\psi_0(Q(h))e^{-ith^{-1}\tilde{Q}(h)}\Big) = O(h^\infty)$$

when $t \in \operatorname{supp}\chi \setminus \Omega$, where Ω is any fixed neighborhood of 0, and that (12.71) holds, where

$$\Gamma_1(\lambda) = (2\pi)^{-n}\psi_0(\frac{T}{2\pi}\lambda)\int_{\frac{T}{2\pi}p_\mu=\frac{T}{2\pi}\lambda} L_{\frac{T}{2\pi}p,\frac{T}{2\pi}\lambda}(dX)$$
$$= (2\pi)^{-n}\psi(\lambda)\frac{2\pi}{T}\int_{p_\mu=\lambda} L_{p_\mu,\lambda}(dX),$$

by virtue of Lemma 12.2.13.

When $n=1$, one has from the Stokes formula (keeping into account orientations) that for any given $E \in (E_1-\varepsilon, E_2+\varepsilon)$

$$J_p(E) - J_p(E_1-\varepsilon) = \int_{\gamma(E)} \xi\, dx - \int_{\gamma(E_1-\varepsilon)} \xi\, dx$$
$$= \int_{E_1-\varepsilon\le p(X)\le E} d\xi\wedge dx = \iint_{E_1-\varepsilon\le p(X)\le E} dX,$$

from which it follows that

$$J_p(E) = \iint_{E_1-\varepsilon\le p(X)\le E} dX + \text{const}.$$

Using the fact that $\nabla_X p_\mu \neq 0$ at the points of $p_\mu^{-1}(E)$ gives

$$dX = L_{p_\mu,E}(dX)dE,$$

so that

$$T_p = J_p'(E) = \int_{p_\mu=E} L_{p_\mu,E}(dX),$$

which then gives $N_{P(h)}(I_k(h)) = 1+O(h)$ for all $(k,h)\in\mathbb{Z}\times(0,h_0]$ with $\frac{2\pi}{T}(k+\frac{1}{2})h+\delta \in [E_1,E_2]$, whence (12.73). □

When the subprincipal symbol $p_{\mu-2}(X)$ is a non-zero *constant* c_0 for $X \in p_\mu^{-1}([E_1-\varepsilon,E_2+\varepsilon])$, one uses the definition (12.54) of β accordingly.

Following Dozias' adaptation of Hörmander [30] to the semiclassical case (see Dozias [8]), we next show how one can replace the condition that $p_{\mu-2}$ be constant with the following condition:

- *(H5′) There exists $c_0\in\mathbb{R}$ such that for all $X \in p_\mu^{-1}([E_1-\varepsilon,E_2+\varepsilon])$ one has*

$$\frac{1}{T}\int_0^T p_{\mu-2}\circ\exp(tH_{p_\mu})(X)dt = c_0.$$

Let hence $p\in S^0_{0,\mathrm{cl}}(m^\mu,g)$ be *real-valued*, $\mu\ge 2$ and $p_\mu \approx m^\mu$. Since we are interested in the spectral properties of $p^{\mathrm{w}}(x,hD)$ in the interval $[E_1-\varepsilon,E_2+\varepsilon]$, we may replace $p^{\mathrm{w}}(x,hD)$ with $\chi(p^{\mathrm{w}}(x,hD))$, where χ is *smooth, bounded, stricly increasing* and $\chi(t) = t$ for $t \in [E_1 - \varepsilon, E_2 + \varepsilon]$, and therefore may assume that $p \in S^0_{\mathrm{cl}}(1)$, with $p \sim p_0 + hp_1 + \dots$. The choice of χ ensures that $p_\mu^{-1}([E_1 - \varepsilon, E_2 + \varepsilon]) = (\chi \circ p_\mu)^{-1}([E_1 - \varepsilon, E_2 + \varepsilon]) = p_0^{-1}([E_1 - \varepsilon, E_2 + \varepsilon])$. The crucial result is contained in the following proposition.

Proposition 12.2.17. *Let $p \in S^0_{\mathrm{cl}}(1)$, $p = p_0 + hp_1 + \dots$. Assume that all the integral trajectories of H_{p_0} in $p_0^{-1}([E_1 - \varepsilon, E_2 + \varepsilon])$ are periodic with the same period $T > 0$ and that $(H5')$ holds for the subprincipal term p_1. Write*

$$P(h) = p_0^{\mathrm{w}}(x, hD) + hp_1^{\mathrm{w}}(x, hD) + h^2 r^{\mathrm{w}}(x, hD) =: P_0(h) + hP_1(h) + h^2 R(h),$$

where $R(h) = O(1)$, for all $h \in (0, h_0]$. Define also

$$\hat{P}_1(h) = \frac{1}{T} \int_0^T e^{ith^{-1}P_0(h)} P_1(h) e^{-ith^{-1}P_0(h)} dt.$$

Then $P_0(h) + hP_1(h)$ and $P_0(h) + h\hat{P}_1(h)$ are unitarily equivalent modulo $O(h^2)$ in the energy band $[E_1 - \varepsilon, E_2 + \varepsilon]$.

Proof. Put

$$P(t) := e^{ith^{-1}P} P_1 e^{-ith^{-1}P}, \quad \text{and} \quad S := -\frac{1}{T} \int_0^T \Big(\int_0^t P(s) ds \Big) dt.$$

Then, by the results in Robert's book [65] (namely, Egorov's Theorem) they are self-adjoint h-pseudodifferential operators with symbols in $S^0_{\mathrm{cl}}(1)$. We now compare

$$e^{iS}\Big(P_0(h) + hP_1(h)\Big)e^{-iS} \quad \text{with} \quad P_0(h) + h\hat{P}_1(h),$$

by computing their principal and subprincipal symbols, which we denote by $\sigma_0(\cdot)$ and $\mathrm{sub}(\cdot)$, respectively. To this end we need the following lemma.

Lemma 12.2.18. *Let $A = A^*$ with symbol in $S^0_{\mathrm{cl}}(1)$. Then $e^{iS}Ae^{-iS}$ is an h-pseudodifferential operator such that*

$$\mathrm{symbol}(e^{iS}Ae^{-iS}) \sim \sum_{j \geq 0} \frac{1}{j!} \mathrm{symbol}\Big((\mathrm{ad}\, iS)^j A\Big), \tag{12.77}$$

where $(\mathrm{ad}\, iS)A = [iS, A]$.

Proof (of the lemma). Since $S = S^*$, by Beals' characterization of h-pseudodifferential operators (Theorem 9.1.11), we have that e^{itS} is an h-pseudodifferential operator with symbol in $S^0_0(1)$, and is unitary. Put

$$A(t) = e^{itS}Ae^{-itS}.$$

Then

$$A'(t) = e^{itS}[iS,A]e^{-itS}, \ A^{(j)}(t) = e^{itS}\left((\mathrm{ad}\, iS)^j A\right)e^{-itS}.$$

Hence

$$A^{(j)}(0) = (\mathrm{ad}\, iS)^j A, \ \forall j \in \mathbb{Z}_+,$$

and therefore, formally,

$$A(1) = \sum_{j\geq 0} \frac{A^{(j)}(0)}{j!}.$$

Now, for all j and all t

$$\|A^{(j)}(t)\|_{L^2\to L^2} \leq C_j h^j,$$

for one sees that the first j terms in the symbol are 0. Using

$$A(1) = \sum_{j=0}^{N} \frac{1}{j!}(\mathrm{ad}\, iS)^j A + \frac{1}{N!}\int_0^1 (1-t)^N A^{(N+1)}(t)dt,$$

we get

$$\|A(1) - \sum_{j=0}^{N} \frac{1}{j!}(\mathrm{ad}\, iS)^j A\|_{L^2\to L^2} \leq C_N' h^{N+1},$$

and this proves the lemma. □

We now use (12.77) and have that on the one hand

$$\sigma_0\left(e^{iS}(P_0 + hP_1)e^{-iS}\right) = p_0,$$
$$\mathrm{sub}\left(e^{iS}(P_0 + hP_1)e^{-iS}\right) = \{\sigma_0(S), p_0\} + p_1,$$

and, on the other,

$$\sigma_0\left(P_0 + h\hat{P}_1\right) = p_0,$$
$$\mathrm{sub}\left(P_0 + h\hat{P}_1\right) = \frac{1}{T}\int_0^T p_1 \circ \exp(tH_{p_0})dt.$$

Hence, equality at the operator-level modulo $O(h^2)$ occurs for $X \in p_0^{-1}([E_1 - \varepsilon, E_2 + \varepsilon])$ if

$$\{\sigma_0(S), p_0\} + p_1 = \frac{1}{T}\int_0^T p_1 \circ \exp(tH_{p_0})dt. \tag{12.78}$$

Notice that since $p_0 \in S_0^0(1)$, the Hamilton vector field H_{p_0} is uniformly bounded on $\mathbb{R}^{2n}$, whence for every $X \in \mathbb{R}^{2n}$ the integral trajectories $t \mapsto \exp(tH_{p_0})(X)$ exist for all $t \in \mathbb{R}$. It follows that

$$\mathbb{R} \times \mathbb{R}^{2n} \ni (t,X) \overset{C^\infty}{\longmapsto} \exp(tH_{p_0})(X),$$

and that

$$\mathbb{R}^{2n} \ni X \longmapsto \frac{1}{T}\int_0^T p_1 \circ \exp(tH_{p_0})(X)dt$$

is well-defined and smooth. Now,

$$\begin{aligned}
\sigma_0(S)(X) &= -\frac{1}{T}\int_0^T \Big(\int_0^t p_1 \circ \exp(sH_{p_0})(X)ds\Big)dt \\
&= -\frac{1}{T}\int_0^T \Big(\int_s^T p_1 \circ \exp(sH_{p_0})(X)dt\Big)ds \\
&= -\frac{1}{T}\int_0^T (T-s)p_1 \circ \exp(sH_{p_0})(X)ds \\
&= -\int_0^T p_1 \circ \exp(sH_{p_0})(X)ds + \frac{1}{T}\int_0^T s p_1 \circ \exp(sH_{p_0})(X)ds \\
&= -Tc_0 + \frac{1}{T}\int_0^T s p_1 \circ \exp(sH_{p_0})(X)ds, \ \forall X \in p_0^{-1}([E_1-\varepsilon, E_2+\varepsilon]).
\end{aligned}$$

Then, for all $X \in p_0^{-1}([E_1-\varepsilon, E_2+\varepsilon])$ we have

$$\begin{aligned}
\{\sigma_0(S), p_0\}(X) &= -\frac{d}{dt}\Big(\sigma_0(S) \circ \exp(tH_{p_0})(X)\Big)\Big|_{t=0} \\
&= -\frac{d}{dt}\frac{1}{T}\int_0^T s p_1 \circ \exp((s+t)H_{p_0})(X)ds\Big|_{t=0} \\
&= -\frac{1}{T}\int_0^T s\frac{d}{dt} p_1 \circ \exp((s+t)H_{p_0})(X)\Big|_{t=0} ds \\
&= -\frac{1}{T}\int_0^T s\frac{d}{ds}\Big(p_1 \circ \exp(sH_{p_0})(X)\Big)ds \\
&= -\frac{1}{T}\int_0^T \frac{d}{ds}\Big(s p_1 \circ \exp(sH_{p_0})(X)\Big)ds + c_0 \\
&= -p_1 \circ \exp(TH_{p_0})(X) + c_0 = -p_1(X) + c_0,
\end{aligned}$$

which shows that (12.78) indeed holds. This concludes the proof of the proposition. □

Proposition 12.2.17 and Corollary 12.2.9 thus yield the following result.

Corollary 12.2.19. *Under the assumptions of Theorem 12.2.4, with $\mu \geq 2$ and with the subprincipal symbol $p_{\mu-2}$ fulfilling hypothesis $(H5')$ instead of hypothesis $(H5)$, one has the same conclusion of Theorem 12.2.4, that is*

$$\mathrm{Spec}(P(h)) \cap [E_1, E_2] \subset \bigcup_{k\in\mathbb{Z}} I_k(h), \ \ \forall h \in (0, h_0],$$

where, this time,

$$I_k(h) = \left[\frac{2\pi}{T}\left(k+\frac{\alpha}{4}\right)h + c_0h + \delta - C_0h^2, \frac{2\pi}{T}\left(k+\frac{\alpha}{4}\right)h + c_0h + \delta + C_0h^2\right],$$

and

$$\beta = A_{p_\mu}(E) + h\frac{1}{T}\int_0^T p_{\mu-2}\circ\exp(tH_{p_\mu})(X)dt = \delta + hc_0.$$

Notice that, again, if h_0 is chosen sufficiently small, then $I_k(h)\cap I_{k'}(h) = \emptyset$, for all $h\in(0,h_0]$ and all $k\neq k'$.

Furthermore, in Theorem 12.2.16, with (H5$'$) replacing (H5), one has the same conclusion, with β as above.

As a consequence of Corollary 12.2.9 and Corollary 12.2.19 we have the following result, which is useful when dealing with h^∞-diagonalizations.

Lemma 12.2.20. *Let $p\in S^0_{0,\mathrm{cl}}(m^\mu,g)$, $\mu\geq 2$, $p\sim p_\mu + hp_{\mu-2}+\dots$, be* **real valued**, *with principal symbol $p_\mu \approx m^\mu$ satisfying hypotheses (H1), (H2) and (H3) for an energy interval $[E_1-\varepsilon, E_2+\varepsilon]$, for some $\varepsilon\in(0,1]$. Suppose in addition that:*

- *(H4) The period $T_{p_\mu}(E)$ is a constant T for all $E\in[E_1-\varepsilon,E_2+\varepsilon]$;*
- *(H5$'$) There exists $c_0\in\mathbb{R}$ such that for all $X\in p_\mu^{-1}([E_1-\varepsilon,E_2+\varepsilon])$ one has*

$$\frac{1}{T}\int_0^T p_{\mu-2}\circ\exp(tH_{p_\mu})(X)dt = c_0.$$

Let $\alpha\in\mathbb{Z}$ be the Maslov index of the trajectories $\gamma_X\subset p_\mu^{-1}(E_1-\varepsilon,E_2+\varepsilon)$. Consider $\tilde{p}\in S^0_{0,\mathrm{cl}}(m^\mu,g)$, $\mu\geq 2$, such that

$$\tilde{p} = p + h\,\mathrm{Im}(f\{\bar{f},p_\mu\}),$$

where $f\colon\mathbb{R}^{2n}\longrightarrow\mathbb{S}^1\subset\mathbb{C}$ belongs to $S(1,g)$. We know from Corollary 12.2.19 that there exist $C_0>0$ and $h_0\in(0,1]$ such that

$$\mathrm{Spec}(P(h))\cap[E_1,E_2]\subset\bigcup_{k\in\mathbb{Z}} I_k(h),$$

for all $h\in(0,h_0]$, where $P(h)=p^{\mathrm{w}}(x,hD;h)$ and

$$I_k(h) = \left[\frac{2\pi}{T}\left(k+\frac{\alpha}{4}\right)h + c_0h + \delta - C_0h^2, \frac{2\pi}{T}\left(k+\frac{\alpha}{4}\right)h + c_0h + \delta + C_0h^2\right].$$

Put $\tilde{P}(h)=\tilde{p}^{\mathrm{w}}(x,hD;h)$. Then also

$$\mathrm{Spec}(\tilde{P}(h))\cap[E_1,E_2]\subset\bigcup_{k\in\mathbb{Z}} I_k(h),\quad \forall h\in(0,h_0].$$

Proof. We need only show that

$$\int_0^T (f\{\bar{f},p_\mu\})\Big(\exp(tH_{p_\mu})(X)\Big)dt = 2\pi i k_0, \tag{12.79}$$

where $k_0 \in \mathbb{Z}$ is *independent of* $X \in p_\mu^{-1}([E_1-\varepsilon,E_2+\varepsilon])$. In fact, one then has

$$\frac{1}{T}\int_0^T \mathrm{Im}\Big((f\{\bar{f},p_\mu\})(\exp(tH_{p_\mu})(X))\Big)dt = \frac{2\pi}{T}k_0,$$

and through Corollary 12.2.9 and Corollary 12.2.19 this shows that

$$\mathrm{Spec}(\tilde{P}(h)) \cap [E_1,E_2] \subset \bigcup_{k'\in\mathbb{Z}} \tilde{I}_{k'}(h), \ \text{ where } \tilde{I}_{k'}(h) = I_{k'+k_0}(h),$$

thus concluding the proof modulo (12.79), that we now prove.

Since $f\bar{f}=1$ we have that

$$0 = \{f\bar{f},p_\mu\} = \bar{f}\{f,p_\mu\} + f\{\bar{f},p_\mu\}, \ \text{ whence } \ f\{\bar{f},p_\mu\} \in i\mathbb{R}.$$

Let $\gamma_X(t) = \exp(tH_{p_\mu})(X)$ for $X \in p_\mu^{-1}([E_1-\varepsilon,E_2+\varepsilon])$. As $f\{\bar{f},p_\mu\} = \bar{f}\{p_\mu,f\}$, we get

$$\begin{aligned}
\int_0^T (f\{\bar{f},p_\mu\})(\gamma_X(t))dt &= \int_0^T (\bar{f}\{p_\mu,f\})(\gamma_X(t))dt \\
&= \int_0^T (\frac{\{p_\mu,f\}}{f} \circ \gamma_X)(t)dt \\
&= \int_0^T \frac{1}{f(\gamma_X(t))}\frac{d}{dt}(f\circ\gamma_X)(t)dt = \int_{f\circ\gamma_X} \frac{dz}{z} \in 2\pi i\mathbb{Z}.
\end{aligned}$$

Since $p_\mu^{-1}([E_1-\varepsilon,E_2+\varepsilon])$ is a *connected* set and

$$p_\mu^{-1}([E_1-\varepsilon,E_2+\varepsilon]) \ni X \longmapsto \int_{f\circ\gamma_X}\frac{dz}{z} \ \text{ is } \textit{continuous},$$

we obtain (12.79) and complete the proof of the corollary. □

Remark 12.2.21. Notice that when $f(X) = e^{i\varphi(X)}$, with $\varphi \in S(1,g)$, then in (12.79) one has $k_0 = 0$. △

Since $P(h)$ and $\tilde{P}(h)$ in Lemma 12.2.20 have the same principal symbol, they have in particular the same Leray-Liouville measure associated with $p_\mu^{-1}(E)$, $E \in [E_1-\varepsilon,E_2+\varepsilon]$. Hence from Theorem 12.2.16 and Corollary 12.2.19 we obtain the following consequence.

Corollary 12.2.22. *Let $k \in \mathbb{Z}$ and $h \in (0,h_0]$ be such that $\frac{2\pi}{T}(k+\frac{\alpha}{4})h + c_0 h + \delta \in [E_1,E_2]$. Let $k' \in \mathbb{Z}$ be such that $k = k' + k_0$, where k_0 is given in (12.79), whence $\tilde{I}_{k'}(h) = I_k(h)$. Then*

$$\mathrm{Spec}(\tilde{P}(h)) \cap I_k(h) \neq \emptyset, \quad \textit{and} \quad \mathrm{Spec}(P(h)) \cap I_k(h) \neq \emptyset,$$

so that

$$\mathrm{dist}\Big(\mathrm{Spec}(\tilde{P}(h)) \cap I_k(h), \mathrm{Spec}(P(h)) \cap I_k(h)\Big) = O(h^2).$$

12.3 Localization and Multiplicity of Spec($A_2^w(x, D)$), with $A_2 \in \mathscr{Q}_2^s$

We are finally in a position to describe in this section some properties of the large eigenvalues of a general 2×2 elliptic positive NCHO in $\mathbb{R}^n$ with symbol $A_2 \in \mathscr{Q}_2^s$, that are linked to properties of the periods of the bicharacteristics of the eigenvalues of the symbol $A_2(X)$. As already mentioned, we shall go through the semiclassical construction.

In the first place we show that assumptions $(H1)$ and $(H2)$ are fulfilled by the eigenvalues of a system $A_2 \in \mathscr{Q}_2$.

Proposition 12.3.1. *Let $\lambda_\pm$ be the eigenvalues of a system $A_2 \in \mathscr{Q}_2$. Then hypotheses $(H1)$ and $(H2)$ hold for $\lambda_\pm$.*

Proof. It is immediate using Proposition 11.1.4, for $\lambda_\pm : \mathbb{R}^{2n} \setminus \{0\} \longrightarrow \mathbb{R}_+$ are positively homogeneous of degree 2. □

We next show that when $A_2 \in \mathscr{Q}_2^s$, we may find h^∞-diagonalizations of the *reference operator* $A_r^w(x,hD)$ for which the subprincipal terms in the diagonal form vanish in the region $\{X \in \mathbb{R}^{2n};\ \lambda_-(X) \geq \varepsilon_1\}$ (see (12.4), (12.13) and Remark 12.1.21).

Proposition 12.3.2. *Suppose that $A_2 \in \mathscr{Q}_2^s$. Then there exists a unitary symbol $e_0 \in S(1,g;\mathsf{M}_2)$, $e_0 e_0^* = e_0^* e_0 = I$, such that for the subprincipal symbol $\Lambda_0 = \mathrm{diag}(\Lambda_0^+,\Lambda_0^-)$ of the h^∞-diagonalization of the reference operator $A_r^w(x,hD)$ we have*

$$\Lambda_0^\pm(X) = 0, \ \forall X \in \{X \in \mathbb{R}^{2n};\ \lambda_-(X) \geq \varepsilon_1\}.$$

By Remark 12.1.21, the proposition is a consequence of the following lemma (related to diagonalizations of systems of *classical* symbols, see Theorem 9.2.3).

Lemma 12.3.3. *When $A_2 \in \mathscr{Q}_2^s$ we may always find a diagonalization whose resulting subprincipal symbol $\Lambda_0 = \mathrm{diag}(\Lambda_0^+,\Lambda_0^-)$ satisfies*

$$\Lambda_0^\pm(X) = 0, \ \forall X \neq 0.$$

Proof. We work for $X \neq 0$. Consider $A_2 \in \mathscr{Q}_2^s$ and write

$$A_2(X) = \begin{bmatrix} a_{11}(X) & \gamma b(X) \\ \bar{\gamma} b(X) & a_{22}(X) \end{bmatrix},$$

where b is a *real-valued homogeneous* polynomial of degree 2 in X, $\gamma = 1$ or $\gamma = i$, and (say) $a_{11}(X) > a_{22}(X)$. Recall that $\lambda_\pm$ are the eigenvalues of A_2 (see (12.4)), and that $\{w_+, w_-\}$ is the canonical basis of $\mathbb{C}^2$, $w_+ = \begin{bmatrix} 1 \\ 0 \end{bmatrix}$, $w_- = \begin{bmatrix} 0 \\ 1 \end{bmatrix}$. Put also (as before)

$$a_\pm = \frac{a_{11} \pm a_{22}}{2}.$$

Then

$$\lambda_\pm = a_+ \pm \sqrt{a_-^2 + b^2} = a_+ \pm \sqrt{\delta}, \text{ where (as before) } \delta = a_-^2 + b^2.$$

Consider therefore (always for $X \neq 0$) the eigenvectors belonging to $\lambda_\pm$, $\pm$-respectively,

$$\tilde{v}_+(X) := \begin{bmatrix} a_-(X) + \sqrt{\delta(X)} \\ \bar{\gamma} b(X) \end{bmatrix}, \quad \tilde{v}_-(X) := \begin{bmatrix} -\gamma b(X) \\ a_-(X) + \sqrt{\delta(X)} \end{bmatrix},$$

and hence the unitary matrix (positively homogeneous of degree 0) whose columns are, in the order, the normalized eigenvectors v_+ and v_-:

$$e_0(X) = [v_+(X) | v_-(X)], \quad v_\pm(X) = \frac{\tilde{v}_\pm(X)}{|\tilde{v}_\pm(X)|_{\mathbb{C}^2}}.$$

It is now crucial to notice that

$$e_0(X) w_+ = v_+(X) \in \mathbb{R} \times \gamma\mathbb{R}, \quad e_0(X) w_- = v_-(X) \in \gamma\mathbb{R} \times \mathbb{R}, \quad \forall X \neq 0, \tag{12.80}$$

whence

$$\partial e_0 w_+ \in \mathbb{R} \times \gamma\mathbb{R}, \quad \partial e_0 w_- \in \gamma\mathbb{R} \times \mathbb{R}, \quad \forall X \neq 0.$$

It follows that

$$\langle \partial e_0 w_\pm, e_0 w_\pm \rangle \in \mathbb{R}, \tag{12.81}$$

so that

$$\langle \partial e_0 w_\pm, e_0 w_\pm \rangle = \langle e_0 w_\pm, \partial e_0 w_\pm \rangle,$$

whence, since $|v_\pm(X)|_{\mathbb{C}^2} = 1$ for all $X \neq 0$,

$$\langle \partial e_0(X) w_\pm, e_0(X) w_\pm \rangle = 0, \quad \forall X \neq 0.$$

We thus have

$$\partial e_0(X)w_\pm \in \mathrm{Span}_{\mathbb{C}}\{e_0(X)w_\pm\}^\perp = \mathrm{Span}_{\mathbb{C}}\{e_0(X)w_\mp\}, \ \forall X \neq 0. \tag{12.82}$$

Recall (see (12.46)) that $\pm$-respectively for all $X \neq 0$

$$\Lambda_0^\pm = \frac{1}{2}\mathrm{Im}\big(\langle\{e_0^*,\lambda_\pm\}e_0w_\pm, w_\pm\rangle\big) + \frac{1}{2}\mathrm{Im}\big(\langle e_0^*\{A_2,e_0\}w_\pm, w_\pm\rangle\big) =: I_1 + I_2.$$

Consider I_1. By (12.81) we have

$$\begin{aligned}\langle\{e_0^*,\lambda_\pm\}e_0w_\pm, w_\pm\rangle &= \sum_{\ell=1}^n\Big(\frac{\partial\lambda_\pm}{\partial x_\ell}\langle\frac{\partial e_0^*}{\partial\xi_\ell}e_0w_\pm, w_\pm\rangle - \frac{\partial\lambda_\pm}{\partial\xi_\ell}\langle\frac{\partial e_0^*}{\partial x_\ell}e_0w_\pm, w_\pm\rangle\Big)\\ &= \sum_{\ell=1}^n\Big(\frac{\partial\lambda_\pm}{\partial x_\ell}\langle e_0w_\pm, \frac{\partial e_0}{\partial\xi_\ell}w_\pm\rangle - \frac{\partial\lambda_\pm}{\partial\xi_\ell}\langle e_0w_\pm, \frac{\partial e_0}{\partial x_\ell}w_\pm\rangle\Big) \in \mathbb{R},\end{aligned}$$

whence

$$I_1 = 0, \ \forall X \neq 0.$$

We next consider I_2. Using (9.42) we have

$$\begin{aligned}\langle\partial A_2\partial' e_0w_\pm, e_0w_\pm\rangle &= \langle\partial' e_0w_\pm, (\partial A_2)e_0w_\pm\rangle\\ &= \langle\partial' e_0w_\pm, (\partial\lambda_\pm)e_0w_\pm - (A_2-\lambda_\pm)\partial e_0w_\pm\rangle,\end{aligned}$$

which, along with (12.81) and the fact that $A_2 - \lambda_\pm$ is Hermitian, gives

$$\begin{aligned}\langle\{A_2,e_0\}w_\pm, e_0w_\pm\rangle = &\underbrace{\sum_{\ell=1}^n\Big(\frac{\partial\lambda_\pm}{\partial\xi_\ell}\langle\frac{\partial e_0}{\partial x_\ell}w_\pm, e_0w_\pm\rangle - \frac{\partial\lambda_\pm}{\partial x_\ell}\langle\frac{\partial e_0}{\partial\xi_\ell}w_\pm, e_0w_\pm\rangle\Big)}_{\in\mathbb{R}}\\ &-\sum_{\ell=1}^n\Big(\langle(A_2-\lambda_\pm)\frac{\partial e_0}{\partial x_\ell}w_\pm, \frac{\partial e_0}{\partial\xi_\ell}w_\pm\rangle\\ &\qquad -\langle(A_2-\lambda_\pm)\frac{\partial e_0}{\partial\xi_\ell}w_\pm, \frac{\partial e_0}{\partial x_\ell}w_\pm\rangle\Big).\end{aligned}$$

Therefore

$$\begin{aligned}I_2 &= -\frac{1}{2}\mathrm{Im}\Big(\sum_{\ell=1}^n\langle(A_2-\lambda_\pm)\frac{\partial e_0}{\partial x_\ell}w_\pm, \frac{\partial e_0}{\partial\xi_\ell}w_\pm\rangle\Big)\\ &\quad+\frac{1}{2}\mathrm{Im}\Big(\sum_{\ell=1}^n\langle(A_2-\lambda_\pm)\frac{\partial e_0}{\partial\xi_\ell}w_\pm, \frac{\partial e_0}{\partial x_\ell}w_\pm\rangle\Big)\\ &= \mathrm{Im}\Big(\sum_{\ell=1}^n\langle(A_2-\lambda_\pm)\frac{\partial e_0}{\partial\xi_\ell}w_\pm, \frac{\partial e_0}{\partial x_\ell}w_\pm\rangle\Big).\end{aligned}$$

Now it is important to notice that from (12.80) it also follows

$$\langle\partial e_0w_\pm, e_0w_\mp\rangle \in \gamma\mathbb{R}.$$

By (12.82) we have

$$\partial e_0 w_\pm = \langle \partial e_0 w_\pm, e_0 w_\mp \rangle e_0 w_\mp,$$

whence, for $1 \le \ell \le n$ and $\pm$-respectively,

$$\langle (A_2 - \lambda_\pm) \frac{\partial e_0}{\partial \xi_\ell} w_\pm, \frac{\partial e_0}{\partial x_\ell} w_\pm \rangle = (\lambda_\mp - \lambda_\pm) \underbrace{\langle \frac{\partial e_0}{\partial \xi_\ell} w_\pm, e_0 w_\mp \rangle}_{\in \gamma \mathbb{R}} \underbrace{\overline{\langle \frac{\partial e_0}{\partial x_\ell} w_\pm, e_0 w_\mp \rangle}}_{\in \gamma \mathbb{R}} \in \mathbb{R}.$$

Thus also $I_2 = 0$, and this concludes the proof of the lemma. □

We now exploit the results of Section 12.2 to obtain at first information on the spectrum of the reference operator $A_r^{\mathrm{w}}(x, hD)$, and then on that of $A^{\mathrm{w}}(x, D)$. In the first place we give the following definition regarding the periods of the eigenvalues of $A_2(X)$, and afterwards define two important sets of integers that we shall use for localizing the eigenvalues.

Definition 12.3.4. Let $A_2 \in \mathscr{Q}_2$ and let $\lambda_\pm$ be its eigenvalues. Let $0 < E_1 < E_2$. We shall say that $A_2 \in \mathscr{Q}_2$ **satisfies hypothesis** $(T_\pm)$ **on the energy interval** $[E_1, E_2]$ if **all** the integral trajectories of the Hamilton vector-fields $H_{\lambda_\pm}$ contained in $\lambda_\pm^{-1}([E_1, E_2])$, $\pm$-respectively, are periodic with periods $T_\pm$, necessarily **independent of** $E \in [E_1, E_2]$ by Lemma 11.1.8. Remark that when $n = 1$, by Proposition 11.1.10, hypothesis $(T_\pm)$ is automatically satisfied on every energy interval $[E_1, E_2] \subset (0, +\infty)$.

Notice that since for the eigenvalues $\lambda_\pm(X)$ of $A_2(X)$ we have $\lambda_-(X) < \lambda_+(X)$ for all $X \neq 0$, it follows from (11.11) of Proposition 11.1.10 that

$$T_- / T_+ > 1. \tag{12.83}$$

Definition 12.3.5. Let $A_2 \in \mathscr{Q}_2$ satisfy hypothesis $(T_\pm)$ on an energy interval $[E_1 - \varepsilon, E_2 + \varepsilon]$, for some $\varepsilon \in (0, 1]$. Let $\alpha_\pm$ be the Maslov indices of the closed trajectories, $\pm$-respectively,

$$\gamma_X^\pm := \left\{ \exp(tH_{\lambda_\pm})(X);\, t \in [0, T_\pm] \right\}, \ X \in \lambda_\pm^{-1}([E_1, E_2]).$$

We define, $\pm$-respectively,

$$\mathbb{Z}^\pm_{E_1,E_2}(h) := \left\{ k \in \mathbb{Z}_+;\ \frac{2\pi}{T_\pm}(k + \frac{\alpha_\pm}{4}) h \in [E_1, E_2] \right\}, \tag{12.84}$$

and, when $n = 1$ (we have $\alpha_\pm = 2$ in this case),

$$\mathbb{Q}_{E_1,E_2}(h) := \left\{ (k, k') \in \mathbb{Z}^+_{E_1,E_2}(h) \times \mathbb{Z}^-_{E_1,E_2}(h);\ \frac{2k+1}{2k'+1} = \frac{T_+}{T_-} \right\}. \tag{12.85}$$

In the first place we have the following result, interesting in its own right, about the large eigenvalue of the reference operator $A_r^{\mathrm{w}}(x, hD)$. As already remarked, this is a fundamental step, to obtain information on the large eigenvalues of $A_2^{\mathrm{w}}(x, hD)$ and hence on those of $A_2^{\mathrm{w}}(x, D)$.

Theorem 12.3.6. *Let $A_2 \in \mathscr{Q}_2^s$ satisfy hypothesis $(T_\pm)$ on an energy interval $[E_1-\varepsilon, E_2+\varepsilon]$, for some $\varepsilon \in (0,1]$, where $E_0 + 10^2 + c_\chi < E_1 < E_2$ (see Theorem 12.1.16). Let A_r be the symbol of the corresponding* **reference** *operator constructed in Section 12.1. We have:*

- *There exists $C_0 > 0$ and $h_0 \in (0,1]$ such that*

$$\mathrm{Spec}(A_r^{\mathrm{w}}(x,hD)) \cap [E_1,E_2] \subset \bigcup_{\pm} \bigcup_{k\in\mathbb{Z}_+} I_k^{\pm}(h),\ \forall h \in (0,h_0], \tag{12.86}$$

where

$$I_k^{\pm}(h) := \Big[\frac{2\pi}{T_\pm}(k+\frac{\alpha_\pm}{4})h - C_0h^2, \frac{2\pi}{T_\pm}(k+\frac{\alpha_\pm}{4})h + C_0h^2\Big].$$

We may choose $h_0 > 0$ so small that $I_k^{\pm}(h) \cap I_{k'}^{\pm}(h) = \emptyset$, $\pm$-respectively, for all $k \neq k'$.

- *Suppose that $T_\pm$ are the* **least** *periods, and suppose there exists $C_* > 0$ such that*

$$\left|\frac{2\pi}{T_+}(k+\frac{\alpha_+}{4}) - \frac{2\pi}{T_-}(k'+\frac{\alpha_-}{4})\right| \geq C_*,\ \forall k,k' \in \mathbb{Z}, \tag{12.87}$$

whence necessarily we must have $\alpha_+/4 \notin \mathbb{Z}$ or $\alpha_-/4 \notin \mathbb{Z}$. It is important to notice that by virtue of (12.87) there is $h_0 \in (0,1]$ such that

$$I_k^j(h) \cap I_{k'}^{j'}(h) = \emptyset,\ \forall k,k' \in \mathbb{Z},\ \forall h \in (0,h_0],\ \forall j,j' = \pm. \tag{12.88}$$

Then for $h_0 > 0$ sufficiently small

$$\mathrm{Spec}(A_r^{\mathrm{w}}(x,hD)) \cap I_k^{\pm}(h) \neq \emptyset,\ \forall k \in \mathbb{Z}_{E_1,E_2}^{\pm}(h),\ \forall h \in (0,h_0], \tag{12.89}$$

$\pm$-respectively. When $n=1$ we have that $\alpha_\pm = 2$, and for any fixed $h \in (0,h_0]$,

$$N_{A_r^{\mathrm{w}}(x,hD)}(I_k^{\pm}(h)) = 1,\ \forall k \in \mathbb{Z}_{E_1,E_2}^{\pm}(h),\ \pm\text{-respectively.} \tag{12.90}$$

Proof. Take an h^∞-diagonalization $\Lambda^{\mathrm{w}}(h) = \mathrm{diag}(\Lambda_+^{\mathrm{w}}(h), \Lambda_-^{\mathrm{w}}(h))$ of the reference operator $A_r^{\mathrm{w}}(x,hD)$ with subprincipal part satisfying Proposition 12.3.2. Hence the action is 0 on $\lambda_\pm^{-1}([E_1-\varepsilon, E_2+\varepsilon])$. From Proposition 11.1.4, Proposition 12.3.1 and the hypotheses, we have that we may use the spectral results of Section 12.2 applied to the scalar operators $\Lambda_\pm^{\mathrm{w}}(h)$. Using (12.43) of Theorem 12.1.19 (and possibly shrinking h_0, depending also on ε) we thus have

$$\mathrm{Spec}(A_r^{\mathrm{w}}(x,hD)) \cap [E_1,E_2] \subset \bigcup_{\tilde\lambda(h)\in \mathrm{Spec}(\Lambda^{\mathrm{w}}(h))\cap[E_1-\varepsilon/2,E_2+\varepsilon/2]} [\tilde\lambda(h)-h^4, \tilde\lambda(h)+h^4]$$

$$= \bigcup_{\pm} \bigcup_{\tilde\lambda_\pm(h)\in \mathrm{Spec}(\Lambda_\pm^{\mathrm{w}}(h))\cap[E_1-\varepsilon/2,E_2+\varepsilon/2]} \Big[\tilde\lambda_\pm(h)-h^4, \tilde\lambda_\pm(h)+h^4\Big],\ \forall h \in (0,h_0],$$

which, in view of Theorem 12.2.4, proves (12.86).

The proof of (12.89) follows from (12.42) and (12.44) of Theorem 12.1.19, from Theorem 12.2.16 and the Minimax. In fact, (12.87) yields, for h_0 sufficiently small and any given $h \in (0, h_0]$,

$$I_k^+(h) \cap I_{k'}^-(h) = \emptyset, \ \forall k \in \mathbb{Z}_{E_1,E_2}^+(h), \ \forall k' \in \mathbb{Z}_{E_1,E_2}^-(h). \tag{12.91}$$

We set

$$J_k^{\pm}(h) = \Big[\frac{2\pi}{T_\pm}(k+\frac{\alpha_\pm}{4})h - \frac{C_0}{2}h^2, \frac{2\pi}{T_\pm}(k+\frac{\alpha_\pm}{4})h + \frac{C_0}{2}h^2\Big] = \text{middle half of } I_k^{\pm}(h). \tag{12.92}$$

Take $k \in \mathbb{Z}_{E_1,E_2}^{\pm}(h)$, $\pm$-respectively. Then by Theorem 12.2.16 for such k

$$\mathrm{Spec}(\Lambda_\pm^{\mathrm{w}}(h)) \cap J_k^{\pm}(h) \neq \emptyset.$$

Let $\tilde{\lambda}(h) \in \mathrm{Spec}(\Lambda^{\mathrm{w}}(h))$ belong to $J_k^+(h)$, resp. $J_k^-(h)$. Since $\tilde{\lambda}(h)$ must be the j-th eigenvalue $\tilde{\lambda}_j(h) \in \mathrm{Spec}(\Lambda^{\mathrm{w}}(h))$, for some j sufficiently large, we have by (12.91) that

$$\tilde{\lambda}_j(h) \in J_k^+(h), \ \ \textbf{resp.} \ \ \tilde{\lambda}_j(h) \in J_k^-(h),$$

and when $n = 1$ this happens exactly with multiplicity 1. Hence from (12.44) and (12.91) we get that, with the same index j,

$$\lambda_j^r(h) \in I_k^+(h), \ \ \textbf{resp.} \ \ \lambda_j^r(h) \in I_k^-(h),$$

which gives (12.89), and when $n = 1$ this must happen (by the Minimax and using (12.91)) with multiplicity 1, which gives (12.90) and concludes the proof. □

We next deal with a case in which condition (12.87) does not hold. We consider cases in which $n = 1$ and $\mathbb{Q}_{E_1,E_2}(h) \neq \emptyset$. We have the following result.

Theorem 12.3.7. *Suppose $n = 1$. There exists $h_0 \in (0, 1]$ so small that for any given $h \in (0, h_0]$, whenever $\mathbb{Q}_{E_1,E_2}(h) \neq \emptyset$ one has*

$$N_{A_r^{\mathrm{w}}(x,hD)}(I_k^+(h)) = 2, \ \forall k \in \mathrm{proj}_1\Big(\mathbb{Q}_{E_1,E_2}(h)\Big), \tag{12.93}$$

where proj$_1$ *denotes the projection onto the first factor.*

Proof. We immediately observe that when $\mathbb{Q}_{E_1,E_2}(h) \neq \emptyset$

$$(k,k'), (k,\tilde{k}') \in \mathbb{Q}_{E_1,E_2}(h) \Longrightarrow k' = \tilde{k}',$$

$$(k,k'), (\tilde{k},k') \in \mathbb{Q}_{E_1,E_2}(h) \Longrightarrow k = \tilde{k},$$

and

$$(k,k') \in \mathbb{Q}_{E_1,E_2}(h) \Longrightarrow I_k^+(h) = I_{k'}^-(h), \ k' = \frac{1}{2}\left(\frac{T_-}{T_+}(2k+1) - 1\right).$$

It follows from (12.83) that

$$(k,k') \in \mathbb{Q}_{E_1,E_2}(h) \implies k \neq k',$$

and, as $I_k^+(h) = I_{k'}^-(h)$,

$$I_k^+(h) \cap I_{k''}^+(h) = I_k^+(h) \cap I_{k''}^-(h) = I_k^+(h) \cap I_k^-(h) = \emptyset, \ \forall k'' \neq k,k'. \tag{12.94}$$

The proof is thus a consequence of (12.42), (12.44), the Minimax, and the fact that the scalar operators $\Lambda_\pm^{\mathrm{w}}(h)$ have spectrum of multiplicity 1 inside each interval $J_k^\pm(h)$, $\pm$-respectively (see (12.92)). In fact, take $(k,k') \in \mathbb{Q}_{E_1,E_2}(h)$. We must then have exactly two eigenvalues $\tilde{\lambda}_j(h)$ and $\tilde{\lambda}_{j'}(h)$ of $\Lambda^{\mathrm{w}}(h)$, with $\tilde{\lambda}_j(h) \in \mathrm{Spec}(\Lambda_+^{\mathrm{w}}(h))$ and $\tilde{\lambda}_{j'}(h) \in \mathrm{Spec}(\Lambda_-^{\mathrm{w}}(h))$, that are contained in $J_k^+(h) = J_{k'}^-(h)$. Hence the eigenvalues $\lambda_j^r(h)$ and $\lambda_{j'}^r(h)$ of $A_r^{\mathrm{w}}(x,hD)$ must be contained in $I_k^+(h) = I_{k'}^-(h)$, and in no other interval, for the other eigenvalues of $A_r^{\mathrm{w}}(x,hD)$ must (by the Minimax) be close to eigenvalues of $\Lambda^{\mathrm{w}}(h)$ that are "far away", by (12.94), from $\tilde{\lambda}_j(h)$ and $\tilde{\lambda}_{j'}(h)$ (in fact, they must be contained in intervals $J_{k''}^\pm(h)$ with $k'' \neq k',k$). This concludes the proof. □

We now use Theorem 12.1.16 and the fact that $A_2^{\mathrm{w}}(x,hD) = hA_2^{\mathrm{w}}(x,D)$ to obtain from Theorems 12.3.6 and 12.3.7 the following spectral information about $A_2^{\mathrm{w}}(x,D)$.

Theorem 12.3.8. *Let $A_2 \in \mathscr{Q}_2^s$ satisfy hypothesis $(T_\pm)$ on an energy interval $[E_1 - \varepsilon, E_2 + \varepsilon]$, for some $\varepsilon \in (0,1]$, where $10(E_0 + 10^2 + c_\chi) < E_1 < E_2$ (see Theorem 12.1.16). We have:*

- *There exists $C_0 > 0$ and $h_0 \in (0,1]$ so small that, for (all)* **fixed** *$h \in (0,h_0]$ one has*

$$\mathrm{Spec}(A_2^{\mathrm{w}}(x,D)) \cap [E_1h^{-1}, E_2h^{-1}] \subset \bigcup_{\pm} \bigcup_{k\in\mathbb{Z}} h^{-1} I_k^\pm(h), \tag{12.95}$$

  where

$$I_k^\pm(h) := \Big[\frac{2\pi}{T_\pm}(k + \frac{\alpha_\pm}{4})h - C_0h^2, \frac{2\pi}{T_\pm}(k + \frac{\alpha_\pm}{4})h + C_0h^2\Big],$$

  and, for h_0 chosen sufficiently small, $I_k^\pm(h) \cap I_{k'}^\pm(h) = \emptyset$, for all $h \in (0,h_0]$, and for all $k \neq k'$, $\pm$-respectively.
- *Suppose there exists $C_* > 0$ such that (12.87) holds. Then there exists $h_0 \in (0,1]$ so small that, for (all)* **fixed** *$h \in (0,h_0]$ one has*

$$\mathrm{Spec}(A_2^{\mathrm{w}}(x,D)) \cap h^{-1} I_k^\pm(h) \neq \emptyset, \ \forall k \in \mathbb{Z}_{E_1,E_2}^\pm(h), \ \pm\text{-respectively}, \tag{12.96}$$

  and when $n = 1$ (so that $\alpha_\pm = 2$)

$$N_{A_2^{\mathrm{w}}(x,D)}(h^{-1} I_k^\pm(h)) = 1, \ \forall k \in \mathbb{Z}_{E_1,E_2}^\pm(h), \ \pm\text{-respectively}. \tag{12.97}$$

Hence, when $n = 1$, $\pm$-respectively,

$$\text{multiplicity}\Big(\text{Spec}(A_2^{\text{w}}(x,D)) \cap h^{-1}I_k^{\pm}(h)\Big) = 1,\ \forall k \in \mathbb{Z}_{E_1,E_2}(h),$$

that is, the eigenvalues of $A_2^{\text{w}}(x,D)$ belonging to the $h^{-1}I_k^{\pm}(h)$, $k \in \mathbb{Z}^{\pm}_{E_1,E_2}(h)$, are all simple.

- *Let $n = 1$. There exists $h_0 \in (0,1]$ so small that, for (all)* **fixed** *$h \in (0,h_0]$, whenever $\mathbb{Q}_{E_1,E_2}(h) \neq \emptyset$ one has*

$$N_{A_2^{\text{w}}(x,D)}(h^{-1}I_k^{+}(h)) = 2,\ \forall k \in \text{proj}_1\Big(\mathbb{Q}_{E_1,E_2}(h)\Big).$$

Proof. Since, by Theorem 12.1.16 and Theorem 12.1.19,

$$\text{Spec}(A_2^{\text{w}}(x,hD)) \cap [E_1,E_2] \subset \bigcup_{\tilde{\lambda}(h)\in \text{Spec}(\Lambda^{\text{w}}(h))\cap[E_1-\varepsilon/2,E_2+\varepsilon/2]} [\tilde{\lambda}(h)-h^4, \tilde{\lambda}(h)+h^4],$$

for all $h \in (0,h_0]$, the proof is a consequence of (12.33), (12.44), the use of Minimax arguments as in the proofs of Theorems 12.3.6 and 12.3.7, and of the fact

$$\text{Spec}(A_2^{\text{w}}(x,hD)) = \big\{\lambda(h) = h\lambda;\ \lambda \in \text{Spec}(A_2^{\text{w}}(x,D))\big\}.$$

□

Remark 12.3.9. By Proposition 9.2.7, Lemma 12.2.20 and Corollary 12.2.22 we have that Theorems 12.3.6, 12.3.7 and 12.3.8 hold regardless the choice of the h^∞-diagonalization of the reference operator $A_r^{\text{w}}(x,hD)$. △

12.4 Localization and Multiplicity of Spec($Q^{\text{w}}_{(\alpha,\beta)}(x, D)$)

In this section, we specialize the results of the previous Section 12.3 to finally give some properties of the large eigenvalues of a NCHO $Q_{(\alpha,\beta)}$, $\alpha \neq \beta$, $\alpha,\beta > 0$ and $\alpha\beta > 1$, that are linked to properties of the periods of the bicharacteristics of the eigenvalues of the symbol $Q_{(\alpha,\beta)}(x,\xi)$. Since $n = 1$, we have that $T_\pm$ are the least periods of the eigenvalues of the symbol of $Q^{\text{w}}_{(\alpha,\beta)}(x,D)$, that do not depend on the energy on the whole energy interval $(0,+\infty)$. Moreover, the Maslov indices $\alpha_\pm$ are both equal to 2.

We decided to explicitly state Theorem 12.4.1 below, because the system $Q^{\text{w}}_{(\alpha,\beta)}(x,D)$ is the basic reference model that led us in our explorations.

Theorem 12.4.1. *Let $\alpha \neq \beta$, with $\alpha,\beta > 0$ and $\alpha\beta > 1$. Let $10(E_0 + 10^2 + c_\chi) \leq E_1 < E_2$. With the notation of Theorems 12.3.6, 12.3.7 and 12.3.8, we have:*

- *There exists* $C_0 > 0$ *and* $h_0 \in (0,1]$ *so small that, for (all)* **fixed** $h \in (0,h_0]$ *one has*

$$\mathrm{Spec}(Q^{\mathrm{w}}_{(\alpha,\beta)}(x,D)) \cap [E_1 h^{-1}, E_2 h^{-1}] \subset \bigcup_{\pm} \bigcup_{k\in\mathbb{Z}_+} h^{-1} I_k^{\pm}(h), \tag{12.98}$$

where

$$I_k^{\pm}(h) := \Big[\frac{2\pi}{T_{\pm}}(k+\frac{1}{2})h - C_0 h^2, \frac{2\pi}{T_{\pm}}(k+\frac{1}{2})h + C_0 h^2\Big].$$

and, for h_0 *chosen sufficiently small,* $I_k^{\pm}(h) \cap I_{k'}^{\pm}(h) = \emptyset$, *for all* $h \in (0,h_0]$, *and for all* $k \neq k'$, $\pm$*-respectively.*

- *Suppose there exists* $C_* > 0$ *such that (12.87) holds. In this case (12.87) may be rewritten as*

$$\left|\frac{T_-}{T_+} - \frac{2k'+1}{2k+1}\right| \geq \frac{C_*}{2k+1}, \quad \forall k,k' \in \mathbb{Z}_+.$$

Then there exists $h_0 \in (0,1]$ *so small that, for (all)* **fixed** $h \in (0,h_0]$ *one has*

$$N_{Q^{\mathrm{w}}_{(\alpha,\beta)}(x,D)}(h^{-1} I_k^{\pm}(h)) = 1, \ \forall k \in \mathbb{Z}^{\pm}_{E_1,E_2}(h), \ \pm\text{-respectively}.$$

Hence, $\pm$*-respectively,*

$$\mathrm{multiplicity}\Big(\mathrm{Spec}(Q^{\mathrm{w}}_{(\alpha,\beta)}(x,D)) \cap h^{-1} I_k^{\pm}(h)\Big) = 1, \ \forall k \in \mathbb{Z}^{\pm}_{E_1,E_2}(h), \tag{12.99}$$

that is, the eigenvalues of $Q^{\mathrm{w}}_{(\alpha,\beta)}(x,D)$ *belonging to the* $h^{-1} I_k^{\pm}(h)$, *for* $k \in \mathbb{Z}^{\pm}_{E_1,E_2}(h)$, *are all simple.*

- *There exists* $h_0 \in (0,1]$ *so small that, for (all)* **fixed** $h \in (0,h_0]$, *whenever* $\mathbb{Q}_{E_1,E_2}(h) \neq \emptyset$ *one has*

$$N_{Q^{\mathrm{w}}_{(\alpha,\beta)}(x,D)}(h^{-1} I_k^{+}(h)) = 2, \forall k \in \mathrm{proj}_1\Big(\mathbb{Q}_{E_1,E_2}(h)\Big).$$

Corollary 12.4.2. *Let* $\alpha \neq \beta$, *with* $\alpha,\beta > 0$, $\alpha\beta > 1$, *and let* $10(E_0 + 10^2 + c_\chi) \leq E_1 < E_2$. *Suppose there exists* $C_* > 0$ *such that (12.87) holds. Then there exists* h_0 *sufficiently small such that for (all)* h **fixed** *in* $(0,h_0]$ *one has*

$$\Sigma_0^{\pm} \cap h^{-1} I_k^j(h) = \Sigma_\infty^{+} \cap \Sigma_\infty^{-} \cap h^{-1} I_k^j(h) = \emptyset, \ \forall k \in \mathbb{Z}^j_{E_1,E_2}(h), \ j = \pm,$$

where $\Sigma_0^{\pm}$ *and* $\Sigma_\infty^{\pm}$ *are the sets introduced in Parmeggiani-Wakayama [58, 59].*

Proof. This follows from the second point in Theorem 12.4.1 above, for in this case we have

$$\mathrm{multiplicity}\Big(\mathrm{Spec}(Q^{\mathrm{w}}(x,D)) \cap h^{-1} I_k^{\pm}(h)\Big) = 1, \ \forall k \in \mathbb{Z}^{\pm}_{E_1,E_2}(h), \ \pm\text{-resp.}$$

Since higher multiplicity eigenvalues must lie in $\Sigma_0^{\pm}$ or $\Sigma_\infty^{+} \cap \Sigma_\infty^{-}$ (see [59]), the result follows. □

As a byproduct of the foregoing corollary, recalling Theorems 2 and 3 of Ochiai [47] we immediately have the following result.

Corollary 12.4.3. *Following Ochiai [46, 47], for suitable real numbers* a', b', c', d' *and complex numbers* a, e' *with* $|a| > 1$, *let*

$$H_\lambda(z,\partial_z) = \partial_z^2 + \Big(\frac{a'}{z} + \frac{b'}{z-1} + \frac{c'}{z-a}\Big)\partial_z + \frac{d'z-e'}{z(z-1)(z-a)},$$

be the Heun operator associated with the eigenvalue λ *of* $Q^{\mathrm{w}}_{(\alpha,\beta)}(x,D)$, *with associated odd eigenfunctions. Let* $\alpha \neq \beta$, *with* $\alpha,\beta > 0$, $\alpha\beta > 1$, *and let* $10(E_0 + 10^2 + c_\chi) \leq E_1 < E_2$. *Suppose there exists* $C_* > 0$ *such that (12.87) holds. Then there exists* h_0 *sufficiently small such that for (all)* h **fixed** *in* $(0, h_0]$, *for all* $k \in \mathbb{Z}^{\pm}_{E_1,E_2}(h)$, $\pm$*-resp., and for* $\lambda \in \mathrm{Spec}(Q^{\mathrm{w}}_{(\alpha,\beta)}(x,D)) \cap h^{-1}I_k^{\pm}(h)$, *one* **cannot** *find non-zero rational functions* $f_1(z), f_2(z)$ *such that*

$$f_1(z), f_2(z)\sqrt{z-a} \in \mathrm{Ker}\, H_\lambda(z,\partial_z) \text{ at the origin.}$$

Proof. By Corollary 12.4.2 we have that $\Sigma_0^- \cap h^{-1}I_k^{\pm}(h) = \emptyset$, for all $k \in \mathbb{Z}^{\pm}_{E_1,E_2}(h)$, $\pm$-resp. Hence the result follows at once from Theorems 2 and 3 of [47]. □

Remark 12.4.4. Notice that (12.99) and Corollary 12.4.2 complement Proposition 3.14 of Parmeggiani [51]. Hence either when condition (12.87) is fulfilled or when $\alpha/\beta \neq 1$ is **not** a ratio of positive odd integers, and $\alpha\beta$ is sufficienly large, then the conclusions of Corollaries 12.4.2 and 12.4.3 hold.

Notice that for $\alpha \neq \beta$, upon putting

$$m_+ := \min\Big\{\sqrt{\frac{\alpha}{\beta}}, \sqrt{\frac{\beta}{\alpha}}\Big\},\quad m_- := \max\Big\{\sqrt{\frac{\alpha}{\beta}}, \sqrt{\frac{\beta}{\alpha}}\Big\},$$

one has

$$T_\pm = \frac{2\pi m_\pm}{\sqrt{\alpha\beta}}(1+o(1)), \quad as \quad \begin{cases} \alpha\beta \longrightarrow +\infty, \\ \alpha/\beta \text{ constant} \neq 1, \end{cases}$$

so that $T_+/T_- \longrightarrow m_+/m_-$. As shown in Proposition 3.14 of Parmeggiani [51], the eigenvalues of $Q^{\mathrm{w}}_{(\alpha,\beta)}(x,D)$ smaller than or equal to any fixed $E > 0$ are still simple for all $\alpha\beta$ sufficiently large with α/β fixed ($\neq 1$), even when m_+/m_- is a ratio of certain positive odd integers (that is, in the case $\alpha < \beta$, say, when $m_+/m_- = \alpha/\beta = (2m_0+1)/(2n_0+1)$ with $m_0 \neq n_0 - 2 - 4k$, any given $k \in \mathbb{Z}_+$). It would therefore be interesting to carry out a refined study of the periods $T_\pm$ as functions of $1/\sqrt{\alpha\beta}$ and α/β in order to get more precise spectral information also in the case $\mathbb{Q}^{\pm}_{E_1,E_2}(h) \neq \emptyset$ or, more generally, when condition (12.87) does not hold. △

12.5 Localization and Multiplicity of Spec$(A_2^{\mathrm{w}}(x, D))$, with $A_2 \in \mathscr{Q}_2 \setminus \mathscr{Q}_2^s$

When $A_2 \in \mathscr{Q}_2 \setminus \mathscr{Q}_2^s$ we no longer have, in general, that for some e_0 the h^∞-diagonalization has no subprincipal term. However, the results at the end of Section 12.2 (for scalar operators with a subprincipal part whose average is constant on the periodic trajectories) and the methods of proof of the last section yield the following theorem.

Theorem 12.5.1. *Let $A_2 \in \mathscr{Q}_2 \setminus \mathscr{Q}_2^s$ satisfy hypothesis $(T_\pm)$ on an energy interval $[E_1-\varepsilon, E_2+\varepsilon]$, for some $\varepsilon \in (0,1]$, where $10(E_0+10^2+c_\chi) < E_1 < E_2$ (see Theorem 12.1.16). Suppose that there exists an h^∞-diagonalization $\Lambda^{\mathrm{w}}(h)$ of the reference operator $A_r^{\mathrm{w}}(x,hD)$ which has subprincipal symbol $\Lambda_0 = \mathrm{diag}(\Lambda_0^+, \Lambda_0^-)$ satisfying, $\pm$-respectively,*

$$\frac{1}{T_\pm}\int_0^{T_\pm} \Lambda_0^\pm \circ \exp(tH_{\lambda_\pm})(X)dt = c_0^\pm, \ \forall X \in \lambda_\pm^{-1}([E_1-\varepsilon, E_2+\varepsilon]), \qquad (12.100)$$

where $c_0^\pm \in \mathbb{R}$ are **constants**. *In this setting define, $\pm$-respectively,*

$$\mathbb{Z}^\pm_{E_1,E_2}(h) := \Big\{k \in \mathbb{Z}_+;\ \frac{2\pi}{T_\pm}(k+\frac{\alpha_\pm}{4})h + c_0^\pm h \in [E_1,E_2]\Big\},$$

and, when $n=1$,

$$\mathbb{Q}_{E_1,E_2}(h) := \Big\{(k,k') \in \mathbb{Z}^+_{E_1,E_2}(h) \times \mathbb{Z}^-_{E_1,E_2}(h);\ \frac{2k+1+T_+c_0^+/\pi}{2k'+1+T_-c_0^-/\pi} = \frac{T_+}{T_-}\Big\}.$$

We have:

- *There exists $C_0>0$ and $h_0 \in (0,1]$ so small that, for (all)* **fixed** *$h \in (0,h_0]$ one has*
$$\mathrm{Spec}(A_2^{\mathrm{w}}(x,D)) \cap [E_1h^{-1}, E_2h^{-1}] \subset \bigcup_{\pm}\bigcup_{k\in\mathbb{Z}} h^{-1} I_k^\pm(h), \qquad (12.101)$$
where
$$I_k^\pm(h) = \Big[\frac{2\pi}{T_\pm}(k+\frac{\alpha_\pm}{4})h + c_0^\pm h - C_0h^2, \frac{2\pi}{T_\pm}(k+\frac{\alpha_\pm}{4})h + c_0^\pm h + C_0h^2\Big],$$
and, for h_0 chosen sufficiently small, $I_k^\pm(h) \cap I_{k'}^\pm(h) = \emptyset$, for all $h \in (0,h_0]$, and for all $k \neq k'$, $\pm$-respectively.
- *Suppose there exists $C_*>0$ such that the analog of (12.87) holds:*
$$\Big|\frac{2\pi}{T_+}(k+\frac{\alpha_+}{4}) + c_0^+ - \Big(\frac{2\pi}{T_-}(k'+\frac{\alpha_-}{4}) + c_0^-\Big)\Big| \geq C_*,\ \forall k,k' \in \mathbb{Z}.$$

Then there exists $h_0 \in (0,1]$ *so small that, for (all)* **fixed** $h \in (0,h_0]$ *one has*

$$\mathrm{Spec}(A_2^{\mathrm{w}}(x,D)) \cap h^{-1}I_k^{\pm}(h) \neq \emptyset, \ \ \forall k \in \mathbb{Z}^{\pm}_{E_1,E_2}(h), \ \pm\text{-respectively}, \tag{12.102}$$

and when $n = 1$ *(so that* $\alpha_{\pm} = 2$*)*

$$N_{A_2^{\mathrm{w}}(x,D)}(h^{-1}I_k^{\pm}(h)) = 1, \ \ \forall k \in \mathbb{Z}^{\pm}_{E_1,E_2}(h), \ \pm\text{-respectively}. \tag{12.103}$$

Hence, when $n = 1$, $\pm$*-respectively,*

$$\mathrm{multiplicity}\Big(\mathrm{Spec}(A_2^{\mathrm{w}}(x,D)) \cap h^{-1}I_k^{\pm}(h)\Big) = 1, \ \forall k \in \mathbb{Z}^{\pm}_{E_1,E_2}(h),$$

that is, the eigenvalues of $A_2^{\mathrm{w}}(x,D)$ *belonging to the* $h^{-1}I_k^{\pm}(h)$, $k \in \mathbb{Z}^{\pm}_{E_1,E_2}(h)$, *are all simple.*

- *Let* $n = 1$. *Suppose that* $c_0^+ = c_0^-$. *Then there exists* $h_0 \in (0,1]$ *so small that, for (all)* **fixed** $h \in (0,h_0]$, *whenever* $\mathbb{Q}_{E_1,E_2}(h) \neq \emptyset$ *one has*

$$N_{A_2^{\mathrm{w}}(x,D)}(h^{-1}I_k^{+}(h)) = 2, \ \ \forall k \in \mathrm{proj}_1\Big(\mathbb{Q}_{E_1,E_2}(h)\Big).$$

Proof. The proof is carried out by following the same lines of the proof of Theorem 12.3.8, and by using Corollary 12.2.19 and Lemma 12.2.20. □

Remark 12.5.2. By Proposition 9.2.7 and the proof of Lemma 12.2.20 we have that if condition (12.100) holds for an h^∞-diagonalization, then it holds for **all** h^∞-diagonalizations (upon shifting the constants $c_0^{\pm}$ by $\frac{2\pi}{T_{\pm}}k_0^{\pm}$, with $k_0^{\pm} \in \mathbb{Z}$, $\pm$-respectively, see (12.79)). Hence Theorem 12.5.1 holds regardless the choice of the h^∞-diagonalization of the reference operator $A_r^{\mathrm{w}}(x,hD)$, provided that (12.100) be satisfied by the subprincipal symbol $\Lambda_0 = \mathrm{diag}(\Lambda_0^+, \Lambda_0^-)$ on the energy interval $[E_1 - \varepsilon, E_2 + \varepsilon]$, for at least one h^∞-diagonalization $\Lambda^{\mathrm{w}}(h)$. △

Remark 12.5.3. Recalling that

$$\Lambda_0^{\pm} = \frac{1}{2}\mathrm{Im}\big(\langle\{e_0^*, \lambda_{\pm}\}e_0 w_{\pm}, w_{\pm}\rangle\big) + \frac{1}{2}\mathrm{Im}\big(\langle e_0^*\{A_2, e_0\}w_{\pm}, w_{\pm}\rangle\big),$$

$\pm$-respectively, where $\{w_+, w_-\}$ is the canonical basis of $\mathbb{C}^2$, condition (12.100) may be thought of as a monodromy condition on the principal symbol e_0 of the operator $e^{\mathrm{w}}(x,hD)$ that h^∞-diagonalizes $A_r^{\mathrm{w}}(x,hD)$. △

12.6 Notes

The results of Section 12.4 were proved in Parmeggiani [55]. In these notes we adapted the method of the proof there, to obtain the localization and multiplicity theorems for the more general case of 2×2 systems belonging to the class $\mathscr{Q}_2$.

Appendix

In this Appendix we collect some useful facts related to the almost-analytic extension of a smooth function and give, following Dimassi-Sjöstrand [7], the proof of the Dyn'kin-Helffer-Sjostrand formula (12.48).

A.1 Almost-Analytic Extension and the Dyn'kin–Helffer–Sjöstrand Formula

Let $x, y \in \mathbb{R}$, and let $z = x + iy \in \mathbb{C}$. We denote by $L(dz) = dxdy$ the Lebesgue measure of $\mathbb{C}$, so that $dz \wedge d\bar{z} = -2idx \wedge dy = -2iL(dz)$, and write $\partial/\partial\bar{z} = (\partial/\partial x + i\partial/\partial y)/2$.

In the first place we recall the *complex Gauss-Green formula*, whose proof is easily deduced from the classical Gauss-Green formula when applied to the real and imaginary part of a complex-valued function.

Lemma A.1.1 (Complex Gauss-Green formula). *Let $D \subset \Omega \subset \mathbb{C}$ be open sets such that D has a C^1 boundary ∂D in Ω. Let $f \in C_0^1(\Omega)$. Then with the proper orientations (that is, with ∂D oriented in such a way that D is kept to the left) one has*

$$\int_{\partial D} f(z)dz = -\iint_D \frac{\partial f}{\partial \bar{z}}(z)dz \wedge d\bar{z} = 2i \iint_D \frac{\partial f}{\partial \bar{z}}(z)L(dz). \tag{A.1}$$

Notice that (A.1) may also be rewritten as

$$\langle \frac{\partial \mathbf{1}_D}{\partial \bar{z}} | f \rangle = \frac{i}{2} \int_{\partial D} f(z)dz, \quad f \in C_0^1(\Omega), \tag{A.2}$$

where the left-hand side denotes the distribution duality, and $\mathbf{1}_D$ is the characteristic function of the set D.

Next, following Hörmander [28], we recall the following result about the fundamental solution of the $\partial/\partial\bar{z}$-operator.

Lemma A.1.2. *With the notation of Lemma A.1.1, let* $\zeta \in D$. *Then, with the proper orientations,*

$$f(\zeta) = -\frac{1}{\pi}\iint_D \frac{\partial f}{\partial \bar{z}}(z)(z-\zeta)^{-1}L(dz) + \frac{1}{2\pi i}\int_{\partial D} f(z)(z-\zeta)^{-1}dz, \forall f \in C_0^1(\Omega). \tag{A.3}$$

In particular, when $D = \Omega$ *there is no curvilinear integral in (A.3), that is,*

$$f(\zeta) = -\frac{1}{\pi}\iint_\Omega \frac{\partial f}{\partial \bar{z}}(z)(z-\zeta)^{-1}L(dz), \ \forall f \in C_0^1(\Omega), \tag{A.4}$$

and, furthermore, considering $(x,y) \longmapsto E_\zeta(x,y) = \pi^{-1}(z-\zeta)^{-1}$, *which is a locally integrable function, gives*

$$\frac{\partial E_\zeta}{\partial \bar{z}} = \delta_\zeta, \tag{A.5}$$

that is, E_ζ *is a fundamental solution of* $\partial/\partial\bar{z}$.

Proof. For $\zeta \in D$, we apply the complex Gauss-Green formula (A.1) to the function $f(z)/(z-\zeta)$, with D replaced by $D \setminus B_\varepsilon$, where B_ε is a disc of radius ε with center at ζ, and ε is picked small. Hence

$$2i\iint_{D\setminus B_\varepsilon} \frac{\partial f}{\partial \bar{z}}(z)(z-\zeta)^{-1}L(dz) = \int_{\partial D} f(z)(z-\zeta)^{-1}dz - \int_{\partial B_\varepsilon} f(z)(z-\zeta)^{-1}dz,$$

with the proper orientations (i.e. ∂B_ε is oriented in the formula as the boundary of B_ε). Since

$$\int_{\partial B_\varepsilon} f(z)(z-\zeta)^{-1}dz = f(\zeta)\int_{\partial B_\varepsilon}(z-\zeta)^{-1}dz + O(\varepsilon) \longrightarrow 2\pi i f(\zeta), \text{ as } \varepsilon \to 0+,$$

letting $\varepsilon \to 0+$ gives (A.3). □

We now pass to the proof of Theorem 12.2.1. In the first place we show that given a function $f \in C_0^2(\mathbb{R})$, it is always possible to find an extension $\tilde{f} \in C_0^1(\mathbb{C})$ such that $\partial\tilde{f}/\partial\bar{z} = O(|\mathrm{Im}\, z|)$, as in the statement of Theorem 12.2.1, and for which formula (A.4) holds true. We have in fact the following lemma.

Lemma A.1.3. *Given any* $f \in C_0^2(\mathbb{R})$ *there exists* $\tilde{f} \in C_0^1(\mathbb{C})$ *such that*

$$\tilde{f}|_{\mathbb{R}} = f, \quad \text{and} \quad \frac{\partial \tilde{f}}{\partial \bar{z}}(z) = O(|\mathrm{Im}\, z|). \tag{A.6}$$

Moreover, one has

$$f(t) = -\frac{1}{\pi}\iint_{\mathbb{C}} \frac{\partial \tilde{f}}{\partial \bar{z}}(z)(z-t)^{-1}L(dz), \ \forall t \in \mathbb{R}. \tag{A.7}$$

Proof. We denote by $g^{(k)}$ the kth-derivative with respect to x or y of a function g. Take $\chi \in C_0^\infty(\mathbb{R})$ with $0 \leq \chi \leq 1$, $\chi\big|_{|y|\leq 1} = 1$, $\chi\big|_{|y|\geq 2} = 0$. Then define

$$\tilde{f}(x+iy) := \Big(f(x) + iyf^{(1)}(x)\Big)\chi(y).$$

It is clear that $\tilde{f} \in C_0^1(\mathbb{C})$. One then computes

$$\frac{\partial \tilde{f}}{\partial x}(x+iy) = \Big(f^{(1)}(x) + iyf^{(2)}(x)\Big)\chi(y),$$
$$\frac{\partial \tilde{f}}{\partial y}(x+iy) = if^{(1)}(x)\chi(y) + \Big(f(x) + iyf^{(1)}(x)\Big)\chi^{(1)}(y),$$

so that

$$\frac{\partial \tilde{f}}{\partial \bar{z}}(z) = \frac{y}{2}\Big(if^{(2)}(x)\chi(y) - f^{(1)}(x)\chi^{(1)}(y)\Big) + if(x)\chi^{(1)}(y). \tag{A.8}$$

Since

$$\operatorname{supp}\chi^{(1)} \subset \{y;\ 1 \leq |y| \leq 2\},$$

we notice that

$$\left|\frac{\chi^{(1)}(y)}{y}\right| \leq |\chi^{(1)}(y)| \leq C,$$

whence we may write

$$\frac{\partial \tilde{f}}{\partial \bar{z}}(z) = \frac{y}{2}\Big(if^{(2)}(x)\chi(y) - f^{(1)}(x)\chi^{(1)}(y)\Big) + iyf(x)\frac{\chi^{(1)}(y)}{y}.$$

which proves (A.6).

The fact that formula (A.7) holds, follows immediately from Lemma A.1.2 by taking $\Omega = \mathbb{C}$ and $\zeta = x \in \mathbb{R}$. □

We next prove Theorem 12.2.1.

Proof (of Theorem 12.2.1). Denote by $Q \in \mathscr{L}(H,H)$ the right-hand side of (12.48). For $u, v \in H$ consider

$$((z-P)^{-1}u, v) = \int (z-t)^{-1}(dE(t)u, v),$$

where $t \mapsto E(t)$ is the spectral family associated with P. It follows that

$$(Qu, v) = -\frac{1}{\pi}\iint_{\mathbb{C}} \frac{\partial \tilde{f}}{\partial \bar{z}}(z)\Big(\int (z-t)^{-1}(dE(t)u, v)\Big)L(dz) =$$

(by Fubini's theorem, using the fact that $\tilde{f}$ is compactly supported and that $\int dE(t) = \mathrm{Id}_H$)

$$= \int \Big(-\frac{1}{\pi} \iint_{\mathbb{C}} \frac{\partial \tilde{f}}{\partial \bar{z}}(z)(z-t)^{-1} L(dz) \Big) (dE(t)u, v) = \int f(t)(dE(t)u, v) = f(P),$$

which concludes the proof. □

However, when P is a pseudodifferential operator and when dealing with the pseudodifferential nature of $f(P)$, as in the proof of Theorem 12.2.2, one does need to consider almost-analytic extensions of a given $f \in C_0^\infty(\mathbb{R})$. The next lemma grants the existence of such almost-analytic extension. (This result also holds, by using a locally-finite partition of unity of $\mathbb{R}$, for functions belonging to $C^\infty(\mathbb{R})$. See Treves [70, Chapter X, Section 2], for more on this.)

Lemma A.1.4. *Let $f \in C_0^\infty(\mathbb{R})$. There exists $\tilde{f} \in C_0^\infty(\mathbb{C})$, called an* **almost-analytic extension** *of f, such that*

$$\tilde{f}\big|_{\mathbb{R}} = f, \text{ and } \forall N \geq 1\ \exists C_N > 0 \text{ such that } \left| \frac{\partial \tilde{f}}{\partial \bar{z}}(z) \right| \leq C_N |\mathrm{Im}\, z|^N, \forall z \in \mathbb{C}, \tag{A.9}$$

$$\operatorname{supp} \tilde{f} \subset \{z = x + iy \in \mathbb{C};\ x \in \operatorname{supp} f,\ |y| \leq C\},$$

where, recall, $\partial/\partial \bar{z} = (\partial/\partial x + i\partial/\partial y)/2$. In addition one has

$$f(t) = -\frac{1}{\pi} \iint_{\mathbb{C}} \frac{\partial \tilde{f}}{\partial \bar{z}}(z)\big(z-t\big)^{-1} L(dz),\ t \in \mathbb{R}. \tag{A.10}$$

Proof. Take $\chi \in C_0^\infty(\mathbb{R})$ as in the proof of Lemma A.1.3. One can then choose a monotone increasing sequence $R_k \nearrow +\infty$ growing sufficiently fast so as to have that the series

$$\tilde{f}(x+iy) := \sum_{k \geq 0} \frac{f^{(k)}(x)}{k!} (iy)^k \chi(R_k y)$$

is uniformly convergent, along with the series of the derivatives to all orders. One now computes

$$\frac{\partial \tilde{f}}{\partial x}(z) = \sum_{k \geq 0} \frac{f^{(k+1)}(x)}{k!} (iy)^k \chi(R_k y),$$

and

$$i \frac{\partial \tilde{f}}{\partial y}(z) = \sum_{k \geq 0} \frac{f^{(k+1)}(x)}{k!} i^{k+2} y^k \chi(R_{k+1} y) + \sum_{k \geq 0} \frac{f^{(k)}(x)}{k!} (iy)^k R_k \chi^{(1)}(R_k y),$$

obtaining

$$\frac{\partial \tilde{f}}{\partial \bar{z}}(z) = \frac{1}{2}\sum_{k\geq 0} \frac{f^{(k+1)}(x)}{k!}(iy)^k\Big(\chi(R_k y) - \chi(R_{k+1}y)\Big)$$
$$+\frac{i}{2}\sum_{k\geq 0} \frac{f^{(k)}(x)}{k!}(iy)^k R_k \chi^{(1)}(R_k y).$$

Since

$$\operatorname{supp}\chi^{(1)}(R_k\cdot) \subset \{y;\, R_k^{-1} \leq |y| \leq 2R_k^{-1}\},$$

and

$$\operatorname{supp}(\chi(R_k\cdot) - \chi(R_{k+1}\cdot)) \subset \{y;\, R_{k+1}^{-1} \leq |y| \leq R_k^{-1}\},$$

one sees that (A.9) holds.

Formula (A.10) follows as before from Lemma A.1.2. This completes the proof. □

Main Notation

B.1 General Notation

- $\mathbb{N} = \{1,2,\ldots\}$, $\mathbb{Z}_+ = \{0,1,2,\ldots\}$, $\mathbb{R}_+ = (0,+\infty)$, and $\overline{\mathbb{R}}_+ = [0,+\infty)$.
- We denote by $\sharp A$ or by $\mathrm{card} A$ the cardinality of the set A.
- As usual, given functions f and g, we write $f \sim g$ *as* $x \to x_0$ when $\lim_{x\to x_0} \dfrac{f(x)}{g(x)} = 1$.
- Let $A, B > 0$. We write $A \lesssim B$ (or equivalently $B \gtrsim A$) when there is a universal constant $C > 0$ such that $A \leq CB$. We therefore write $A \approx B$ whenever $A \lesssim B$ and $B \lesssim A$.
- Given sequences $\{A_j\}_j$, $\{B_j\}_j \subset \mathbb{R}$, we write

  $$A_j \lesssim B_j, \text{ as } j \to +\infty$$

  (or equivalently $B_j \gtrsim A_j$ *as* $j \to +\infty$) if there are constants $C > 0$ and $j_0 \in \mathbb{N}$ such that

  $$A_j \leq CB_j, \ \forall j \geq j_0.$$

  We hence write

  $$A_j \approx B_j, \text{ as } j \to +\infty,$$

  when $A_j \lesssim B_j$ and $B_j \lesssim A_j$ as $j \to +\infty$.
- The $N \times N$ identity matrix is denoted by I, regardless N, or by I_N, or by $I_{\mathbb{C}^N}$.
- The norm of a vector v belonging to $\mathbb{R}^N$, resp. $\mathbb{C}^N$, is denoted by $|v|$, resp. $|v|_{\mathbb{C}^N}$. The inner product in $\mathbb{R}^N$, resp. the Hermitian product in $\mathbb{C}^N$, of vectors v, v', is denoted by $\langle v, v'\rangle$, resp. $\langle v, v'\rangle_{\mathbb{C}^N}$.
- The canonical inner product in L^2 is denoted by $(\cdot,\cdot)$, the induced norm by $\|\cdot\|_0$.
- The ring of $N \times N$ complex matrices is always denoted by M_N.
- An **excision** function is a C^∞ function χ with $0 \leq \chi \leq 1$, supported away from the origin and $\equiv 1$ outside a large compact set. The standard one used in these notes is such that $\chi \equiv 0$ for $|X| \leq 1/2$ and $\chi \equiv 1$ for $|X| \geq 1$.
- The duality between $\mathscr{S}$ and $\mathscr{S}'$ is denoted by $\langle u|\varphi\rangle_{\mathscr{S}',\mathscr{S}}$, where $u \in \mathscr{S}'$ and $\varphi \in \mathscr{S}$.

- The set of linear continuous maps between Hilbert (or Banach, or Fréchet) spaces H_1 and H_2 is denoted by $\mathscr{L}(H_1,H_2)$.
- We write Tr for the operator-trace, and Tr for the $N \times N$ matrix-trace.
- The symbol of the standard harmonic oscillator $p_0(x,\xi) = (|\xi|^2 + |x|^2)/2$, also written $p_0(X) = |X|^2/2$, with $X = (x,\xi)$.
- The matrices J and K:

$$J = \begin{bmatrix} 0 & -1 \\ 1 & 0 \end{bmatrix}, \quad K = \begin{bmatrix} 0 & 1 \\ 1 & 0 \end{bmatrix}.$$

- We say that $f(h) = O(h^{N_0})$ for some $N_0 \in \mathbb{Z}_+$ if there exists $C_{N_0} > 0$ such that $|f(h)| \leq C_{N_0} h^{N_0}$. We say that $f(h) = O(h^\infty)$ if for any given $N_0 \in \mathbb{Z}_+$ one has $f(h) = O(h^{N_0})$.
- The characteristic function of a set V is denoted by $\mathbf{1}_V$.
- $\hookrightarrow\hookrightarrow$ denotes compact embedding.

B.2 Symbol, Function and Operator Spaces

- The symbol space $S(m,g)$ in the Weyl-Hörmander calculus, see Definition 3.1.6.
- The global weight-function $m(X) = (1+|X|^2)^{1/2}$ and the global metric $g_X = |dX|^2/m(X)^2$, $X \in \mathbb{R}^{2n}$, see (3.13).
- The symbol space $S_{\mathrm{cl}}(m^\mu,g)$, see Definition 3.2.3.
- The global polynomial differential (GPD) symbols, see Definition 3.2.6 and also Definition 3.2.9.
- The smoothing symbol class $S(m^{-\infty},g)$, see Definition 3.2.14.
- The function space B^s, see Definition 3.2.25 and Proposition 3.2.26.
- The symbol space $S(\mu,r)$, see Definition 6.1.1.
- The set $\mathrm{OP}S_{\mathrm{cl}}(\mu,r)$ of pseudodifferential operators, see Definition 6.1.2.
- The semiclassical symbol space $S^k_\delta(m^\mu,g)$, see (9.1) of Definition 9.1.1.
- The semiclassical symbol space $S^k_\delta(m^\mu)$, see (9.2) of Definition 9.1.1.
- The smoothing semiclassical symbol class $S^{-\infty}(m^{-\infty},g)$, see (9.7).
- The smoothing semiclassical symbol class $S^{-\infty}(m^\mu)$, see (9.8).
- The class of classical semiclassical symbols $S^k_{\mathrm{cl}}(m^\mu)$, see Point 1. of Definition 9.1.9.
- The class of classical semiclassical symbols $S^k_{0,\mathrm{cl}}(m^\mu,g)$, see Point 2. of Definition 9.1.9.
- The class of semiclassical GPD symbols, see Definition 9.4.1.
- The isometry of L^2, also automorphism of $\mathscr{S}'$ and $\mathscr{S}$ $(E>0)$

$$U_E\colon u(x) \longmapsto E^{-n/4}u(x/\sqrt{E}),$$

see (9.4).

B.3 The Spectral Counting Function and the Spectral ζ-Function

- The spectral counting functions $N(\lambda)$ and $N_0(\lambda)$, see (4.4).
- The spectral ζ-function $\zeta_A(\lambda)$ associated with a positive operator A with a discrete spectrum, see (4.10).

B.4 Dynamical Quantities and Assumptions

- Hypotheses $(H1)$-$(H3)$, see Assumption 11.1.1.
- Hypotheses $(H4)$ and $(H5)$, see Theorem 12.2.4.
- Hypothesis $(H5')$, see Proposition 12.2.17 (and just a few lines above it).
- Hypothesis $(H6)$, see Theorem 12.2.16.
- The functions $J_p(E)$ and $T_p(E)$ associated with a periodic Hamiltonian trajectory of a symbol p at energy E, see Lemma 11.1.2.
- The averaged action-integral $A_p(E)$, see Definition 11.1.5.
- The Leray-Liouville measure $L_{f,\lambda}(dX)$ associated with the hypersurface $f^{-1}(\lambda)$, see Definition 12.2.11, Proposition 12.2.14 and Corollary 12.2.15.

B.5 Classes of Systems

- Non-commutative harmonic oscillator (for short NCHO), see Definition 3.2.11.
- The class $\mathscr{Q}_2$, see Definition 12.1.1.
- The class $\mathscr{Q}_2^s$, see Definition 12.1.2.

References

[1] J. Aramaki. Complex powers of vector valued operators and their application to asymptotic behavior of eigenvalues. J. Funct. Anal. **87** (1989), 294–320.

[2] V. I. Arnold. On a characteristic class entering into conditions of quantization. Funct. Anal. Appl. **1** (1967), 1–13.

[3] J.-M. Bony and N. Lerner. Quantification asymptotique et microlocalisations d'ordre supérieur, I. Ann. Sci. École Norm. Sup. **22** (1989), 377–433.

[4] F. Cardoso and R. Mendoza. The spectral distribution of a globally elliptic operator. Ann. Scuola Norm. Sup. Pisa Cl. Sci. (4) **14** (1987), 143–163.

[5] J. Chazarain. Spectre d'un Hamiltonien quantique et Mécanique Classique. Comm. Partial Differential Equations **5** (1980), 595–644.

[6] J. Chazarain and A. Piriou. *Introduction á la Théorie des Équations aux Dérivées Partielles Linéaires*. Gauthier-Villars, Paris, 1981.

[7] M. Dimassi and J. Sjöstrand. *Spectral Asymptotics in the Semi-Classical Limit*. London Math. Soc. Lecture Note Ser. **268**. Cambridge University Press, 1999.

[8] S. Dozias. Clustering for the spectrum of h-pseudodifferential operators with periodic flow on an energy surface. J. Funct. Anal. **145** (1997), 296–311.

[9] J. J. Duistermaat and L. Hörmander. Fourier integral operators. II. Acta Math. **128** (1972), 183–269.

[10] J. J. Duistermaat. Oscillatory integrals, Lagrange immersions and unfolding of singularities. Comm. Pure Appl. Math. **27** (1974), 207–281.

[11] J. J. Duistermaat. On the Morse index in variational calculus. Advances in Math. **21** (1976), 173–195.

[12] J. J. Duistermaat. *Fourier Integral Operators*. Progr. in Mathematics **130**. Birkhuser Boston, Inc., Boston, MA, 1996.

[13] N. Dunford and J. T. Schwartz. *Linear Operators, Vol. II*. John Wiley and Sons, 1988.

[14] J.-P. Eckmann and R. Sénéor. The Maslov-WKB Method for the (an)Harmonic Oscillator. Arch. Rational Mech. Anal. **61** (1976), 153–173.

[15] L. C. Evans and M. Zworski. *Lectures on Semiclassical Analysis*. Notes of the course, UC Berkeley, http://math.berkeley.edu/∼zworski, 2006.

[16] P. R. Halmos. *Measure Theory*. Graduate Texts in Mathematics **18** Springer-Verlag Berlin, 2nd ed., 1976.

[17] B. Helffer. *Théorie Spectrale Pour Des Opérateurs Globalement Elliptiques*. Astérisque **112**, Soc. Math. de France, Paris, 1984.

[18] B. Helffer and D. Robert. Comportement semi-classique du spectre des hamiltoniens quantiques elliptiques. Ann. Inst. Fourier **31** (1981), 169–223.

[19] B. Helffer and D. Robert. Propriétés asymptotiques du spectre d'opérateurs pseudodifferentiels sur $\mathbb{R}^n$. Comm. Partial Differential Equations **7** (1982), 795–882.

[20] B. Helffer and D. Robert. Puits de potentiel généralisés at asymptotique semi-classique. Ann. Inst. H. Poincar Phys. Théor. **41** (1984), 291–331.

[21] B. Helffer and J. Sjöstrand. Analyse semic-classique pour l'équation de Harper II. Comportement semi-classique près d'un rationnel. Mém. Soc. Math. France (N.S.) No. 40 (1990), 139 pp.
[22] M. Hirokawa. The Dicke-type crossing among eigenvalues of differential operators in a class of non-commutative oscillators. Indiana Univ. Math. J. **58** (2009), 1493–1536.
[23] M. Hitrik and I. Polterovich. Regularized traces and Taylor expansions for the heat semigroup. J. London Math. Soc. (2) **68** (2003), 402–418.
[24] M. Hitrik and I. Polterovich. Resolvent expansions and trace regularizations for Schrödinger operators. Advances in differential equations and mathematical physics (Birmingham, AL, 2002), 161–173, Contemp. Math. **327**, Amer. Math. Soc., Providence, RI, 2003.
[25] R. Howe and E. C. Tan. *Non-Abelian Harmonic Analysis. Applications of* $SL(2,\mathbb{R})$. Springer-Verlag, 1992.
[26] L. Hörmander. Fourier integral operators. I. Acta Math. **127** (1971), 79–183.
[27] L. Hörmander. The Weyl calculus of pseudo-differential operators. Comm. Pure Appl. Math. **32** (1979), 359–443.
[28] L. Hörmander. *The Analysis of Linear Partial Differential Operators - I.* Grundlehren der mathematischen Wissenschaften 256, Springer-Verlag, 1983.
[29] L. Hörmander. *The Analysis of Linear Partial Differential Operators - III.* Grundlehren der mathematischen Wissenschaften 274, Springer-Verlag, 1985.
[30] L. Hörmander. *The Analysis of Linear Partial Differential Operators - IV.* Grundlehren der mathematischen Wissenschaften 275, Springer-Verlag, 2nd Printing, 1994.
[31] T. Ichinose and M. Wakayama. Zeta functions for the spectrum of the non-commutative harmonic oscillators. Comm. Math. Phys. **258** (2005), 697–739.
[32] T. Ichinose and M. Wakayama. Special values of the spectral zeta function of the non-commutative harmonic oscillator and confluent Heun equations. Kyushu J. Math. **59** (2005), 39–100.
[33] T. Ichinose and M. Wakayama. On the spectral zeta function for the non-commutative harmonic oscillator. Rep. Math. Phys. **59** (2007), 421–432.
[34] V. Ivrii. *Microlocal Analysis and Precise Spectral Asymptotics.* Springer Monographs in Mathematics. Springer-Verlag, 1998.
[35] T. Kato. *Perturbation Theory for Linear Operators.* Second edition, Springer-Verlag, Berlin, 1976.
[36] K. Kimoto and M. Wakayama. Apéry-like numbers arising from special values of spectral zeta functions for non-commutative harmonic oscillators. Kyushu J. Math. **60** (2006), 383–404.
[37] K. Kimoto and M. Wakayama. Elliptic curves arising from the spectral zeta function for non-commutative harmonic oscillators and $\Gamma_0(4)$-modular forms. In "The Conference on L-Functions", 201–218, World Sci. Publ., Hackensack, NJ, 2007.
[38] K. Kimoto and Y. Yamasaki. A partial alternating multiple zeta value. Proc. Amer. Math. Soc. **137** (2009), 2503–2515.
[39] W. Lay and S. Slavyanov. *Special functions. A unified theory based on singularities*, Oxford Mathematical Monographs. Oxford Science Publications. Oxford University Press, Oxford, 2000.
[40] A. Martinez. *An introduction to semiclassical and microlocal analysis.* Universitext. Springer-Verlag, New York, 2002.
[41] J. E. Marsden and A. Weinstein. Review to the books: *Geometric asymptotics* and *Symplectic geometry and Fourier analysis*. Bull. Amer. Math. Soc. (N.S.) **1** (1979), 545–553.
[42] V. P. Maslov. *Théorie des perturbations et méthodes asymptotiques*. Dudnod, Paris, 1972.
[43] D. McDuff, D. Salamon. *Introduction to symplectic topology. Second edition.* Oxford Mathematical Monographs. The Clarendon Press, Oxford University Press, New York, 1998.
[44] A. S. Mishchenko, V. E. Shatalov, and B. Yu. Sternin. *Lagrangian manifolds and the Maslov operator*. Springer Series in Soviet Mathematics. Springer-Verlag, Berlin, 1990.
[45] K. Nagatou, M. T. Nakao and M. Wakayama. Verified numerical computations for eigenvalues of non-commutative harmonic oscillators. Numer. Funct. Anal. Optim. **23** (2002), 633–650.

[46] H. Ochiai. Non-commutative harmonic oscillators and Fuchsian ordinary differential operators. Comm. Math. Phys. **217** (2001), 357–373.
[47] H. Ochiai. Non-commutative harmonic oscillators and the connection problem for the Heun differential equation. Lett. Math. Phys. **70** (2004), 133–139.
[48] H. Ochiai. A special value of the spectral zeta function of the non-commutative harmonic oscillators. Ramanujan J. **15** (2008), 31–36.
[49] C. Parenti and A. Parmeggiani. Lower bounds for systems with double characteristics. J. Anal. Math. **86** (2002), 49–91.
[50] C. Parenti and A. Parmeggiani. A Lyapunov Lemma for elliptic systems. Ann. Glob. Anal. Geom. **25** (2004), 27–41.
[51] A. Parmeggiani. On the spectrum and the lowest eigenvalue of certain non-commutative harmonic oscillators. Kyushu J. Math. **58** (2004), 277–322.
[52] A. Parmeggiani. On the spectrum of certain noncommutative harmonic oscillators. Proceedings of the conference "Around Hyperbolic Problems - in memory of Stefano"; Ann. Univ. Ferrara Sez. VII Sci. Mat. **52**, (2006), 431–456.
[53] A. Parmeggiani. On the Fefferman-Phong inequality for systems of PDEs. In "Phase Space Analysis of Partial Differential Equations". Birkhäuser Verlag, Progress in Nonlinear Differential Equations and their Applications **69** (2006), 247–266.
[54] A. Parmeggiani. On positivity of certain systems of partial differential equations. Proc. Natl. Acad. Sci. USA **104** (2007), 723–726.
[55] A. Parmeggiani. On the spectrum of certain non-commutative harmonic oscillators and semiclassical analysis. Comm. Math. Phys. **279**, 285–308 (2008).
[56] A. Parmeggiani. A remark on the semiclassical Fefferman-Phong inequality for certain systems of PDEs. Rend. Lincei Mat. Appl. **19** (2008), 339–359.
[57] A. Parmeggiani. A remark on the Fefferman-Phong inequality for 2×2 systems. Pure and Applied Mathematics Quarterly **6** (Special Issue in Honor of Joseph J. Kohn) (2010), 1081–1103.
[58] A. Parmeggiani and M. Wakayama. Oscillator representations and systems of ordinary differential equations. Proc. Natl. Acad. Sci. USA **98**, (2001), 26–30.
[59] A. Parmeggiani and M. Wakayama. Non-commutative harmonic oscillators-I,-II, Corrigenda and Remarks to I. Forum Math. **14** (2002), 539–604, 669–690, ibid. **15** (2003), 955–963.
[60] A. Parmeggiani and M. Wakayama. A remark on systems of differential equations associated with representations of $\mathfrak{sl}_2(\mathbb{R})$ and their perturbations. Kodai Math. J. **25** (2002), 254–277.
[61] M. Reed and B. Simon. *Methods of Modern Mathematical Pyisics IV: Analysis of Operators*. Academic Press, New York-London, 1978.
[62] J. Robbin, D. Salamon. The Maslov index for paths. Topology **32** (1993), 827–844.
[63] D. Robert. Propriétés spectrales d'opérateurs pseudodifferentiels. Comm. Partial Differential Equations **3** (1978), 755–826.
[64] D. Robert. Calcul fonctionnel sur les opérateurs admissibles et application. J. Funct. Anal. **45** (1982), 74–94.
[65] D. Robert. *Autour de l'Approximation Semi-Classique*. Progress in Mathematics **68**, Birkhäuser, 1987.
[66] W. Rossmann. *Lie groups. An introduction through linear groups*. Oxford Graduate Texts in Mathematics, **5**. Oxford University Press, Oxford, 2002.
[67] M. A. Shubin. *Pseudodifferential Operators and Spectral Theory*, 2nd edition. Springer-Verlag, 2001.
[68] S. Taniguchi. The heat semigroup and kernel associated with certain non-commutative harmonic oscillators. Kyushu J. Math. **62** (2008), 63–68.
[69] M. Taylor. *Pseudodifferential Operators*. Princeton University Press, 1981.
[70] F. Treves. *Introduction to pseudodifferential and Fourier integral operators, Vol. 2. Fourier integral operators*. The University Series in Mathematics. Plenum Press, New York-London, 1980.
[71] A. Voros. An algebra of pseudodifferential operators and the asymptotics of quantum mechanics. J. Funct. Anal. **29** (1978), 104–132.

Index

Lecture Notes in Mathematics

For information about earlier volumes
please contact your bookseller or Springer
LNM Online archive: springerlink.com

Vol. 1809: O. Steinbach, Stability Estimates for Hybrid Coupled Domain Decomposition Methods (2003)
Vol. 1810: J. Wengenroth, Derived Functors in Functional Analysis (2003)
Vol. 1811: J. Stevens, Deformations of Singularities (2003)
Vol. 1812: L. Ambrosio, K. Deckelnick, G. Dziuk, M. Mimura, V. A. Solonnikov, H. M. Soner, Mathematical Aspects of Evolving Interfaces. Madeira, Funchal, Portugal 2000. Editors: P. Colli, J. F. Rodrigues (2003)
Vol. 1813: L. Ambrosio, L. A. Caffarelli, Y. Brenier, G. Buttazzo, C. Villani, Optimal Transportation and its Applications. Martina Franca, Italy 2001. Editors: L. A. Caffarelli, S. Salsa (2003)
Vol. 1814: P. Bank, F. Baudoin, H. Föllmer, L.C.G. Rogers, M. Soner, N. Touzi, Paris-Princeton Lectures on Mathematical Finance 2002 (2003)
Vol. 1815: A. M. Vershik (Ed.), Asymptotic Combinatorics with Applications to Mathematical Physics. St. Petersburg, Russia 2001 (2003)
Vol. 1816: S. Albeverio, W. Schachermayer, M. Talagrand, Lectures on Probability Theory and Statistics. Ecole d'Eté de Probabilités de Saint-Flour XXX-2000. Editor: P. Bernard (2003)
Vol. 1817: E. Koelink, W. Van Assche (Eds.), Orthogonal Polynomials and Special Functions. Leuven 2002 (2003)
Vol. 1818: M. Bildhauer, Convex Variational Problems with Linear, nearly Linear and/or Anisotropic Growth Conditions (2003)
Vol. 1819: D. Masser, Yu. V. Nesterenko, H. P. Schlickewei, W. M. Schmidt, M. Waldschmidt, Diophantine Approximation. Cetraro, Italy 2000. Editors: F. Amoroso, U. Zannier (2003)
Vol. 1820: F. Hiai, H. Kosaki, Means of Hilbert Space Operators (2003)
Vol. 1821: S. Teufel, Adiabatic Perturbation Theory in Quantum Dynamics (2003)
Vol. 1822: S.-N. Chow, R. Conti, R. Johnson, J. Mallet-Paret, R. Nussbaum, Dynamical Systems. Cetraro, Italy 2000. Editors: J. W. Macki, P. Zecca (2003)
Vol. 1823: A. M. Anile, W. Allegretto, C. Ringhofer, Mathematical Problems in Semiconductor Physics. Cetraro, Italy 1998. Editor: A. M. Anile (2003)
Vol. 1824: J. A. Navarro González, J. B. Sancho de Salas, $\mathscr{C}^\infty$ – Differentiable Spaces (2003)
Vol. 1825: J. H. Bramble, A. Cohen, W. Dahmen, Multiscale Problems and Methods in Numerical Simulations, Martina Franca, Italy 2001. Editor: C. Canuto (2003)
Vol. 1826: K. Dohmen, Improved Bonferroni Inequalities via Abstract Tubes. Inequalities and Identities of Inclusion-Exclusion Type. VIII, 113 p, 2003.
Vol. 1827: K. M. Pilgrim, Combinations of Complex Dynamical Systems. IX, 118 p, 2003.
Vol. 1828: D. J. Green, Gröbner Bases and the Computation of Group Cohomology. XII, 138 p, 2003.
Vol. 1829: E. Altman, B. Gaujal, A. Hordijk, Discrete-Event Control of Stochastic Networks: Multimodularity and Regularity. XIV, 313 p, 2003.
Vol. 1830: M. I. Gil', Operator Functions and Localization of Spectra. XIV, 256 p, 2003.
Vol. 1831: A. Connes, J. Cuntz, E. Guentner, N. Higson, J. E. Kaminker, Noncommutative Geometry, Martina Franca, Italy 2002. Editors: S. Doplicher, L. Longo (2004)
Vol. 1832: J. Azéma, M. Émery, M. Ledoux, M. Yor (Eds.), Séminaire de Probabilités XXXVII (2003)
Vol. 1833: D.-Q. Jiang, M. Qian, M.-P. Qian, Mathematical Theory of Nonequilibrium Steady States. On the Frontier of Probability and Dynamical Systems. IX, 280 p, 2004.
Vol. 1834: Yo. Yomdin, G. Comte, Tame Geometry with Application in Smooth Analysis. VIII, 186 p, 2004.
Vol. 1835: O.T. Izhboldin, B. Kahn, N.A. Karpenko, A. Vishik, Geometric Methods in the Algebraic Theory of Quadratic Forms. Summer School, Lens, 2000. Editor: J.-P. Tignol (2004)
Vol. 1836: C. Năstăsescu, F. Van Oystaeyen, Methods of Graded Rings. XIII, 304 p, 2004.
Vol. 1837: S. Tavaré, O. Zeitouni, Lectures on Probability Theory and Statistics. Ecole d'Eté de Probabilités de Saint-Flour XXXI-2001. Editor: J. Picard (2004)
Vol. 1838: A.J. Ganesh, N.W. O'Connell, D.J. Wischik, Big Queues. XII, 254 p, 2004.
Vol. 1839: R. Gohm, Noncommutative Stationary Processes. VIII, 170 p, 2004.
Vol. 1840: B. Tsirelson, W. Werner, Lectures on Probability Theory and Statistics. Ecole d'Eté de Probabilités de Saint-Flour XXXII-2002. Editor: J. Picard (2004)
Vol. 1841: W. Reichel, Uniqueness Theorems for Variational Problems by the Method of Transformation Groups (2004)
Vol. 1842: T. Johnsen, A. L. Knutsen, K_3 Projective Models in Scrolls (2004)
Vol. 1843: B. Jefferies, Spectral Properties of Noncommuting Operators (2004)
Vol. 1844: K.F. Siburg, The Principle of Least Action in Geometry and Dynamics (2004)
Vol. 1845: Min Ho Lee, Mixed Automorphic Forms, Torus Bundles, and Jacobi Forms (2004)
Vol. 1846: H. Ammari, H. Kang, Reconstruction of Small Inhomogeneities from Boundary Measurements (2004)
Vol. 1847: T.R. Bielecki, T. Björk, M. Jeanblanc, M. Rutkowski, J.A. Scheinkman, W. Xiong, Paris-Princeton Lectures on Mathematical Finance 2003 (2004)
Vol. 1848: M. Abate, J. E. Fornaess, X. Huang, J. P. Rosay, A. Tumanov, Real Methods in Complex and CR Geometry, Martina Franca, Italy 2002. Editors: D. Zaitsev, G. Zampieri (2004)
Vol. 1849: Martin L. Brown, Heegner Modules and Elliptic Curves (2004)

Vol. 1850: V. D. Milman, G. Schechtman (Eds.), Geometric Aspects of Functional Analysis. Israel Seminar 2002-2003 (2004)
Vol. 1851: O. Catoni, Statistical Learning Theory and Stochastic Optimization (2004)
Vol. 1852: A.S. Kechris, B.D. Miller, Topics in Orbit Equivalence (2004)
Vol. 1853: Ch. Favre, M. Jonsson, The Valuative Tree (2004)
Vol. 1854: O. Saeki, Topology of Singular Fibers of Differential Maps (2004)
Vol. 1855: G. Da Prato, P.C. Kunstmann, I. Lasiecka, A. Lunardi, R. Schnaubelt, L. Weis, Functional Analytic Methods for Evolution Equations. Editors: M. Iannelli, R. Nagel, S. Piazzera (2004)
Vol. 1856: K. Back, T.R. Bielecki, C. Hipp, S. Peng, W. Schachermayer, Stochastic Methods in Finance, Bressanone/Brixen, Italy, 2003. Editors: M. Fritelli, W. Runggaldier (2004)
Vol. 1857: M. Émery, M. Ledoux, M. Yor (Eds.), Séminaire de Probabilités XXXVIII (2005)
Vol. 1858: A.S. Cherny, H.-J. Engelbert, Singular Stochastic Differential Equations (2005)
Vol. 1859: E. Letellier, Fourier Transforms of Invariant Functions on Finite Reductive Lie Algebras (2005)
Vol. 1860: A. Borisyuk, G.B. Ermentrout, A. Friedman, D. Terman, Tutorials in Mathematical Biosciences I. Mathematical Neurosciences (2005)
Vol. 1861: G. Benettin, J. Henrard, S. Kuksin, Hamiltonian Dynamics – Theory and Applications, Cetraro, Italy, 1999. Editor: A. Giorgilli (2005)
Vol. 1862: B. Helffer, F. Nier, Hypoelliptic Estimates and Spectral Theory for Fokker-Planck Operators and Witten Laplacians (2005)
Vol. 1863: H. Führ, Abstract Harmonic Analysis of Continuous Wavelet Transforms (2005)
Vol. 1864: K. Efstathiou, Metamorphoses of Hamiltonian Systems with Symmetries (2005)
Vol. 1865: D. Applebaum, B.V. R. Bhat, J. Kustermans, J. M. Lindsay, Quantum Independent Increment Processes I. From Classical Probability to Quantum Stochastic Calculus. Editors: M. Schürmann, U. Franz (2005)
Vol. 1866: O.E. Barndorff-Nielsen, U. Franz, R. Gohm, B. Kümmerer, S. Thorbjønsen, Quantum Independent Increment Processes II. Structure of Quantum Lévy Processes, Classical Probability, and Physics. Editors: M. Schürmann, U. Franz, (2005)
Vol. 1867: J. Sneyd (Ed.), Tutorials in Mathematical Biosciences II. Mathematical Modeling of Calcium Dynamics and Signal Transduction. (2005)
Vol. 1868: J. Jorgenson, S. Lang, $\mathrm{Pos}_n(\mathrm{R})$ and Eisenstein Series. (2005)
Vol. 1869: A. Dembo, T. Funaki, Lectures on Probability Theory and Statistics. Ecole d'Eté de Probabilités de Saint-Flour XXXIII-2003. Editor: J. Picard (2005)
Vol. 1870: V.I. Gurariy, W. Lusky, Geometry of Müntz Spaces and Related Questions. (2005)
Vol. 1871: P. Constantin, G. Gallavotti, A.V. Kazhikhov, Y. Meyer, S. Ukai, Mathematical Foundation of Turbulent Viscous Flows, Martina Franca, Italy, 2003. Editors: M. Cannone, T. Miyakawa (2006)
Vol. 1872: A. Friedman (Ed.), Tutorials in Mathematical Biosciences III. Cell Cycle, Proliferation, and Cancer (2006)
Vol. 1873: R. Mansuy, M. Yor, Random Times and Enlargements of Filtrations in a Brownian Setting (2006)
Vol. 1874: M. Yor, M. Émery (Eds.), In Memoriam Paul-André Meyer - Séminaire de Probabilités XXXIX (2006)
Vol. 1875: J. Pitman, Combinatorial Stochastic Processes. Ecole d'Eté de Probabilités de Saint-Flour XXXII-2002. Editor: J. Picard (2006)
Vol. 1876: H. Herrlich, Axiom of Choice (2006)
Vol. 1877: J. Steuding, Value Distributions of L-Functions (2007)
Vol. 1878: R. Cerf, The Wulff Crystal in Ising and Percolation Models, Ecole d'Eté de Probabilités de Saint-Flour XXXIV-2004. Editor: Jean Picard (2006)
Vol. 1879: G. Slade, The Lace Expansion and its Applications, Ecole d'Eté de Probabilités de Saint-Flour XXXIV-2004. Editor: Jean Picard (2006)
Vol. 1880: S. Attal, A. Joye, C.-A. Pillet, Open Quantum Systems I, The Hamiltonian Approach (2006)
Vol. 1881: S. Attal, A. Joye, C.-A. Pillet, Open Quantum Systems II, The Markovian Approach (2006)
Vol. 1882: S. Attal, A. Joye, C.-A. Pillet, Open Quantum Systems III, Recent Developments (2006)
Vol. 1883: W. Van Assche, F. Marcellàn (Eds.), Orthogonal Polynomials and Special Functions, Computation and Application (2006)
Vol. 1884: N. Hayashi, E.I. Kaikina, P.I. Naumkin, I.A. Shishmarev, Asymptotics for Dissipative Nonlinear Equations (2006)
Vol. 1885: A. Telcs, The Art of Random Walks (2006)
Vol. 1886: S. Takamura, Splitting Deformations of Degenerations of Complex Curves (2006)
Vol. 1887: K. Habermann, L. Habermann, Introduction to Symplectic Dirac Operators (2006)
Vol. 1888: J. van der Hoeven, Transseries and Real Differential Algebra (2006)
Vol. 1889: G. Osipenko, Dynamical Systems, Graphs, and Algorithms (2006)
Vol. 1890: M. Bunge, J. Funk, Singular Coverings of Toposes (2006)
Vol. 1891: J.B. Friedlander, D.R. Heath-Brown, H. Iwaniec, J. Kaczorowski, Analytic Number Theory, Cetraro, Italy, 2002. Editors: A. Perelli, C. Viola (2006)
Vol. 1892: A. Baddeley, I. Bárány, R. Schneider, W. Weil, Stochastic Geometry, Martina Franca, Italy, 2004. Editor: W. Weil (2007)
Vol. 1893: H. Hanßmann, Local and Semi-Local Bifurcations in Hamiltonian Dynamical Systems, Results and Examples (2007)
Vol. 1894: C.W. Groetsch, Stable Approximate Evaluation of Unbounded Operators (2007)
Vol. 1895: L. Molnár, Selected Preserver Problems on Algebraic Structures of Linear Operators and on Function Spaces (2007)
Vol. 1896: P. Massart, Concentration Inequalities and Model Selection, Ecole d'Été de Probabilités de Saint-Flour XXXIII-2003. Editor: J. Picard (2007)
Vol. 1897: R. Doney, Fluctuation Theory for Lévy Processes, Ecole d'Été de Probabilités de Saint-Flour XXXV-2005. Editor: J. Picard (2007)
Vol. 1898: H.R. Beyer, Beyond Partial Differential Equations, On linear and Quasi-Linear Abstract Hyperbolic Evolution Equations (2007)
Vol. 1899: Séminaire de Probabilités XL. Editors: C. Donati-Martin, M. Émery, A. Rouault, C. Stricker (2007)
Vol. 1900: E. Bolthausen, A. Bovier (Eds.), Spin Glasses (2007)

Vol. 1901: O. Wittenberg, Intersections de deux quadriques et pinceaux de courbes de genre 1, Intersections of Two Quadrics and Pencils of Curves of Genus 1 (2007)

Vol. 1902: A. Isaev, Lectures on the Automorphism Groups of Kobayashi-Hyperbolic Manifolds (2007)

Vol. 1903: G. Kresin, V. Maz'ya, Sharp Real-Part Theorems (2007)

Vol. 1904: P. Giesl, Construction of Global Lyapunov Functions Using Radial Basis Functions (2007)

Vol. 1905: C. Prévôt, M. Röckner, A Concise Course on Stochastic Partial Differential Equations (2007)

Vol. 1906: T. Schuster, The Method of Approximate Inverse: Theory and Applications (2007)

Vol. 1907: M. Rasmussen, Attractivity and Bifurcation for Nonautonomous Dynamical Systems (2007)

Vol. 1908: T.J. Lyons, M. Caruana, T. Lévy, Differential Equations Driven by Rough Paths, Ecole d'Été de Probabilités de Saint-Flour XXXIV-2004 (2007)

Vol. 1909: H. Akiyoshi, M. Sakuma, M. Wada, Y. Yamashita, Punctured Torus Groups and 2-Bridge Knot Groups (I) (2007)

Vol. 1910: V.D. Milman, G. Schechtman (Eds.), Geometric Aspects of Functional Analysis. Israel Seminar 2004-2005 (2007)

Vol. 1911: A. Bressan, D. Serre, M. Williams, K. Zumbrun, Hyperbolic Systems of Balance Laws. Cetraro, Italy 2003. Editor: P. Marcati (2007)

Vol. 1912: V. Berinde, Iterative Approximation of Fixed Points (2007)

Vol. 1913: J.E. Marsden, G. Misiołek, J.-P. Ortega, M. Perlmutter, T.S. Ratiu, Hamiltonian Reduction by Stages (2007)

Vol. 1914: G. Kutyniok, Affine Density in Wavelet Analysis (2007)

Vol. 1915: T. Bıyıkoğlu, J. Leydold, P.F. Stadler, Laplacian Eigenvectors of Graphs. Perron-Frobenius and Faber-Krahn Type Theorems (2007)

Vol. 1916: C. Villani, F. Rezakhanlou, Entropy Methods for the Boltzmann Equation. Editors: F. Golse, S. Olla (2008)

Vol. 1917: I. Veselić, Existence and Regularity Properties of the Integrated Density of States of Random Schrödinger (2008)

Vol. 1918: B. Roberts, R. Schmidt, Local Newforms for GSp(4) (2007)

Vol. 1919: R.A. Carmona, I. Ekeland, A. Kohatsu-Higa, J.-M. Lasry, P.-L. Lions, H. Pham, E. Taflin, Paris-Princeton Lectures on Mathematical Finance 2004. Editors: R.A. Carmona, E. Çinlar, I. Ekeland, E. Jouini, J.A. Scheinkman, N. Touzi (2007)

Vol. 1920: S.N. Evans, Probability and Real Trees. Ecole d'Été de Probabilités de Saint-Flour XXXV-2005 (2008)

Vol. 1921: J.P. Tian, Evolution Algebras and their Applications (2008)

Vol. 1922: A. Friedman (Ed.), Tutorials in Mathematical BioSciences IV. Evolution and Ecology (2008)

Vol. 1923: J.P.N. Bishwal, Parameter Estimation in Stochastic Differential Equations (2008)

Vol. 1924: M. Wilson, Littlewood-Paley Theory and Exponential-Square Integrability (2008)

Vol. 1925: M. du Sautoy, L. Woodward, Zeta Functions of Groups and Rings (2008)

Vol. 1926: L. Barreira, V. Claudia, Stability of Nonautonomous Differential Equations (2008)

Vol. 1927: L. Ambrosio, L. Caffarelli, M.G. Crandall, L.C. Evans, N. Fusco, Calculus of Variations and Non-Linear Partial Differential Equations. Cetraro, Italy 2005. Editors: B. Dacorogna, P. Marcellini (2008)

Vol. 1928: J. Jonsson, Simplicial Complexes of Graphs (2008)

Vol. 1929: Y. Mishura, Stochastic Calculus for Fractional Brownian Motion and Related Processes (2008)

Vol. 1930: J.M. Urbano, The Method of Intrinsic Scaling. A Systematic Approach to Regularity for Degenerate and Singular PDEs (2008)

Vol.1931:M.Cowling,E.Frenkel,M.Kashiwara,A.Valette,D.A.Vogan,Jr.,N.R.Wallach, RepresentationTheoryandComplexAnalysis.Venice,Italy2004.Editors:E.C.Tarabusi,A. D'Agnolo,M.Picardello(2008)

Vol. 1932: A.A. Agrachev, A.S. Morse, E.D. Sontag, H.J. Sussmann, V.I. Utkin, Nonlinear and Optimal Control Theory. Cetraro, Italy 2004. Editors: P. Nistri, G. Stefani (2008)

Vol. 1933: M. Petkovic, Point Estimation of Root Finding Methods (2008)

Vol. 1934: C. Donati-Martin, M. Émery, A. Rouault, C. Stricker (Eds.), Séminaire de Probabilités XLI (2008)

Vol. 1935: A. Unterberger, Alternative Pseudodifferential Analysis (2008)

Vol. 1936: P. Magal, S. Ruan (Eds.), Structured Population Models in Biology and Epidemiology (2008)

Vol. 1937: G. Capriz, P. Giovine, P.M. Mariano (Eds.), Mathematical Models of Granular Matter (2008)

Vol. 1938: D. Auroux, F. Catanese, M. Manetti, P. Seidel, B. Siebert, I. Smith, G. Tian, Symplectic 4-Manifolds and Algebraic Surfaces. Cetraro, Italy 2003. Editors: F. Catanese, G. Tian (2008)

Vol. 1939: D. Boffi, F. Brezzi, L. Demkowicz, R.G. Durán, R.S. Falk, M. Fortin, Mixed Finite Elements, Compatibility Conditions, and Applications. Cetraro, Italy 2006. Editors: D. Boffi, L. Gastaldi (2008)

Vol. 1940: J. Banasiak, V. Capasso, M.A.J. Chaplain, M. Lachowicz, J. Miękisz, Multiscale Problems in the Life Sciences. From Microscopic to Macroscopic. Będlewo, Poland 2006. Editors: V. Capasso, M. Lachowicz (2008)

Vol. 1941: S.M.J. Haran, Arithmetical Investigations. Representation Theory, Orthogonal Polynomials, and Quantum Interpolations (2008)

Vol. 1942: S. Albeverio, F. Flandoli, Y.G. Sinai, SPDE in Hydrodynamic. Recent Progress and Prospects. Cetraro, Italy 2005. Editors: G. Da Prato, M. Röckner (2008)

Vol. 1943: L.L. Bonilla (Ed.), Inverse Problems and Imaging. Martina Franca, Italy 2002 (2008)

Vol. 1944: A. Di Bartolo, G. Falcone, P. Plaumann, K. Strambach, Algebraic Groups and Lie Groups with Few Factors (2008)

Vol. 1945: F. Brauer, P. van den Driessche, J. Wu (Eds.), Mathematical Epidemiology (2008)

Vol. 1946: G. Allaire, A. Arnold, P. Degond, T.Y. Hou, Quantum Transport. Modelling, Analysis and Asymptotics. Cetraro, Italy 2006. Editors: N.B. Abdallah, G. Frosali (2008)

Vol. 1947: D. Abramovich, M. Mariño, M. Thaddeus, R. Vakil, Enumerative Invariants in Algebraic Geometry and String Theory. Cetraro, Italy 2005. Editors: K. Behrend, M. Manetti (2008)

Vol. 1948: F. Cao, J-L. Lisani, J-M. Morel, P. Musé, F. Sur, A Theory of Shape Identification (2008)

Vol. 1949: H.G. Feichtinger, B. Helffer, M.P. Lamoureux, N. Lerner, J. Toft, Pseudo-Differential Operators. Quantization and Signals. Cetraro, Italy 2006. Editors: L. Rodino, M.W. Wong (2008)
Vol. 1950: M. Bramson, Stability of Queueing Networks, Ecole d'Eté de Probabilités de Saint-Flour XXXVI-2006 (2008)
Vol. 1951: A. Moltó, J. Orihuela, S. Troyanski, M. Valdivia, A Non Linear Transfer Technique for Renorming (2009)
Vol. 1952: R. Mikhailov, I.B.S. Passi, Lower Central and Dimension Series of Groups (2009)
Vol. 1953: K. Arwini, C.T.J. Dodson, Information Geometry (2008)
Vol. 1954: P. Biane, L. Bouten, F. Cipriani, N. Konno, N. Privault, Q. Xu, Quantum Potential Theory. Editors: U. Franz, M. Schuermann (2008)
Vol. 1955: M. Bernot, V. Caselles, J.-M. Morel, Optimal Transportation Networks (2008)
Vol. 1956: C.H. Chu, Matrix Convolution Operators on Groups (2008)
Vol. 1957: A. Guionnet, On Random Matrices: Macroscopic Asymptotics, Ecole d'Eté de Probabilités de Saint-Flour XXXVI-2006 (2009)
Vol. 1958: M.C. Olsson, Compactifying Moduli Spaces for Abelian Varieties (2008)
Vol. 1959: Y. Nakkajima, A. Shiho, Weight Filtrations on Log Crystalline Cohomologies of Families of Open Smooth Varieties (2008)
Vol. 1960: J. Lipman, M. Hashimoto, Foundations of Grothendieck Duality for Diagrams of Schemes (2009)
Vol. 1961: G. Buttazzo, A. Pratelli, S. Solimini, E. Stepanov, Optimal Urban Networks via Mass Transportation (2009)
Vol. 1962: R. Dalang, D. Khoshnevisan, C. Mueller, D. Nualart, Y. Xiao, A Minicourse on Stochastic Partial Differential Equations (2009)
Vol. 1963: W. Siegert, Local Lyapunov Exponents (2009)
Vol. 1964: W. Roth, Operator-valued Measures and Integrals for Cone-valued Functions and Integrals for Cone-valued Functions (2009)
Vol. 1965: C. Chidume, Geometric Properties of Banach Spaces and Nonlinear Iterations (2009)
Vol. 1966: D. Deng, Y. Han, Harmonic Analysis on Spaces of Homogeneous Type (2009)
Vol. 1967: B. Fresse, Modules over Operads and Functors (2009)
Vol. 1968: R. Weissauer, Endoscopy for GSP(4) and the Cohomology of Siegel Modular Threefolds (2009)
Vol. 1969: B. Roynette, M. Yor, Penalising Brownian Paths (2009)
Vol. 1970: M. Biskup, A. Bovier, F. den Hollander, D. Ioffe, F. Martinelli, K. Netočný, F. Toninelli, Methods of Contemporary Mathematical Statistical Physics. Editor: R. Kotecký (2009)
Vol. 1971: L. Saint-Raymond, Hydrodynamic Limits of the Boltzmann Equation (2009)
Vol. 1972: T. Mochizuki, Donaldson Type Invariants for Algebraic Surfaces (2009)
Vol. 1973: M.A. Berger, L.H. Kauffmann, B. Khesin, H.K. Moffatt, R.L. Ricca, De W. Sumners, Lectures on Topological Fluid Mechanics. Cetraro, Italy 2001. Editor: R.L. Ricca (2009)
Vol. 1974: F. den Hollander, Random Polymers: École d'Été de Probabilités de Saint-Flour XXXVII – 2007 (2009)
Vol. 1975: J.C. Rohde, Cyclic Coverings, Calabi-Yau Manifolds and Complex Multiplication (2009)
Vol. 1976: N. Ginoux, The Dirac Spectrum (2009)
Vol. 1977: M.J. Gursky, E. Lanconelli, A. Malchiodi, G. Tarantello, X.-J. Wang, P.C. Yang, Geometric Analysis and PDEs. Cetraro, Italy 2001. Editors: A. Ambrosetti, S.-Y.A. Chang, A. Malchiodi (2009)
Vol. 1978: M. Qian, J.-S. Xie, S. Zhu, Smooth Ergodic Theory for Endomorphisms (2009)
Vol. 1979: C. Donati-Martin, M. Émery, A. Rouault, C. Stricker (Eds.), Śeminaire de Probablitiés XLII (2009)
Vol. 1980: P. Graczyk, A. Stos (Eds.), Potential Analysis of Stable Processes and its Extensions (2009)
Vol. 1981: M. Chlouveraki, Blocks and Families for Cyclotomic Hecke Algebras (2009)
Vol. 1982: N. Privault, Stochastic Analysis in Discrete and Continuous Settings. With Normal Martingales (2009)
Vol. 1983: H. Ammari (Ed.), Mathematical Modeling in Biomedical Imaging I. Electrical and Ultrasound Tomographies, Anomaly Detection, and Brain Imaging (2009)
Vol. 1984: V. Caselles, P. Monasse, Geometric Description of Images as Topographic Maps (2010)
Vol. 1985: T. Linß, Layer-Adapted Meshes for Reaction-Convection-Diffusion Problems (2010)
Vol. 1986: J.-P. Antoine, C. Trapani, Partial Inner Product Spaces. Theory and Applications (2009)
Vol. 1987: J.-P. Brasselet, J. Seade, T. Suwa, Vector Fields on Singular Varieties (2010)
Vol. 1988: M. Broué, Introduction to Complex Reflection Groups and Their Braid Groups (2010)
Vol. 1989: I.M. Bomze, V. Demyanov, Nonlinear Optimization. Cetraro, Italy 2007. Editors: G. di Pillo, F. Schoen (2010)
Vol. 1990: S. Bouc, Biset Functors for Finite Groups (2010)
Vol. 1991: F. Gazzola, H.-C. Grunau, G. Sweers, Polyharmonic Boundary Value Problems (2010)
Vol. 1992: A. Parmeggiani, Spectral Theory of Non-Commutative Harmonic Oscillators: An Introduction (2010)

Recent Reprints and New Editions

Vol. 1702: J. Ma, J. Yong, Forward-Backward Stochastic Differential Equations and their Applications. 1999 – Corr. 3rd printing (2007)
Vol. 830: J.A. Green, Polynomial Representations of GL_n, with an Appendix on Schensted Correspondence and Littelmann Paths by K. Erdmann, J.A. Green and M. Schoker 1980 – 2nd corr. and augmented edition (2007)
Vol. 1693: S. Simons, From Hahn-Banach to Monotonicity (Minimax and Monotonicity 1998) – 2nd exp. edition (2008)
Vol. 470: R.E. Bowen, Equilibrium States and the Ergodic Theory of Anosov Diffeomorphisms. With a preface by D. Ruelle. Edited by J.-R. Chazottes. 1975 – 2nd rev. edition (2008)
Vol. 523: S.A. Albeverio, R.J. Høegh-Krohn, S. Mazzucchi, Mathematical Theory of Feynman Path Integral. 1976 – 2nd corr. and enlarged edition (2008)
Vol. 1764: A. Cannas da Silva, Lectures on Symplectic Geometry 2001 – Corr. 2nd printing (2008)

LECTURE NOTES IN MATHEMATICS Springer

Edited by J.-M. Morel, F. Takens, B. Teissier, P.K. Maini

Editorial Policy (for the publication of monographs)

1. Lecture Notes aim to report new developments in all areas of mathematics and their applications - quickly, informally and at a high level. Mathematical texts analysing new developments in modelling and numerical simulation are welcome.

   Monograph manuscripts should be reasonably self-contained and rounded off. Thus they may, and often will, present not only results of the author but also related work by other people. They may be based on specialised lecture courses. Furthermore, the manuscripts should provide sufficient motivation, examples and applications. This clearly distinguishes Lecture Notes from journal articles or technical reports which normally are very concise. Articles intended for a journal but too long to be accepted by most journals, usually do not have this "lecture notes" character. For similar reasons it is unusual for doctoral theses to be accepted for the Lecture Notes series, though habilitation theses may be appropriate.

2. Manuscripts should be submitted either online at www.editorialmanager.com/lnm to Springer's mathematics editorial in Heidelberg, or to one of the series editors. In general, manuscripts will be sent out to 2 external referees for evaluation. If a decision cannot yet be reached on the basis of the first 2 reports, further referees may be contacted: The author will be informed of this. A final decision to publish can be made only on the basis of the complete manuscript, however a refereeing process leading to a preliminary decision can be based on a pre-final or incomplete manuscript. The strict minimum amount of material that will be considered should include a detailed outline describing the planned contents of each chapter, a bibliography and several sample chapters.

   Authors should be aware that incomplete or insufficiently close to final manuscripts almost always result in longer refereeing times and nevertheless unclear referees' recommendations, making further refereeing of a final draft necessary.

   Authors should also be aware that parallel submission of their manuscript to another publisher while under consideration for LNM will in general lead to immediate rejection.

3. Manuscripts should in general be submitted in English. Final manuscripts should contain at least 100 pages of mathematical text and should always include

   – a table of contents;
   – an informative introduction, with adequate motivation and perhaps some historical remarks: it should be accessible to a reader not intimately familiar with the topic treated;
   – a subject index: as a rule this is genuinely helpful for the reader.

   For evaluation purposes, manuscripts may be submitted in print or electronic form (print form is still preferred by most referees), in the latter case preferably as pdf- or zipped ps-files. Lecture Notes volumes are, as a rule, printed digitally from the authors' files. To ensure best results, authors are asked to use the LaTeX2e style files available from Springer's web-server at:

   ftp://ftp.springer.de/pub/tex/latex/svmonot1/ (for monographs) and
   ftp://ftp.springer.de/pub/tex/latex/svmultt1/ (for summer schools/tutorials).

Additional technical instructions, if necessary, are available on request from: lnm@springer.com.

4. Careful preparation of the manuscripts will help keep production time short besides ensuring satisfactory appearance of the finished book in print and online. After acceptance of the manuscript authors will be asked to prepare the final LaTeX source files and also the corresponding dvi-, pdf- or zipped ps-file. The LaTeX source files are essential for producing the full-text online version of the book (see http://www.springerlink.com/openurl.asp?genre=journal&issn=0075-8434 for the existing online volumes of LNM).

   The actual production of a Lecture Notes volume takes approximately 12 weeks.

5. Authors receive a total of 50 free copies of their volume, but no royalties. They are entitled to a discount of 33.3% on the price of Springer books purchased for their personal use, if ordering directly from Springer.

6. Commitment to publish is made by letter of intent rather than by signing a formal contract. Springer-Verlag secures the copyright for each volume. Authors are free to reuse material contained in their LNM volumes in later publications: a brief written (or e-mail) request for formal permission is sufficient.

Addresses:

Professor J.-M. Morel, CMLA,
École Normale Supérieure de Cachan,
61 Avenue du Président Wilson, 94235 Cachan Cedex, France
E-mail: Jean-Michel.Morel@cmla.ens-cachan.fr

Professor F. Takens, Mathematisch Instituut,
Rijksuniversiteit Groningen, Postbus 800,
9700 AV Groningen, The Netherlands
E-mail: F.Takens@rug.nl

Professor B. Teissier, Institut Mathématique de Jussieu,
UMR 7586 du CNRS, Équipe "Géométrie et Dynamique",
175 rue du Chevaleret,
75013 Paris, France
E-mail: teissier@math.jussieu.fr

For the "Mathematical Biosciences Subseries" of LNM:

Professor P.K. Maini, Center for Mathematical Biology,
Mathematical Institute, 24-29 St Giles,
Oxford OX1 3LP, UK
E-mail: maini@maths.ox.ac.uk

Springer, Mathematics Editorial, Tiergartenstr. 17,
69121 Heidelberg, Germany,
Tel.: +49 (6221) 487-259
Fax: +49 (6221) 4876-8259
E-mail: lnm@springer.com